Metasolutions of Parabolic Equations in Population Dynamics

Metasolutions of Parabolic Equations in Population Dynamics

Julián López-Gómez

Universidad Complutense de Madrid

Spain

CRC Press
Taylor & Francis Group
Boca Raton London New York

CRC Press is an imprint of the
Taylor & Francis Group, an **informa** business

A CHAPMAN & HALL BOOK

CRC Press
Taylor & Francis Group
6000 Broken Sound Parkway NW, Suite 300
Boca Raton, FL 33487-2742

First issued in paperback 2019

ISBN-13: 978-1-4822-3898-3 (hbk)
ISBN-13: 978-0-367-37731-1 (pbk)

Visit the Taylor & Francis Web site at
http://www.taylorandfrancis.com

and the CRC Press Web site at
http://www.crcpress.com

To Rosa Gómez-González
my mother,
with love and gratitude

Contents

List of Figures

Preface

This book studies the dynamics of a generalized prototype of the semilinear parabolic logistic problem

$$
\begin{cases}
\frac{\partial u}{\partial t} - d\Delta u = \lambda u + a(x)u^2 & \text{in } \Omega \times (0, \infty), \\
u = 0 & \text{on } \partial\Omega \times (0, \infty), \\
u(\cdot, 0) = u_0 > 0 & \text{in } \Omega,
\end{cases} \tag{P.1}
$$

where $\Omega \subset \mathbb{R}^N$, with $N \geq 1$, is a smooth bounded domain, $t > 0$ stands for the time, and $a(x)$ is an arbitrary continuous function such that $a(x) \leq 0$ for all $x \in \Omega$ but $a \neq 0$. So, (P.1) is a parabolic boundary value problem for the *degenerate diffusive logistic equation*

$$
\frac{\partial u}{\partial t} - d\Delta u = \lambda u + a(x)u^2 \tag{P.2}
$$

in Ω. It is said to be *degenerate* because $a(x)$ can vanish on some patches of Ω, in contrast to the classical case when $a(x) < 0$ for all $x \in \bar{\Omega}$.

In the context of population dynamics, $N \leq 3$, Ω is the inhabiting area where the individuals of a species, u, disperse randomly at a constant rate measured by $d > 0$; $u(x, t)$ is the density of the individuals of the species at the location $x \in \Omega$ after time $t > 0$; λ is the *intrinsic rate of natural increase* of the species; u_0 is the initial distribution of the species in Ω; and

$$
K(x) \equiv -\frac{\lambda}{a(x)}, \qquad x \in \Omega, \tag{P.3}
$$

is the *carrying capacity* of Ω at each location $x \in \Omega$. As we are imposing homogeneous Dirichlet boundary conditions on $\partial\Omega$, the surroundings of Ω are assumed to be hostile for the species u. So, no individual of the species can survive on the habitat edges. However, this assumption is far from necessary for the validity of most of the results discussed in this book. Two classic books on population dynamics from the perspective of reaction diffusion equations are by J. D. Murray [197] and A. Okubo and S. A. Levin [200].

Although it is folklore that the classical non-spatial logistic equation

$$
u'(t) = \lambda u(t) + au^2(t)
$$

where a is a negative constant goes back to P. F. Verhulst [230] (1838), it is less known that the diffusive logistic equation (P.2) was introduced by A. N.

Kolmogorov I. G. Petrovsky and N. S. Piskunov [124], and independently by R. A. Fisher [88], in 1937, to study some problems of a biological nature. In the classical context, $a(x)$ is a continuous function such that $a(x) < 0$ for all $x \in \bar{\Omega}$. The analysis of the degenerate parabolic problem when $a \le 0$ in Ω but $a \equiv 0$ on some subset of Ω with non-empty interior goes back to J. M. Fraile et al. [90] (1996). An elliptic counterpart of these degenerate models had been previously analyzed by H. Brézis and L. Oswald [31], and by T. Ouyang [203], [204], as part of his PhD thesis under the supervision of W. M. Ni.

Naturally, in spatially heterogeneous environments, the carrying capacity, $K(x)$, might suffer dramatic variations according to the location of the individuals of the species on the territory, $x \in \Omega$. Indeed, although $K(x)$ might be very small on some patches of the territory as an effect of harsh environmental conditions or abiotic stress, in benign areas, natural refuges or special protected zones, $K(x)$ might reach huge values.

From the mathematical point of view, a rather reasonable methodology to deal with huge variations of the carrying capacity $K(x)$ in the territory Ω is assuming that $K = \infty$, or equivalently $a \equiv 0$, in the 'protected areas,' while it is finite in less favorable zones. This strategy also makes sense from the biological point of view, as it is equivalent to combining, simultaneously, within the same territory, the Mathus and the Verhulst laws regulating the growth of the species. A further perturbation analysis should reveal the complete list of admissible limiting distribution patterns of the population as time passes in general diffusive spatially heterogeneous logistic problems.

In the region where $a(x) < 0$ the temporal evolution of species u is assumed to be governed by a logistic growth, while in the region where $a(x) \equiv 0$ the species u increases according to an exponential growth. The main goal of this book is predicting the time evolution of the species u in Ω under such circumstances. Should the species exhibit a genuine logistic behavior in Ω, or, on the contrary, should it exhibit an exponential growth? There is the possibility that u grows according to the Malthus law on some areas of Ω, while it simultaneously inherits a limited growth on others.

In an effort to summarize the contents of this book in this short general preliminary presentation, suppose $a(x)$ has a nodal behavior of the type described in Figure P.1, where the territory Ω contains ten *protected zones*, $\Omega_{0,1}^1$, $\Omega_{0,1}^2$, which are two balls, or discs if $N = 2$, with the same radius R_1, $\Omega_{0,2}^i$, $1 \le i \le 4$, which are four balls with radius $R_2 < R_1$, and $\Omega_{0,3}^i$, $1 \le i \le 4$, which are four balls with radius $R_3 < R_2$. The weight function $a(x)$ is assumed to vanish in all these *refuges*, or protected zones, while it is negative on their complement, the shadow region of Figure P.1, denoted by Ω_-.

To describe the main findings of this book for this special configuration of the territory we need to introduce some notation. Given any nice open connected subset, D, of Ω we will denote by $\lambda_1[-\Delta, D]$ the lowest eigenvalue of the linear eigenvalue problem

$$
\begin{cases}
-\Delta u = \lambda u & \text{in } D, \\
u = 0 & \text{on } \partial D.
\end{cases}
\tag{P.4}
$$

As will be discussed in Chapter 1, from a biological perspective, $d\lambda_1[-\Delta, D]$ measures the critical size of the rate of natural increase, λ, so that the inhabited area D can maintain the species u dispersing at the rate d in the patch D, in the sense that u is driven to extinction if $\lambda < d\lambda_1[-\Delta, D]$, while it is permanent if $\lambda > d\lambda_1[-\Delta, D]$. So, the condition $d\lambda_1[-\Delta, D] < \lambda$ measures the necessary geometrical properties and size of the patch D to maintain the species dispersing at the rate d in D with an intrinsic rate of natural increase λ. When D is a ball of radius R, a simple change of scale reveals that

$$\lambda_1[-\Delta, D] = \frac{\lambda_1[-\Delta, B_1]}{R^2}$$

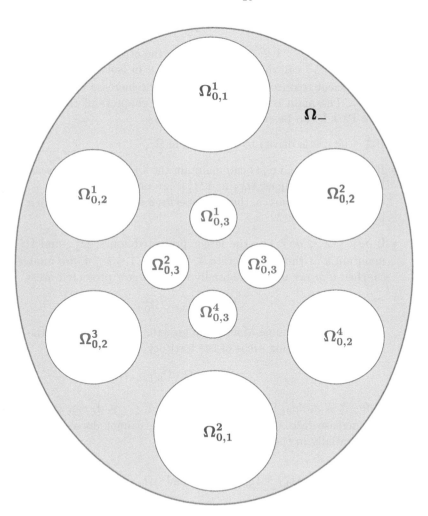

FIGURE P.1: An admissible nodal configuration for $a(x)$.

where B_1 is the unit ball of $\mathbb{R}^N$. In particular, the larger the protected zone the smaller the principal eigenvalue. Consequently, setting

$$\sigma_0 = \lambda_1[-\Delta, \Omega], \qquad \sigma_j := \lambda_1[-\Delta, \Omega_{0,j}^1], \qquad 1 \le j \le 3,$$

it becomes apparent that

$$d\sigma_0 = d\lambda_1[-\Delta, \Omega] < d\sigma_1 = d\lambda_1[-\Delta, \Omega_{0,1}^i] = \frac{d\lambda_1[-\Delta, B_1]}{R_1^2} \quad (1 \le i \le 2)$$

$$< d\sigma_2 = d\lambda_1[-\Delta, \Omega_{0,2}^i] = \frac{d\lambda_1[-\Delta, B_1]}{R_2^2} \qquad (1 \le i \le 4)$$

$$< d\sigma_3 = d\lambda_1[-\Delta, \Omega_{0,3}^i] = \frac{d\lambda_1[-\Delta, B_1]}{R_3^2} \qquad (1 \le i \le 4)$$

Naturally, for each $j = 1, 2, 3$, $d\sigma_j$ measures the critical size of λ so that the protected zone $\Omega_{0,j}^i$ can maintain the species u in isolation. In other words, $\Omega_{0,j}^i$ has sufficient resources to maintain u at the increase rate λ if, and only if, $\lambda > d\sigma_j$. The main results of the first five chapters of this book, which constitute Part I, can be summarized as follows:

- If $\lambda \le d\sigma_0$, u is driven to extinction in Ω.

- If $d\sigma_0 < \lambda < d\sigma_1$, i.e., Ω can maintain the species at the increase rate λ, but the larger refuges, $\Omega_{0,1}^1$ and $\Omega_{0,1}^2$, are unable to maintain it, by e.g., a shortage of resources, then the species u grows according to a logistic law everywhere in Ω.

- If $d\sigma_1 \le \lambda < d\sigma_2$, i.e., the larger protected zones, $\Omega_{0,1}^1$ and $\Omega_{0,1}^2$, can maintain u at the increase rate λ, but $\Omega_{0,2}^i$, $1 \le i \le 4$, are unable to do so, then u grows up exponentially in the largest protected areas

$$\Omega_{0,1} \equiv \Omega_{0,1}^1 \cup \Omega_{0,1}^2,$$

according to a genuine Malthusian growth, but according to the logistic law in the remaining areas of the territory

$$\Omega_1 \equiv \Omega \setminus \bar{\Omega}_{0,1}.$$

- If $d\sigma_2 \le \lambda < d\sigma_3$, i.e., the refuges $\Omega_{0,2}^i$, $1 \le i \le 4$, can maintain u at the increase rate λ, but $\Omega_{0,3}^i$, $1 \le i \le 4$, cannot do so, then u grows exponentially in the protected areas

$$\Omega_{0,1} \equiv \Omega_{0,1}^1 \cup \Omega_{0,1}^2 \quad \text{and} \quad \Omega_{0,2} \equiv \bigcup_{i=1}^{4} \Omega_{0,1}^i$$

whereas it has a limited logistic increase in

$$\Omega_2 \equiv \Omega \setminus (\bar{\Omega}_{0,1} \cup \bar{\Omega}_{0,2}).$$

- If $\lambda \geq d\sigma_3$, i.e., any refuge is able to maintain u at the increase rate λ, then u grows exponentially in all the protected zones

$$\Omega_{0,1} \equiv \Omega_{0,1}^1 \cup \Omega_{0,1}^2, \quad \Omega_{0,2} \equiv \bigcup_{i=1}^{4} \Omega_{0,2}^i \quad \text{and} \quad \Omega_{0,3} \equiv \bigcup_{i=1}^{4} \Omega_{0,3}^i$$

but according to a logistic law in the region

$$\Omega_3 \equiv \Omega \setminus (\bar{\Omega}_{0,1} \cup \bar{\Omega}_{0,2} \cup \bar{\Omega}_{0,3}) = \Omega_-.$$

Consequently, if, for instance, $d\sigma_1 < \lambda < d\sigma_2$, and we denote by $u(x,t)$ the unique solution of (P.1), then, as a consequence of the analysis in Part I, we find that

$$\lim_{t \uparrow \infty} u(x,t) = \infty \quad \text{for all} \quad x \in \Omega_{0,1} = \Omega_{0,1}^1 \cup \Omega_{0,1}^2$$

whereas, in the region $\Omega \setminus \bar{\Omega}_{0,1}$,

$$L_\lambda^{\min} \leq \liminf_{t \to \infty} u(\cdot, t) \leq \limsup_{t \to \infty} u(\cdot, t) \leq L_\lambda^{\max} \tag{P.5}$$

where $L_\lambda^{\min}$ and $L_\lambda^{\max}$ stand for the minimal and the maximal positive solutions of the singular problem

$$\begin{cases} -d\Delta L = \lambda L + a(x)L^2 & \text{in } \Omega \setminus \bar{\Omega}_{0,1}, \\ L = \infty & \text{on } \partial\Omega_{0,1}^1 \cup \partial\Omega_{0,1}^2, \\ L = 0 & \text{on } \partial\Omega. \end{cases} \tag{P.6}$$

Therefore, the limiting profile of $u(x,t)$ as time $t \uparrow \infty$ becomes infinity in the larger refuges, $\bar{\Omega}_{0,1}^1$ and $\bar{\Omega}_{0,1}^2$, while it remains bounded in the complement. These limiting profiles are referred to in this book as *metasolutions* supported in the complement of the largest protected zones, Ω_1, because Ω_1 is the portion of the inhabiting area where the growth of u inherits a genuine logistic character and hence it is limited. It should be noted that the smaller refuges cannot support the species u in isolation if $\lambda < d\sigma_2$. The formal concept metasolution was coined in [109], submitted for publication in September 1998. Then, it was incorporated into the PhD thesis of R. Gómez-Reñasco [105], under the supervision of the author and defended at the University of La Laguna (Tenerife, Spain) in early May 1999.

For those readers not familiarized yet with the most recent advances in the theory of nonlinear parabolic problems, possibly under the influence of the established (wrong) paradigm that the Harnack inequality is one of the driving forces of the theory of nonlinear partial differential equations, the emergence of such *metasolutions* in the context of population dynamics might be slightly shocking, as large solutions and metasolutions provide us with uncontestable evidence that the Harnack inequality is a technical device of a linear nature of doubtful interest in analyzing global nonlinear problems, as will become apparent in Section 4.9.

This might possibly explain the reaction of an anonymous reviewer of [195] who noted that a series of classical solutions and metasolutions were computed in the disc of radius 1 centered at the origin, B_1, with the choices

$$\Omega = B_1(0) = \{x \in \mathbb{R}^2 \; : \; |x| < 1\},$$
$$\Omega_{0,1} = A(0.5, 1) = \{x \in \mathbb{R}^2 \; : \; 0.5 < |x| < 1\},$$
$$\Omega_{0,2} = B_{0.3}(0) = \{x \in \mathbb{R}^2 \; : \; |x| < 0.3\},$$
$$\Omega_- = A(0.3, 0.5) = \{x \in \mathbb{R}^2 \; : \; 0.3 < |x| < 0.5\},$$

by using pseudo-spectral methods.

What the heck is a "metasolution"? Please provide a formal definition. Okay, one is provided in (4.8), but "metasolutions" is used in the abstract and intro; definition needs to be earlier. The definition puzzles me. The "large" solution would seem to be very difficult to compute because of the singularity on the boundary. And what is the use or point of a solution that is infinite everywhere on another subdomain? Metasolutions are wierd...

I am alarmed by the references to "blow up" and "approach infinity on the boundary". Spectral methods are notoriously sensitive to singularities of the solution including singularities on the boundaries...

The serious problem with the paper is that the discontinuities of slope in coefficients of partial differential equation and the infinities on the boundary both makes the solution of partial differential equation singular within the domain.

I hate their $B_r(0)$, $A(R_0, R_1)$ notation for what are simply the disk of radius R and the annulus bounded in radius by R_0 and R_1. For goodness' sake, use conventional notation and wording: "disk of radius R, $r \in [0, R]$,"...

I am further bothered that their coefficient function $a(x)$ is nonzero only for $r \in [0.3, 0.5]$ for a problem in the unit disk. The PDE thus has a coefficient with a slope discontinuity. The function $u(r, \theta)$ will be singular on the lines $r = 0.3$ and $r = 0.5$. The usual spectral strategy would be to split the domain into three and solve the linear Helmholtz equation on $r \in [0, 0.3]$ and $r \in [0.5, 1]$, the nonlinear PDE on $[0.3, 0.5]$ and carefully match the pieces taking account of the singularities. Instead the authors blithely ignored the singularities entirely...

Although the strategy proposed by the reviewer in the previous paragraphs is the most natural one when dealing with linear problems where the Harnack inequality applies, it is of no help in dealing with *singular boundary value problems* such as those treated in [195] and in this book. Contrary to what happens in most 'academic problems,' real problems might be highly nonlinear and hence can develop internal interfaces whose numerical treatment is a top level challenge.

It is the hope of the author that the readers of this book will not be 'alarmed' by the large solutions and the metasolutions as much as the reviewer of [195] was. Although, at first glance, metasolutions might be slightly hard to digest because of the number of technicalities involved in their study, during the last two decades they have proven to be categorical imperatives to describe the dynamics of wide classes of parabolic equations and systems in the presence of spatial heterogeneities.

It should be noted that (P.5) does not fully characterize the asymptotic behavior of the solutions of (P.1) unless,

$$L_\lambda^{\min} = L_\lambda^{\max}.$$

Consequently, to characterize the exact asymptotic profiles as $t \uparrow \infty$ of the solutions of (P.1), one must face the problem of the uniqueness of the solutions to the singular problem (P.6) and some other closely related singular problems that the reader will find in Chapter 4. This is the main bulk of Part II, consisting of Chapters 6, 7 and 8, where a series of very sharp optimal uniqueness results found by the author and his coworkers will be analyzed in a self-contained way.

Finally, the main goal of Part III, formed by the last two chapters, is to reinforce the evidence that metasolutions also are categorical imperatives to describe the dynamics of huge classes of spatially heterogeneous semilinear parabolic problems. Precisely, Chapter 9 analyzes (P.1) in the more general case when $a(x)$ changes sign, giving a rather complete account of some of the most relevant recent advances in the theory of *superlinear indefinite problems*, and Chapter 10 studies a paradigmatic competing species model with a protected zone for one of the species to illustrate how large solutions and metasolutions play a pivotal role in describing the dynamics of spatially heterogeneous systems.

This book grew from the monograph [160] and the lecture notes of the Metasolutions course delivered by the author at the National Center for Theoretical Sciences, Tsing Hua University, Hsinchu (Taiwan), during July and August of 2009. The author is delighted to thank Professor Sze-Bi Hsu for his kind invitation to deliver it, as well as for his brilliant questions and sharp comments during these lectures. The time spent in Taiwan by the author was certainly unforgettable, both personally and professionally.

To complete this book, the author has been supported by Research Grant Ref: MTM2012-30669 of the Ministry of Economy and Competitiveness of Spain.

Madrid

J. López-Gómez

Part I

Large solutions and metasolutions: Dynamics

Chapter 1

Introduction: Preliminaries

This book focuses attention into the problem of ascertaining the asymptotic behavior of the solutions of the parabolic problem

$$\begin{cases} \frac{\partial u}{\partial t} - d\Delta u = \lambda u + a(x)f(x,u)u & \text{in} \quad \Omega \times (0,\infty), \\ u = 0 & \text{on} \quad \partial\Omega \times (0,\infty), \\ u(\cdot,0) = u_0 > 0 & \text{in} \quad \Omega, \end{cases} \qquad (1.1)$$

where $d > 0$ is a constant, Δ stands for the Laplace operator in $\mathbb{R}^N$,

$$\Delta := \sum_{j=1}^{N} \frac{\partial^2}{\partial x_j^2}, \qquad x = (x_1,...,x_N) \in \mathbb{R}^N,$$

Ω is a bounded domain of $\mathbb{R}^N$, $N \geq 1$, with smooth boundary $\partial\Omega$ of class $\mathcal{C}^{2+\nu}$, for some $\nu \in (0,1]$, $\lambda \in \mathbb{R}$ is regarded as a parameter, and

$$a \in \mathcal{C}^\nu(\bar{\Omega}), \qquad f \in \mathcal{C}^{\nu,1+\nu}(\bar{\Omega} \times [0,\infty)),$$

although these regularity requirements can be substantially relaxed to assume $\partial\Omega$ is of class $\mathcal{C}^2$, $a \in L^\infty(\Omega)$ and $f \in \mathcal{C}^{0,1}(\bar{\Omega} \times [0,\infty))$. However the first eight chapters of this book will study the special situation when

$$a < 0 \quad (a \leq 0, \ a \neq 0),$$

which is usually referred to as the *sublinear* case if $f \geq 0$, because in such circumstances we have

$$\lambda u + a(x)f(x,u)u \leq \lambda u \qquad \text{for all} \ u \geq 0,$$

the general case when $a(x)$ changes sign will be dealt with in Chapter 9.

Throughout this book, given a $\mathcal{C}^2$-subdomain D of Ω and $V \in L^\infty(D)$, we will denote by $\lambda_1[-d\Delta + V, D]$ the principal eigenvalue of $-d\Delta + V$ in D under homogeneous Dirichlet boundary conditions on $\partial\Omega$, i.e.,

$$\lambda_1[-d\Delta + V, D] = \inf_{\psi \in \mathcal{C}_0^1(D) \setminus \{0\}} \frac{\int_D \left(d|\nabla\psi|^2 + V\psi^2\right)}{\int_D \psi^2}. \tag{1.2}$$

As the nodal behavior of $a(x)$ might be rather involved without further assumptions, in most of this book we will assume that $a(x)$ satisfies the next structural hypothesis:

Hypothesis (Ha)

The open sets

$$\Omega_- := \{\, x \in \Omega \;:\; a(x) < 0 \,\}, \qquad \Omega_0 := \Omega \setminus \bar{\Omega}_-,$$

are of class $\mathcal{C}^{2+\nu}$ and consist of finitely many components

$$\Omega_{-,j} \subset \Omega_-, \quad 1 \le j \le q_-, \qquad \Omega_{0,j}^i \subset \Omega_0, \quad 1 \le j \le q_0, \quad 1 \le i \le m_j,$$

such that

$$\Omega_- = \bigcup_{j=1}^{q_-} \Omega_{-,j}, \qquad \bar{\Omega}_{-,j} \cap \bar{\Omega}_{-,\tilde{j}} = \emptyset \quad \text{if } j \ne \tilde{j},$$

$$\Omega_0 = \bigcup_{j=1}^{q_0} \bigcup_{i=1}^{m_j} \Omega_{0,j}^i, \qquad \bar{\Omega}_{0,j}^i \cap \bar{\Omega}_{0,\tilde{j}}^{\tilde{i}} = \emptyset \quad \text{if } (j,i) \ne (\tilde{j},\tilde{i}),$$

for some integers $q_- \ge 1$, $q_0 \ge 1$ and $m_j \ge 1$, $1 \le j \le q_0$. Moreover,

- *If Γ_0 is a component of $\partial\Omega_0$ such that $\Gamma_0 \cap \partial\Omega \ne \emptyset$, then Γ_0 is a component of $\partial\Omega$.*

- *If Γ_- is a component of $\partial\Omega_-$ such that $\Gamma_- \cap \partial\Omega \ne \emptyset$, then Γ_- is a component of $\partial\Omega$.*

Without loss of generality, throughout this book we will assume that the labeling of these components has been already carried out so that

$$\begin{aligned} \sigma_j &:= \lambda_1[-\Delta, \Omega_{0,j}^i], & 1 \le i \le m_j, \; 1 \le j \le q_0, \\ \sigma_j &< \sigma_{j+1}, & 1 \le j \le q_0 - 1. \end{aligned} \tag{1.3}$$

Also, for every $1 \le j \le q_0$, we will set

$$\Omega_{0,j} := \bigcup_{i=1}^{m_j} \Omega_{0,j}^i, \qquad \lambda_1[-\Delta, \Omega_{0,j}] := \sigma_j = \lambda_1[-\Delta, \Omega_{0,j}^i], \; 1 \le i \le m_j.$$

Figure 1.1 illustrates a typical situation where assumption (Ha) is fulfilled with $q_0 = 2$ and $q_- = 1$, where we have denoted

$$\Gamma = \partial\Omega, \quad \gamma_1 = \partial\Omega_{0,1} \setminus \Gamma, \quad \gamma_2 := \partial\Omega_{0,2}, \quad \partial\Omega_- = \gamma_1 \cup \gamma_2.$$

In this example, (1.3) becomes

$$\sigma_1 := \lambda_1[-\Delta, \Omega_{0,1}] < \sigma_2 := \lambda_1[-\Delta, \Omega_{0,2}]. \tag{1.4}$$

Thanks to the Faber–Krahn inequality (cf. C. Faber [83] and E. Krahn [126]), (1.4) holds if $\Omega_{0,2}$ has sufficiently small Lebesgue measure (e.g., [144, Section 5]). Indeed, according to these results, among all domains D of $\mathbb{R}^N$ with a fixed Lebesgue measure, $|D|$, the ball has the smallest principal eigenvalue. Consequently, setting

$$B_\varrho := \{x \in \mathbb{R}^N \ : \ |x| < \varrho\}, \quad \varrho > 0,$$

and

$$R := (|D|/|B_1|)^{1/N},$$

we have that $|D| = |B_R|$ and hence,

$$\lambda_1[-\Delta, D] \geq \lambda_1[-\Delta, B_R] = \lambda_1[-\Delta, B_1]R^{-2} = \lambda_1[-\Delta, B_1]|B_1|^{2/N}|D|^{-2/N}.$$

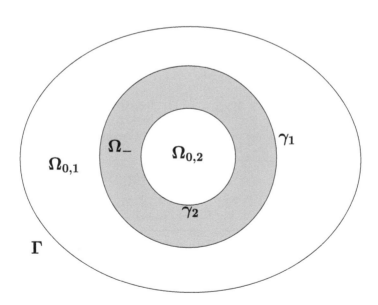

FIGURE 1.1: An admissible nodal configuration for $a(x)$.

Therefore,

$$\lim_{|D|\downarrow 0} \lambda_1[-\Delta, D] = \infty.$$

Although one might think of (1.3) as a sort of hierarchical ordering between the components of Ω_0 establishing that $\Omega_{0,j}$ is larger than $\Omega_{0,j+1}$ for all $1 \leq j \leq q_0 - 1$, one should not forget that $\lambda_1[-\Delta, D]$ also depends on certain (hidden) geometrical properties of D. Setting

$$\sigma_0 := \lambda_1[-\Delta, \Omega],$$

(1.3) implies that

$$\sigma_0 < \sigma_1 < \sigma_2 < \cdots < \sigma_{q_0}, \tag{1.5}$$

by the monotonicity of the principal eigenvalue with respect to the domain. Although most of the properties of $\lambda_1[-\Delta, D]$ invoked in this book can be easily inferred from the variational characterization (1.2), the reader is sent to Chapters 8 and 9 of [163], if necessary, where all the properties are collected in a much more general setting.

As for the function $f(x, u)$, in most of this book we will assume that it satisfies the following structural assumption:

Hypothesis (Hf)

$f \in \mathcal{C}^{\nu, 1+\nu}(\bar{\Omega} \times [0, \infty))$ *satisfies* $f(x, 0) = 0$ *and* $\partial_u f(x, u) > 0$ *for all* $x \in \bar{\Omega}$ *and* $u > 0$.

Obviously, the function

$$f(x, u) = b(x)u^p, \qquad (x, u) \in \bar{\Omega} \times [0, \infty), \tag{1.6}$$

satisfies (Hf) for all $p \geq 1$ and $b \in \mathcal{C}^{\nu}(\bar{\Omega})$ such that $b(x) > 0$ for all $x \in \bar{\Omega}$. In case (1.6), the function $b(x)$ can be glued by $a(x)$ as soon as $b(x) > 0$ for all $x \in \Omega_-$. In many circumstances, f will be also required to satisfy

---

Hypothesis (Hg)

There exists $g \in \mathcal{C}^{1+\nu}[0, \infty)$ *with* $f(\cdot, u) \geq g(u)$ *for all* $u \geq 0$ *such that* $g(0) = 0$, $g'(u) > 0$ *for all* $u > 0$ *and* $\lim_{u\uparrow\infty} g(u) = \infty$, *where* $' = d/du$.

Note that (Hg) implies

$$\lim_{u\uparrow\infty} f(x, u) = \infty \quad \text{uniformly in} \quad x \in \bar{\Omega},$$

and that the special choice (1.6) satisfies (Hg) with

$$g(u) = u^p, \qquad u \geq 0.$$

Under these structural assumptions on $a(x)$ and $f(x,u)$, it will be established in Chapter 4 that the set of positive steady states of (1.1), which are the positive solutions of

$$\begin{cases} -d\Delta u = \lambda u + a(x)f(x,u)u & \text{in } \Omega, \\ u = 0 & \text{on } \partial\Omega, \end{cases} \qquad (1.7)$$

consist of a curve of class $\mathcal{C}^1$, as a function of the parameter λ, that bifurcates from $u = 0$ at $\lambda = d\sigma_0$ and blows up to infinity in $\Omega_{0,1}$ as $\lambda \uparrow d\sigma_1$. It turns out that whether the positive solutions of (1.7) stay bounded in $\Omega \setminus \bar{\Omega}_{0,1}$ as $\lambda \uparrow d\sigma_1$ depends on whether f, or the function g of (Hg), satisfies a further condition reminiscent of J. B. Keller [123] and R. Osserman [202], which can be stated as follows:

Hypothesis (KO)

f *satisfies* (Hf), (Hg), *and, for every* $\alpha > 0$, *there exists* $u^* > g^{-1}(1/\alpha)$ *such that*

$$I(u) := \int_1^\infty \frac{d\theta}{\sqrt{\int_1^\theta [\alpha\, g(ut) - 1]\, t\, dt}} < \infty \qquad (1.8)$$

for every $u \geq u^*$, *and*

$$\lim_{u\uparrow\infty} I(u) = 0. \qquad (1.9)$$

As

$$I(u) \leq I(v) \qquad \text{if} \quad u > v > g^{-1}(1/\alpha),$$

it is apparent that $I(u^*) < \infty$ implies (1.8) for all $u \geq u^*$ and, consequently, the hypothesis (KO) can be shortly expressed as follows

---

Abbreviated hypothesis (KO)

f *satisfies* (Hf), (Hg), *and* $\lim_{u\uparrow\infty} I(u) = 0$ *for all* $\alpha > 0$.

Throughout this book, (KO) will be referred to as the *Keller–Osserman condition*. It is an imperative condition to get uniform a priori estimates in Ω_- for the positive solutions of the equation

$$-d\Delta u = \lambda u + a(x)f(x,u)u. \qquad (1.10)$$

It should be noted that (KO) holds if there exist $p > 0$ and $C > 0$ such that

$$g(u) \geq Cu^p \qquad \text{for all} \quad u \geq 0. \qquad (1.11)$$

Indeed, (1.11) implies that

$$
\int_1^\theta [\alpha\, g(ut) - 1]\, t\, dt \geq \int_1^\theta (\alpha\, C u^p t^p - 1)\, t\, dt
$$
$$
= \frac{\alpha\, C u^p}{p+2} (\theta^{p+2} - 1) - \frac{1}{2}(\theta^2 - 1)
$$
$$
> \frac{1}{2}(\theta^{p+2} - \theta^2) = \frac{\theta^2}{2}(\theta^p - 1) > 0
$$

provided

$$
\alpha > 0, \qquad u > \left(\frac{p+2}{2\alpha C}\right)^{\frac{1}{p}}, \qquad \theta > 1.
$$

Thus,

$$
I(u) \leq \int_1^\infty \left[\frac{\alpha\, C u^p}{p+2}(\theta^{p+2} - 1) - \frac{1}{2}(\theta^2 - 1)\right]^{-1/2} d\theta < \infty, \qquad (1.12)
$$

because the function

$$
R(\theta) := \frac{\alpha\, C u^p}{p+2}(\theta^{p+2} - 1) - \frac{1}{2}(\theta^2 - 1), \qquad \theta \geq 1,
$$

satisfies

$$
R(1) = 0, \qquad R'(1) = \alpha\, C u^p - 1 > 0, \qquad \lim_{\theta\uparrow\infty} \frac{R(\theta)}{\theta^{p+2}} = \frac{\alpha\, C u^p}{p+2} > 0,
$$

and, since $p > 0$,

$$
\int_1^\infty \theta^{-\frac{p+2}{2}} d\theta = \frac{2}{p} < \infty.
$$

Moreover, letting $u \uparrow \infty$ in (1.12) yields (1.9). Therefore, (KO) indeed holds under condition (1.11).

1.1   The meaning of the Keller–Osserman condition

The condition (KO) is imposed so that, for every $\lambda > 0$, $A > 0$ and $L > 0$, there is a unique value of

$$
x := x(\lambda, A, L) > 0
$$

for which the singular one-dimensional problem

$$
\begin{cases} -du'' = \lambda u - Ag(u)u & \text{in } (0, L), \\ u(0) = x, \quad u'(0) = 0, \quad u(L) = \infty, \end{cases} \qquad (1.13)
$$

possesses a positive solution. Consequently, by reflection about 0, the diffusive logistic equation

$$-du'' = \lambda u - Ag(u)u \tag{1.14}$$

has a *large*, or *explosive*, solution in $(-L, L)$, i.e., a positive solution u such that

$$\lim_{t \downarrow -L} u(t) = \lim_{t \uparrow L} u(t) = +\infty.$$

In such case, the large solution is unique.

As will become apparent in Chapter 3, under conditions (Hf) and (Hg) the existence of these one-dimensional solutions allows the construction of uniform a priori bounds for all the positive solutions of (1.14), even in the general multidimensional problem, because large solutions are above any other positive solution.

Multiplying the differential equation by $v = u'$ yields

$$dvv' + \lambda uu' - Ag(u)uu' = 0,$$

or, equivalently,

$$\frac{d}{2}v^2 + \frac{\lambda}{2}u^2 - A\int_0^u g(s)s\,ds = \frac{\lambda}{2}x^2 - A\int_0^x g(s)s\,ds. \tag{1.15}$$

The potential energy of the associated (u, v)–system is given by

$$P(u) := \frac{\lambda}{2}u^2 - A\int_0^u g(s)s\,ds, \qquad u \in \mathbb{R}.$$

As under condition (Hg) we have that $g(0) = 0$, $g'(u) > 0$ for all $u > 0$ and $\lim_{u \uparrow \infty} g(u) = \infty$, there exists a unique $\omega > 0$ such that $g(\omega) = \lambda/A$. Thus, 0 and ω are the unique equilibria of (1.14). Obviously,

$$P'(u) = \lambda u - Ag(u)u, \qquad P''(u) = \lambda - Ag'(u)u - Ag(u),$$

and hence,

$$P''(0) = \lambda > 0, \qquad P''(\omega) = -Ag'(\omega)\omega < 0.$$

Thus, 0 is a center and ω is a saddle point. Moreover, $P'(u) > 0$ for all $u \in (0, \omega)$ and

$$g(u) > g(\omega) = \lambda/A \quad \text{for all } u > \omega.$$

Consequently, $\lambda - Ag(u) < 0$ and

$$P''(u) < -Ag'(u)u < 0 \quad \text{for all } u > \omega.$$

Therefore, there exists a unique $u_0 > \omega$ such that $P(u_0) = 0$, and

$$\lim_{u \uparrow \infty} P(u) = -\infty.$$

So, the underlying phase portrait of the positive solutions of (1.14) looks like Figure 1.2

In order to solve (1.13) we should find $x > \omega$ for which the half-upper trajectory is run exactly in a time L. Rearranging (1.15) we find that

$$
\begin{aligned}
dv^2 &= \lambda(x^2 - u^2) + 2A\left(\int_0^u g(s)s\,ds - \int_0^x g(s)s\,ds\right) \\
&= -2\lambda \int_x^u s\,ds + 2A\int_x^u g(s)s\,ds = 2\lambda\left(\frac{A}{\lambda}\int_x^u g(s)s\,ds - \int_x^u s\,ds\right) \\
&= 2\lambda \int_x^u \left(\frac{A}{\lambda}g(s) - 1\right)s\,ds.
\end{aligned}
$$

Consequently, performing the change of variable $s = xt$ in the previous integral yields

$$
v^2 = \frac{2\lambda}{d}x^2 \int_1^{u/x} \left(\frac{A}{\lambda}g(xt) - 1\right)t\,dt.
$$

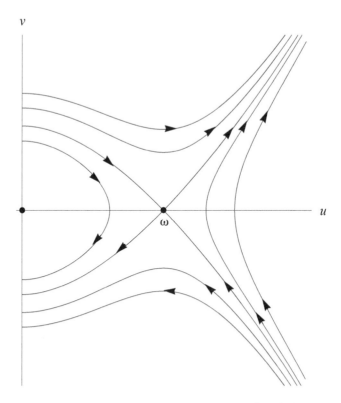

FIGURE 1.2: The phase plane of equation (1.14) for $u > 0$.

Therefore, the solution of the singular problem (1.13) is characterized through

$$L = \int_x^\infty \frac{du}{\sqrt{\frac{2\lambda}{d}x}\sqrt{\int_1^{u/x}\left(\frac{A}{\lambda}g(xt)-1\right)t\,dt}},$$

or, equivalently, after performing the change of variable $u = x\theta$,

$$\sqrt{\frac{2\lambda}{d}}L = \int_1^\infty \frac{d\theta}{\sqrt{\int_1^\theta\left(\frac{A}{\lambda}g(xt)-1\right)t\,dt}} = I(x),$$

where $I(x)$ is the function defined in (1.8) with $\alpha = A/\lambda$. As g is increasing, I is decreasing. Moreover, since $(w, 0)$ is an equilibrium, by continuous dependence it becomes apparent that

$$\lim_{x\downarrow w} I(x) = \infty.$$

Thus, if we impose (1.9), i.e.,

$$\lim_{x\uparrow\infty} I(x) = 0,$$

then there is a unique $x > 0$ such that

$$I(x) = \sqrt{\frac{2\lambda}{d}}L.$$

Therefore, (1.13) admits a unique positive solution. Consequently, in this context, the condition (KO) is equivalent to the existence of a unique $x > 0$ for which (1.13) admits a positive solution, independently of the size of the positive constants λ, A and L.

1.2 Model in population dynamics

Throughout this book, for a given $D \subset \Omega$ of class $\mathcal{C}^1$ and a function $h \in \mathcal{C}(\bar{D})$, it is said that h is positive in D, $h > 0$, if $h \geq 0$ but $h \neq 0$. Obviously, $h < 0$ if $-h > 0$. Similarly, given $h \in \mathcal{C}^1(\bar{D})$, it is said that h is strongly positive in D, $h \gg 0$, if $h(x) > 0$ for all $x \in D$ and $\frac{\partial h}{\partial n}(x) < 0$ for all $x \in \partial D$ with $h(x) = 0$, where n stands for the outward unit normal vector field along ∂D. Finally, it is said that $h \ll 0$ if $-h \gg 0$, and, given $h_1, h_2 \in \mathcal{C}(\bar{D})$, we write $h_1 > h_2$ if $h_1 - h_2 > 0$ and $h_1 \gg h_2$ if $h_1 - h_2 \gg 0$.

In the context of population dynamics, the parabolic problem (1.1) models the evolution of a single species $u(x, t)$ randomly dispersed in the inhabiting area Ω, with constant dispersion rate $d > 0$. In such models, λ stands for the intrinsic growth rate of u and $a(x)$ measures the crowding effects of the population in Ω_-, while in $\Omega_0 = \text{int } a^{-1}(0)$ the species increases according to the Malthus law of population dynamics, because $a = 0$ there in. In our setting, the territory Ω is fully surrounded by completely hostile regions, because we are imposing $u = 0$ on $\partial\Omega$. The function $u_0 \in \mathcal{C}(\bar{\Omega})$, $u_0 > 0$ represents the initial population distribution. So, (1.1) may be regarded as a prototype model linking the Malthus and the logistic laws of population dynamics within the same territory. Indeed, if $f(x, u) = u$, $\Omega_- = \Omega$ and $n := -a$ satisfies $n(x) > 0$ for all $x \in \bar{\Omega}$, then (1.1) provides us with the classical diffusive logistic problem

$$\begin{cases} \frac{\partial u}{\partial t} - d\Delta u = \lambda u - n(x)u^2 & \text{in} \quad \Omega \times (0, \infty), \\ u = 0 & \text{on} \quad \partial\Omega \times (0, \infty), \\ u(\cdot, 0) = u_0 > 0 & \text{in} \quad \Omega, \end{cases} \qquad (1.16)$$

whereas it provides us with the (linear) diffusive Malthusian problem

$$\begin{cases} \frac{\partial u}{\partial t} - d\Delta u = \lambda u & \text{in} \quad \Omega \times (0, \infty), \\ u = 0 & \text{on} \quad \partial\Omega \times (0, \infty), \\ u(\cdot, 0) = u_0 > 0 & \text{in} \quad \Omega, \end{cases} \qquad (1.17)$$

when $\Omega_0 = \Omega$. The main goal of this book is ascertaining the interplay between these two angular laws of population dynamics when they arise simultaneously in a heterogeneous environment.

As the unique solution of (1.17) is given through the heat semigroup by the formula

$$u(\cdot, t; u_0) := e^{t(\lambda + d\Delta)} u_0,$$

it is apparent that

$$\lim_{t\uparrow\infty} u(\cdot, t; u_0) = 0 \quad \text{uniformly in } \Omega$$

if $\lambda < d\sigma_0$, while

$$\lim_{t\uparrow\infty} u(\cdot, t; u_0) = \infty$$

uniformly exponentially in compact subsets of Ω, if $\lambda > d\sigma_0$. Indeed, let $\varphi \gg 0$ be a principal eigenfunction associated with σ_0. By the parabolic maximum principle (e.g., L. Nirenberg [199]), we have that

$$u(\cdot, t; u_0) := e^{t(\lambda + d\Delta)} u_0 \gg 0$$

for all $t > 0$ and hence, there exists $0 < \varepsilon < C$ such that

$$\varepsilon\varphi < e^{\lambda + d\Delta} u_0 < C\varphi.$$

Thus,

$$u(\cdot, t; \varepsilon\varphi) \ll e^{(t+1)(\lambda + d\Delta)} u_0 \ll u(\cdot, t; C\varphi)$$

for all $t > 0$. On the other hand, for every $\xi \in \mathbb{R}$, we have that

$$u(\cdot, t; \xi\varphi) = \xi e^{t(\lambda + d\Delta)} \varphi = \xi e^{(\lambda - d\sigma_0)t} \varphi$$

and hence,

$$\varepsilon e^{(\lambda-d\sigma_0)t}\varphi \ll e^{(t+1)(\lambda+d\Delta)}u_0 \ll Ce^{(\lambda-d\sigma_0)t}\varphi.$$

Consequently,

$$\lim_{t\uparrow\infty} u = 0 \quad \text{with decay rate} \quad e^{(\lambda-d\sigma_0)t} \quad \text{if} \quad \lambda < d\sigma_0,$$

while

$$\lim_{t\to\infty} u = \infty \quad \text{with exponential growth rate} \quad e^{(\lambda-d\sigma_0)t} \quad \text{if} \quad \lambda > d\sigma_0.$$

In the intermediate case when $\lambda = d\sigma_0$, (1.17) possesses an entire half line of positive steady-state solutions. Namely,

$$\mathcal{E}_+ := \{\xi\varphi \, : \, \xi \geq 0\}.$$

Figure 1.3 represents the corresponding dynamics. The arrows provide us with a scheme of the flow of (1.17). The thick half line at $\lambda = d\sigma_0$ represents the set of positive equilibria $\mathcal{E}_+$.

The dynamics of the classical diffusive logistic problem (1.16) change drastically because, as a result of the theory developed in Chapter 2, for every $\lambda > d\sigma_0$ the problem (1.16) possesses a unique positive steady state, which

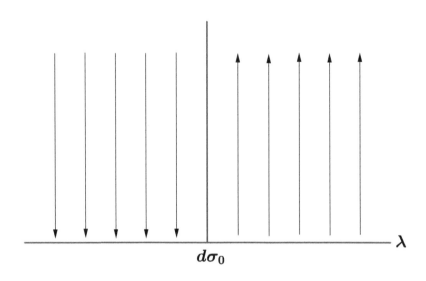

FIGURE 1.3: The dynamics of u according to the Malthus law.

will be denoted by $\theta_{[\lambda,\Omega]}$ throughout this book, while zero is the unique non-negative steady state for each $\lambda \leq d\sigma_0$. Moreover, 0 is a global attractor for all the positive solutions of (1.16) if $\lambda \leq d\sigma_0$, while $\theta_{[\lambda,\Omega]}$ is a global attractor for the positive solutions if $\lambda > d\sigma_0$. Figure 1.4 represents the corresponding dynamics. As a consequence from the results of Chapter 2, it will become apparent that the curve $\lambda \mapsto \theta_{[\lambda,\Omega]}$ is point-wise increasing and smooth, as sketched in Figure 1.4. As in all the forthcoming dynamical schemes, the sense of the arrows provides us with the direction of the flow of (1.16).

In the context of population dynamics, the previous results can be stated as follows. Suppose the species has an intrinsic growth rate $\lambda > 0$ and the territory Ω is sufficiently large, or enjoys the appropriate geometric properties, so that $d\sigma_0 < \lambda$. Then, u increases to infinity, exponentially, if it has Malthusian growth, while it approximates the limiting profile $\theta_{[\lambda,\Omega]}$ if it governed by the logistic law. Moreover, independently of the governing growth law, if the territory is sufficiently small so that $\lambda \leq d\sigma_0$, then the species is driven to extinction inexorably.

Although in the classical cases that we have just described the dynamics of (1.1) are governed by the non-negative steady states, or by ∞ if $a = 0$ and $\lambda > d\sigma_0$, in the general setting of this book, under Hypothesis (Ha) the classical steady states of (1.1), given by the elliptic problem (1.7), cannot describe the dynamics of (1.1) if $\lambda \geq d\sigma_1$, because in such regimes $u(\cdot, t; u_0)$ must approximate a *metasolution*. The metasolutions of (1.1) are the extensions by infinity of the *large* (or *explosive*) solutions of equation (1.10) in the

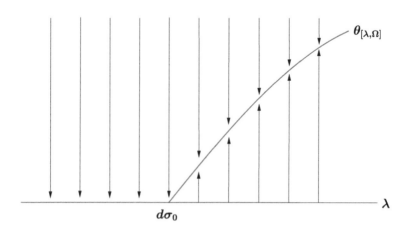

FIGURE 1.4: The dynamics of u according to the logistic law.

components of

$$\left\{ \Omega \setminus \left(\bar{\Omega}_{0,1} \cup \cdots \cup \bar{\Omega}_{0,j} \right) \ : \ 1 \le j \le q_0 \right\}.$$

Note that

$$\Omega_- = \Omega \setminus \left(\bar{\Omega}_{0,1} \cup \cdots \cup \bar{\Omega}_{0,q_0} \right).$$

Under Hypotheses (Ha), (Hf), (Hg) and (KO), the main findings of the first five chapters of this book, which constitute Part I, can be summarized as follows:

- If $\lambda \le d\sigma_0$, then the inhabiting region Ω cannot support the species u.

- If $d\sigma_0 < \lambda < d\sigma_1$, then the species u exhibits a logistic growth in Ω.

- If $d\sigma_1 \le \lambda < d\sigma_2$, then the species u has Malthusian growth in $\bar{\Omega}_{0,1} \setminus \partial\Omega$; however its growth in $\Omega \setminus \bar{\Omega}_{0,1}$ is of logistic type.

- If $d\sigma_j \le \lambda < d\sigma_{j+1}$ for some $j \in \{2, ..., q_0 - 1\}$, then u has Malthusian growth in
$$\left(\bar{\Omega}_{0,1} \cup \cdots \cup \bar{\Omega}_{0,j} \right) \setminus \partial\Omega,$$
but logistic growth in $\Omega \setminus \left(\bar{\Omega}_{0,1} \cup \cdots \cup \bar{\Omega}_{0,j} \right)$.

- If $\lambda \ge d\sigma_{q_0}$, then the species u exhibits Malthusian growth in $\Omega \setminus \bar{\Omega}_-$, but logistic growth in Ω_-.

In particular, both growth laws can coexist within a spatially heterogeneous territory, where the asymptotic behavior of the species can be seriously affected by the geometric and the distribution of the several protection zones of Ω_0, whose components might be viewed as sort of natural protected areas for the species u.

The next sections of this chapter contain some mathematical preliminaries necessary to reach the main goals of Part I.

1.3 Characterization of the maximum principle

Subsequently, for a sufficiently smooth subdomain D of Ω and a given bounded and measurable *potential* $V \in L^\infty(D)$, we consider the linear eigenvalue problem

$$\begin{cases} (-d\Delta + V)u = \lambda u & \text{in } D, \\ u = 0 & \text{on } \partial D. \end{cases} \tag{1.18}$$

It is a very classical result (e.g., R. Courant and D. Hilbert [61]) that any eigenvalue of (1.18) must be real and that (1.18) has an unbounded sequence of eigenvalues

$$\lambda_1 < \lambda_2 \le \cdots \le \lambda_n \le \cdots$$

whose associated eigenfunctions are total in $L^2(D)$. Throughout this book, the *lowest*, or *principal*, eigenvalue, λ_1, will be denoted by $\lambda_1[-d\Delta + V, D]$. It is well known that it satisfies the variational characterization (1.2). Moreover, from (1.2) one can easily infer the following monotonicity properties:

- If $V_1 < V_2$, then $\lambda_1[-d\Delta + V_1, D] < \lambda_1[-d\Delta + V_2, D]$.

- If $D_0 \subsetneq D$, then $\lambda_1[-d\Delta + V, D] < \lambda_1[-d\Delta + V, D_0]$.

The first property is usually referred to as the *monotonicity of the principal eigenvalue with respect to the potential*, and the second one is known as the *monotonicity with respect to the domain*. From the first one, the continuity of the map

$$
\begin{array}{ccc}
L^\infty(D) & \longrightarrow & \mathbb{R} \\
V & \mapsto & \lambda_1[-d\Delta + V, D]
\end{array}
$$

readily follows, which will be refereed to as the *continuity of the principal eigenvalue with respect to the potential*. In particular, the real function

$$\lambda \mapsto \lambda_1[-d\Delta + \lambda V, D]$$

is continuous.

A most sophisticated property, going back to R. Courant and D. Hilbert [61] for general self-adjoint operators, establishes that if $\{D_n\}_{n\geq 1}$ is a sequence of smooth subdomains of D approximating a nice subdomain $D_0 \subset D$ as $n \uparrow \infty$, in the appropriate sense, then

$$\lim_{n\to\infty} \lambda_1[-d\Delta + V, D_n] = \lambda_1[-d\Delta + V, D_0]. \tag{1.19}$$

This property will be called the *continuity of the principal eigenvalue with respect to the domain*. This result was later refined by E. N. Dancer [64], J. López-Gómez [144] and S. Cano-Casanova and J. López-Gómez [35, 37, 38] to cover a general class of operators under rather general mixed boundary conditions (see [163] for further details).

An important property of $\lambda_1[-d\Delta + V, D]$ is the fact that it admits a positive eigenfunction, φ, unique up to a multiplicative constant, and that it is algebraically simple. Moreover, $\varphi \gg 0$ as discussed in Section 1.2, and, since $-d\Delta + V$ defines a self-adjoint operator in $L^2(D)$, it is easy to see that

$$\int_D \varphi\varphi_n = 0, \qquad n \geq 2,$$

for any eigenfunction φ_n associated to λ_n. Consequently, the principal eigenvalue is the unique one which admits a positive eigenfunction. The associated eigenfunction, φ, will be referred to as the *principal eigenfunction* of (1.18).

These results admit very general counterparts valid for wide classes of linear second order elliptic operators, not necessarily selfadjoint, under general mixed boundary conditions (see J. López-Gómez [163]).

The next characterization of the maximum principle provides us with a pivotal property of the principle eigenvalue. It will be used very often throughout the remainder of this book.

Theorem 1.1 *Suppose D is an open subdomain of $\mathbb{R}^N$, $N \geq 1$, of class $C^{2+\nu}$, for some $\nu \in (0,1]$, and $V \in C^\nu(\bar{D})$. Then, for every $d > 0$, the following assertions are equivalent.*

(a) *$\lambda_1[-d\Delta + V, D] > 0$.*

(b) *There exists a function $h \in C^2(D) \cap C^1(\bar{D})$ such that $h > 0$ in D,*

$$(-d\Delta + V)h \geq 0 \qquad in \quad D,$$

*and either $h|_{\partial D} > 0$, or $(-d\Delta + V)h > 0$ in D. Such a function is called a **positive strict supersolution** of $-d\Delta + V$ in D (under Dirichlet boundary conditions).*

(c) *The operator $-d\Delta + V$ satisfies the **strong maximum principle** in D, in the sense that, for every $f \in C^\nu(\bar{D})$ and $b \in C^{2+\nu}(\partial D)$, with $f \geq 0$, $b \geq 0$ and $(f, b) \neq (0,0)$, and any $u \in C^{2+\nu}(\bar{D})$ satisfying*

$$\begin{cases} (-d\Delta + V)u = f & in \ D, \\ u = b & on \ \partial D, \end{cases}$$

necessarily $u \gg 0$ in D, i.e.,

$$u(x) > 0 \quad \forall \ x \in D \quad and \quad \frac{\partial u}{\partial n}(x) < 0 \quad \forall \ x \in u^{-1}(0) \cap \partial D.$$

Theorem 1.1 goes back to J. López-Gómez and M. Molina-Meyer [165]. Completely self-contained proofs for the scalar problem can be found in [144], [152]. The monograph of J. López-Gómez [163] focuses on a refinement of this result going back to H. Amann and J. López-Gómez [13].

1.4   Existence through subsolutions and supersolutions

Throughout this section, D stands for a subdomain of class $C^{2+\nu}$ of Ω such that $a < 0$ in D, i.e.,

$$D \cap \Omega_- \neq \emptyset. \tag{1.20}$$

As an immediate consequence, from H. Amann [11] the next result holds.

Theorem 1.2 *Suppose $f \in C^{\nu,\nu}(\bar{\Omega} \times [0, \infty))$, $b \in C^{2+\nu}(\partial D)$, $b \geq 0$, and the problem*

$$\begin{cases} -d\Delta u = \lambda u + af(\cdot, u)u & in \ D, \\ u = b & on \ \partial D, \end{cases} \tag{1.21}$$

possesses a subsolution $\underline{u} \in \mathcal{C}^{2+\nu}(\bar{D})$ *and a supersolution* $\bar{u} \in \mathcal{C}^{2+\nu}(\bar{D})$ *with* $0 \leq \underline{u} \leq \bar{u}$ *in* D. *Then,* (1.21) *possesses a solution* $u \in \mathcal{C}^{2+\nu}(\bar{D})$ *such that*

$$0 \leq \underline{u} \leq u \leq \bar{u}. \tag{1.22}$$

Moreover, (1.21) *has a minimal and a maximal solution in* $[\underline{u}, \bar{u}]$, *in the sense that there exist* $u_*, u^* \in \mathcal{C}^{2+\nu}(\bar{D})$ *such that:*

(i) u_* *and* u^* *solve* (1.21) *and satisfy* $\underline{u} \leq u_* \leq u^* \leq \bar{u}$.

(ii) *Any solution* $u \in \mathcal{C}^{2+\nu}(\bar{D})$ *of* (1.21) *with* $\underline{u} \leq u \leq \bar{u}$ *satisfies*

$$u_* \leq u \leq u^*.$$

Remark 1.3 If $\underline{u}$ (resp. $\bar{u}$) is a strict subsolution (resp. supersolution) of (1.21), then any solution $u \in [\underline{u}, \bar{u}]$ satisfies $\underline{u} < u \leq \bar{u}$ (resp. $\underline{u} \leq u < \bar{u}$).

As a consequence of Theorem 1.2, the next result holds.

Theorem 1.4 *Suppose* $f \in \mathcal{C}^{\nu,\nu}(\bar{\Omega} \times [0,\infty))$ *and* $\underline{u}, \bar{u} \in \mathcal{C}^{2+\nu}(D)$ *satisfy* $0 \leq \underline{u} \leq \bar{u}$ *in* D *and*

$$\begin{cases} -d\Delta\underline{u} \leq \lambda\underline{u} + af(\cdot, \underline{u})\underline{u} \\ -d\Delta\bar{u} \geq \lambda\bar{u} + af(\cdot, \bar{u})\bar{u} \end{cases} \quad in \ \ D, \qquad \lim_{\text{dist}\,(x,\partial D)\downarrow 0} \underline{u}(x) = \infty.$$

Then, the singular boundary value problem

$$\begin{cases} -d\Delta u = \lambda u + af(\cdot, u)u & in \ \ D, \\ u = \infty & on \ \ \partial D, \end{cases} \tag{1.23}$$

possesses a solution $u \in \mathcal{C}^{2+\nu}(D)$ *such that*

$$\underline{u} \leq u \leq \bar{u}.$$

Remark 1.5 By a solution of (1.23) it is meant any function $u \in \mathcal{C}^{2+\nu}(D)$ solving (1.10) in D with

$$\lim_{\text{dist}\,(x,\partial D)\downarrow 0} u(x) = \infty.$$

In this book, these solutions are called *large*, or *explosive*, solutions of (1.10) in D. Naturally, $\underline{u}$ (resp. $\bar{u}$) is said to be a *large subsolution* (resp. *large supersolution*) of (1.10) in D.

Proof of Theorem 1.4: For sufficiently large $n \geq 1$, say $n \geq n_0$, consider

$$D_n := \{x \in D \ : \ \text{dist}\,(x, \partial D) > 1/n\}.$$

The integer $n_0 \geq 1$ should be chosen so that ∂D_n inherits the same regularity properties as ∂D. Since

$$\underline{u} \leq \frac{\underline{u} + \bar{u}}{2} \leq \bar{u},$$

we find from Theorem 1.2 that, for every $n \geq n_0$, the auxiliary problem

$$\begin{cases} -d\Delta u = \lambda u + af(\cdot, u)u & \text{in} \quad D_n, \\ u = (\underline{u} + \bar{u})/2 & \text{on} \quad \partial D_n, \end{cases}$$

possesses a solution $u_n \in \mathcal{C}^{2+\nu}(\bar{D}_n)$ such that

$$\underline{u}|_{D_n} \leq u_n \leq \bar{u}|_{D_n} \quad \text{in} \quad D_n. \tag{1.24}$$

Now, pick an integer $k \geq n_0 + 1$. As $D_k \subset D_n$ for all $n \geq k$, (1.24) implies

$$\underline{u}|_{D_k} \leq u_n \leq \bar{u}|_{D_k} \quad \text{in} \quad D_k \quad \text{for all} \quad n \geq k.$$

Consequently, thanks to the Schauder interior estimates (see Section 6.1 of D. Gilbarg and N. S. Trudinger [103]), there exists a constant $C = C(k)$ such that

$$\|u_n\|_{\mathcal{C}^{2+\nu}(\bar{D}_{k-1})} \leq C(k) \quad \text{for all} \quad n \geq k.$$

Thus, since the injection

$$\mathcal{C}^{2+\nu}(\bar{D}_{k-1}) \hookrightarrow \mathcal{C}^2(\bar{D}_{k-1})$$

is compact (e.g., J. López-Gómez [163, p. 197]), there exists a subsequence, $\{u_{n_m}\}_{m\geq 1}$, of $\{u_n\}_{n\geq k}$ such that

$$\lim_{m\to\infty} \|u_{n_m} - \tilde{u}_{k-1}\|_{\mathcal{C}^2(\bar{D}_{k-1})} = 0$$

for some function $\tilde{u}_{k-1} \in \mathcal{C}^{2+\nu}(\bar{D}_{k-1})$ such that

$$-d\Delta \tilde{u}_{k-1} = \lambda \tilde{u}_{k-1} + af(\cdot, \tilde{u}_{k-1})\tilde{u}_{k-1} \quad \text{in} \quad D_{k-1}.$$

Next, consider the sequence

$$\{u_{n_m}|_{D_{k+1}}\}_{m\geq m_0} \tag{1.25}$$

for a sufficiently large $m_0 \geq 1$ such that $n_{m_0} \geq k+1$. Arguing as above, there exists a subsequence of (1.25), relabeled by n_m, such that

$$\lim_{m\to\infty} \|u_{n_m} - \tilde{u}_k\|_{\mathcal{C}^2(\bar{D}_k)} = 0$$

for some function $\tilde{u}_k \in \mathcal{C}^{2+\nu}(\bar{D}_k)$ such that

$$-d\Delta \tilde{u}_k = \lambda \tilde{u}_k + af(\cdot, \tilde{u}_k)\tilde{u}_k \quad \text{in} \quad D_k.$$

By construction,

$$\tilde{u}_k|_{D_{k-1}} = \tilde{u}_{k-1} \quad \text{in} \quad D_{k-1} \quad \text{for all} \quad k \geq n_0 + 1. \tag{1.26}$$

Initializing this iterative scheme at $k = n_0 + 1$, it is apparent that the limit

$$u := \lim_{k\to\infty} \tilde{u}_k$$

provides us with the desired solution of (1.23). Note that (1.24) implies

$$\underline{u} \leq \tilde{u}_k \leq \bar{u} \qquad \text{in } D_k$$

for all $k \geq n_0 + 1$ and, consequently,

$$\underline{u} \leq u \leq \bar{u} \qquad \text{in } D.$$

The proof is complete. $\square$

1.5   Some abstract pivotal results

The next result collects some important properties that are going to be used throughout this book. As in the previous section, D is a subdomain of Ω of class $\mathcal{C}^{2+\nu}$ satisfying (1.20). So, $a < 0$ in D in the sense that $a \leq 0$ but $a \neq 0$.

Lemma 1.6 *Suppose f satisfies* (Hf), *$b \in \mathcal{C}^{2+\nu}(\partial D)$, $b \geq 0$, and $\bar{u} \in \mathcal{C}^{2+\nu}(\bar{D})$, $\bar{u} > 0$, is a supersolution of* (1.21). *Then, $\bar{u} \gg 0$. In particular, any positive solution u of* (1.21) *satisfies $u \gg 0$. Moreover,*

$$\lambda_1[-d\Delta - \lambda - af(\cdot, \bar{u}), D] \geq 0$$

and, actually,

$$\lambda = \lambda_1[-d\Delta - af(\cdot, \bar{u}), D] \tag{1.27}$$

if, in addition, $b = 0$ and $\bar{u}$ solves (1.21). *Furthermore, $\kappa\bar{u}$ is a supersolution of* (1.21) *for all $\kappa \geq 1$.*

Proof: Since $\bar{u}|_{\partial D} \geq b \geq 0$ and

$$(-d\Delta - \lambda - af(\cdot, \bar{u})) \bar{u} \geq 0 \qquad \text{in } D,$$

$\bar{u} > 0$ is a positive supersolution of $-d\Delta - \lambda - af(\cdot, \bar{u})$ in D under Dirichlet boundary conditions. Two different situations may arise.

(i) If either $\bar{u} > 0$ on ∂D, or

$$-d\Delta\bar{u} - \lambda\bar{u} - af(\cdot, \bar{u})\bar{u} > 0 \qquad \text{in } D,$$

then, it follows from Theorem 1.1 that

$$\lambda_1[-d\Delta - \lambda - af(\cdot, \bar{u}), D] > 0 \tag{1.28}$$

and that $\bar{u} \gg 0$.

(ii) If $\bar{u} = 0$ on ∂D and

$$-d\Delta\bar{u} - \lambda\bar{u} - af(\cdot,\bar{u})\bar{u} = 0 \qquad \text{in } D,$$

then $\bar{u} > 0$ solves

$$\begin{cases} -d\Delta u = \lambda u + af(\cdot,u)u & \text{in } D, \\ u = 0 & \text{on } \partial D. \end{cases} \qquad (1.29)$$

Thus, (1.27) holds, because

$$(-d\Delta - af(\cdot,\bar{u}))\bar{u} = \lambda\bar{u},$$

and hence, λ must be the principal eigenvalue of $-d\Delta - af(\cdot,\bar{u})$. Moreover, since $\bar{u}$ is a principal eigenfunction, necessarily $\bar{u} \gg 0$. Indeed, let $\omega > 0$ sufficiently large so that $\lambda + \omega > 0$. Then,

$$(-d\Delta - af(\cdot,\bar{u}) + \omega)\bar{u} = (\lambda + \omega)\bar{u} > 0 \qquad \text{in } D$$

and $\bar{u} = 0$ on ∂D. Consequently, by Theorem 1.1(c), $\bar{u} \gg 0$.

Finally, pick $\kappa \geq 1$. Then,

$$b \leq \bar{u}|_{\partial D} \leq \kappa\bar{u}|_{\partial D},$$

whereas in D, we have

$$-d\kappa\Delta\bar{u} \geq \lambda\kappa\bar{u} + af(\cdot,\bar{u})\kappa\bar{u} \geq \lambda\kappa\bar{u} + af(\cdot,\kappa\bar{u})\kappa\bar{u},$$

because $a < 0$ in D and, owing to (Hf),

$$f(\cdot,\kappa\bar{u}) \geq f(\cdot,\bar{u}) \qquad \text{for all } \kappa \geq 1.$$

This concludes the proof. $\square$

The next theorem will simplify extraordinarily the mathematical analysis of this book.

Theorem 1.7 *Suppose f satisfies* (Hf), *$b \in C^{2+\nu}(\partial D)$ satisfies $b \geq 0$, (1.21) possesses a supersolution $\bar{u} > 0$ and, in addition, $\lambda > \lambda_1[-d\Delta, D]$ if $b = 0$. Then, (1.21) has a unique positive solution, which will be throughout denoted by $\theta_{[\lambda,D,b]}$. Moreover, the following properties are satisfied:*

(a) *For every positive subsolution (resp. supersolution) $\underline{u}$ (resp. $\bar{u}$) of (1.21),*

$$\underline{u} \leq \theta_{[\lambda,D,b]} \qquad (resp. \quad \theta_{[\lambda,D,b]} \leq \bar{u}).$$

(b) *For every $u_0 \in C(\bar{D})$, $u_0 > 0$,*

$$\lim_{t\uparrow\infty} \|u_{[\lambda,D,b]}(\cdot,t;u_0) - \theta_{[\lambda,D,b]}\|_{C(\bar{D})} = 0, \qquad (1.30)$$

where $u_{[\lambda,D,b]}(x,t;u_0)$ stands for the unique solution of

$$\begin{cases} \frac{\partial u}{\partial t} - d\Delta u = \lambda u + af(\cdot,u)u & \text{in } D \times (0,\infty), \\ u = b & \text{on } \partial D \times (0,\infty), \\ u(\cdot,0) = u_0 & \text{in } D. \end{cases} \qquad (1.31)$$

Furthermore, if f satisfies (Hf), *(1.21) admits a supersolution $\bar{u} > 0$, $b = 0$ and $\lambda \leq \lambda_1[-d\Delta, D]$, then (1.21) cannot admit a positive subsolution and*

$$\lim_{t \uparrow \infty} \|u_{[\lambda, D, b]}(\cdot, t; u_0)\|_{C(\bar{D})} = 0 \qquad \text{for all } u_0 > 0. \qquad (1.32)$$

In particular, (1.21) cannot possess a positive solution.

Proof: Suppose $b > 0$. Then, $\underline{u} := 0$ is a strict subsolution of (1.21) and hence, $(0, \bar{u})$ is an ordered sub-supersolution pair. Thus, by Theorem 1.2 and Remark 1.3, (1.21) possesses a solution, u, such that $0 < u \leq \bar{u}$. By Lemma 1.6, $u \gg 0$ and $\kappa \bar{u} \gg 0$ is a supersolution of (1.21) for all $\kappa \geq 1$.

Suppose $b = 0$ and $\lambda > \lambda_1[-d\Delta, D]$. Let $\varphi \gg 0$ be a principal eigenfunction associated with $\lambda_1[-d\Delta, D]$. Then, for sufficiently small $\varepsilon > 0$, $\underline{u} := \varepsilon\varphi$ is a positive strict subsolution of (1.21). Indeed, $\varepsilon\varphi|_{\partial D} = 0$ and

$$-d\Delta(\varepsilon\varphi) = \varepsilon\lambda_1[-d\Delta, D]\varphi < \lambda\varepsilon\varphi + af(\cdot, \varepsilon\varphi)\varepsilon\varphi \qquad \text{in} \quad D$$

for sufficiently small $\varepsilon > 0$, because $\lambda_1[-d\Delta, D] < \lambda$ and, thanks to (Hf),

$$\lim_{\varepsilon \downarrow 0} \|af(\cdot, \varepsilon\varphi)\|_{C(\bar{D})} = 0.$$

Fix one of these ε's and observe that $\underline{u} = \varepsilon\varphi \gg 0$. Since $\bar{u} \gg 0$, there exists $\kappa > 1$ such that $\varepsilon\varphi < \kappa\bar{u}$. Thus, $(\varepsilon\varphi, \kappa\bar{u})$ provides us with an ordered sub-supersolution pair of (1.21). Consequently, by Theorem 1.2 and Remark 1.3, (1.21) possesses a solution, u, such that

$$\varepsilon\varphi < u \leq \kappa\bar{u}.$$

Moreover, by Lemma 1.6, $u \gg 0$. This completes the proof of the existence of a positive solution for (1.21).

To prove the uniqueness we proceed by contradiction. Suppose (1.21) has two different positive solutions

$$u_1 \neq u_2.$$

Then, by Lemma 1.6, $u_1 \gg 0$, $u_2 \gg 0$ and, thanks to the previous analysis, there exist $\varepsilon > 0$, $\kappa > 1$ and a strict subsolution

$$\underline{u} \in \{0, \varepsilon\varphi\}$$

such that

$$\underline{u} < \min\{u_1, u_2\} < \max\{u_1, u_2\} \leq \kappa\bar{u}.$$

According to Theorem 1.2, the problem (1.21) possesses a minimal positive solution, u_*, and a maximal positive solution, u^*, in the order interval $[\underline{u}, \kappa\bar{u}]$. Necessarily,

$$\underline{u} < u_* \leq \min\{u_1, u_2\} < \max\{u_1, u_2\} \leq u^* \leq \kappa\bar{u}$$

and therefore, (1.21) admits two ordered positive solutions, $u_* < u^*$. By Lemma 1.6, $u^* \gg 0$ and $u_* \gg 0$. Subsequently, we set

$$w := u^* - u_* > 0.$$

By construction, $w|_{\partial D} = 0$ and

$$(-d\Delta - \lambda)w = af(\cdot, u^*)u^* - af(\cdot, u_*)u_* \quad \text{in } D.$$

Moreover, setting

$$\psi(t) := f(\cdot, tu^* + (1-t)u_*)(tu^* + (1-t)u_*), \qquad t \in [0,1],$$

we have that

$$f(\cdot, u^*)u^* - f(\cdot, u_*)u_* = \psi(1) - \psi(0) = \int_0^1 \frac{d\psi}{dt}(t)\, dt$$

$$= \int_0^1 \frac{\partial f}{\partial u}(\cdot, tu^* + (1-t)u_*)(tu^* + (1-t)u_*)\, dt\, w$$

$$+ \int_0^1 f(\cdot, tu^* + (1-t)u_*)\, dt\, w.$$

Consequently, w solves the problem

$$\begin{cases} (-d\Delta - \lambda + V)w = 0 & \text{in } D, \\ w = 0 & \text{on } \partial D, \end{cases} \tag{1.33}$$

where V is the potential defined by

$$V := -a \int_0^1 \frac{\partial f}{\partial u}(\cdot, tu^* + (1-t)u_*)(tu^* + (1-t)u_*)\, dt$$

$$- a \int_0^1 f(\cdot, tu^* + (1-t)u_*)\, dt. \tag{1.34}$$

Necessarily $w \gg 0$, as it is a principal eigenfunction associated to

$$\lambda_1[-d\Delta - \lambda + V, D] = 0.$$

Thus, since $u^* \gg u_* \gg 0$ and $-a > 0$ in D, we find from (Hf) that

$$V > -a \int_0^1 f(\cdot, tu^* + (1-t)u_*)\, dt > -af(\cdot, u_*).$$

Therefore, by the monotonicity of the principal eigenvalue with respect to the potential, we find from (1.27) that

$$\lambda_1[-d\Delta - \lambda + V, D] > \lambda_1[-d\Delta - \lambda - af(\cdot, u_*), D] = 0,$$

which is impossible. This contradiction ends the proof of the uniqueness. Throughout this book, we will denote by $\theta_{[\lambda,D,b]}$ the unique positive solution of (1.21).

Suppose $\underline{u} > 0$ is a subsolution of (1.21). Then, since $\bar{u} \gg 0$, for sufficiently large $\kappa > 1$ we have that $\underline{u} < \kappa\bar{u}$, and hence, by Lemma 1.6 and Theorem 1.2, (1.21) has a positive solution in the order interval $[\underline{u}, \kappa\bar{u}]$. By the uniqueness,

$$\underline{u} \leq \theta_{[\lambda,D,b]} \leq \kappa\bar{u}.$$

Now, suppose $b > 0$. Then, $\underline{u} = 0$ is a strict subsolution of (1.21) and, due to Theorem 1.2, (1.21) has a positive solution u in the order interval $[0, \bar{u}]$. By uniqueness,

$$\theta_{[\lambda,D,b]} \leq \bar{u}, \tag{1.35}$$

as claimed by Part (a). Similarly, when $b = 0$, the function $\underline{u} := \varepsilon\varphi$ is a strict subsolution of (1.21) for sufficiently small $\varepsilon > 0$. Thus, if ε is chosen so that $\varepsilon\varphi < \bar{u}$, by Theorem 1.2, (1.21) has a positive solution, u, in $[\varepsilon\varphi, \bar{u}]$. Therefore, again by uniqueness, (1.35) holds, which ends the proof of Part (a).

Next, we will prove (1.30). First, we assume that

$$b = 0 \qquad \text{and} \qquad \lambda > \lambda_1[-d\Delta, D].$$

Since $u_0 > 0$, by the parabolic maximum principle, we have that

$$u_{[\lambda,D,b]}(\cdot, t; u_0) \gg 0 \qquad \text{for all } t > 0.$$

Now, pick a sufficiently small $\varepsilon > 0$ and a sufficiently large $\kappa > 1$ such that $(\varepsilon\varphi, \kappa\bar{u})$ is an ordered sub-supersolution pair of (1.21) with

$$\varepsilon\varphi < u_{[\lambda,D,b]}(\cdot, 1; u_0) < \kappa\bar{u}.$$

Enlarging $\kappa > 1$, if necessary, we can also assume that

$$\kappa\bar{u} > \theta_{[\lambda,D,b]}.$$

By Lemma 1.6 and the uniqueness of the positive solution, $\kappa\bar{u}$ must be a positive strict supersolution of (1.21). Thanks again to the parabolic maximum principle, we find from the semigroup property that

$$\begin{aligned} u_{[\lambda,D,b]}(\cdot, t; \varepsilon\varphi) &\leq u_{[\lambda,D,b]}(\cdot, t; u_{[\lambda,D,b]}(\cdot, 1; u_0)) \\ &= u_{[\lambda,D,b]}(\cdot, t + 1; u_0) \leq u_{[\lambda,D,b]}(\cdot, t; \kappa\bar{u}) \end{aligned} \tag{1.36}$$

for all $t > 0$. Moreover, according to D. Sattinger [220], since $\varepsilon\varphi$ is a strict subsolution of (1.21), the solution $u_{[\lambda,D,b]}(\cdot, t; \varepsilon\varphi)$ increases approximating in $\mathcal{C}(\bar{D})$ the minimal positive solution of (1.21) in $[\varepsilon\varphi, \kappa\bar{u}]$ as $t \uparrow \infty$. Similarly, since $\kappa\bar{u}$ is a strict supersolution of (1.21), $u_{[\lambda,D,b]}(\cdot, t; \kappa\bar{u})$ decreases approximating in $\mathcal{C}(\bar{D})$ the maximal positive solution of (1.21) in $[\varepsilon\varphi, \kappa\bar{u}]$ as $t \uparrow \infty$.

Since $\theta_{[\lambda,D,b]}$ is the unique positive solution of (1.21), letting $t \uparrow \infty$ in (1.36) yields

$$\lim_{t \uparrow \infty} u_{[\lambda,D,b]}(\cdot,t;u_0) = \theta_{[\lambda,D,b]} \quad \text{in} \quad \mathcal{C}(\bar{D}),$$

which ends the proof of (1.30) when $b = 0$.

Now, suppose $b > 0$ and pick a sufficiently large $\kappa > 1$ such that

$$\kappa\bar{u} > \theta_{[\lambda,D,b]} \quad \text{and} \quad \underline{u} := 0 < u_{[\lambda,D,b]}(\cdot,1;u_0) < \kappa\bar{u}.$$

Then, arguing as above, we find that

$$u_{[\lambda,D,b]}(\cdot,t;0) \le u_{[\lambda,D,b]}(\cdot,t+1;u_0) \le u_{[\lambda,D,b]}(\cdot,t;\kappa\bar{u}) \tag{1.37}$$

for all $t > 0$. Thus, letting $t \uparrow \infty$ in (1.37), (1.30) holds.

Finally, suppose

$$b = 0 \quad \text{and} \quad \lambda \le \lambda_1[-d\Delta, D]. \tag{1.38}$$

To prove that under condition (1.38) the problem (1.21) cannot admit a positive subsolution and hence, cannot admit a positive solution neither, we proceed by contradiction. So, suppose (1.21) has a positive subsolution $\underline{u} > 0$. As $\bar{u} \gg 0$, there exists $\kappa > 1$ such that $\underline{u} < \kappa\bar{u}$ and therefore, by Theorem 1.2, (1.21) possesses a positive solution, u. Thanks to Lemma 1.6, $u \gg 0$ and

$$\lambda = \lambda_1[-d\Delta - af(\cdot,u), D].$$

Moreover, since $u \gg 0$ and $a < 0$ in D, we find from (Hf) that

$$-af(\cdot,u) > 0 \quad \text{in} \quad D$$

and hence, by the monotonicity of the principal eigenvalue with respect to the potential,

$$\lambda = \lambda_1[-d\Delta - af(\cdot,u), D] > \lambda_1[-d\Delta, D],$$

which contradicts (1.38). Therefore, (1.21) cannot admit a positive subsolution under condition (1.38).

To complete the proof, it remains to show that (1.38) implies (1.32). Let $\kappa > 1$ be such that

$$0 < u_{[\lambda,D,b]}(\cdot,1;u_0) < \kappa\bar{u}.$$

Then, for every $t > 0$,

$$0 < u_{[\lambda,D,b]}(\cdot,t+1;u_0) < u_{[\lambda,D,b]}(\cdot,t;\kappa\bar{u}). \tag{1.39}$$

Since $\kappa\bar{u}$ is a supersolution of (1.21), $u_{[\lambda,D,b]}(\cdot,t;\kappa\bar{u})$ decreases approximating the maximal non-negative solution of (1.21) in $[0,\kappa\bar{u}]$ as $t \uparrow \infty$. As 0 is the maximal non-negative solution of (1.21) in $[0,\kappa\bar{u}]$, letting $t \uparrow \infty$ in (1.39) shows (1.32) and completes the proof. □

The following strong comparison holds from Theorem 1.7.

Lemma 1.8 *Suppose f satisfies* (Hf), *$b \in C^{2+\nu}(\partial D)$ satisfies $b \geq 0$, (1.21) possesses a supersolution $\bar{u} > 0$ and $\lambda > \lambda_1[-d\Delta, D]$ if $b = 0$. Then, for every positive strict subsolution (resp. supersolution) $\underline{u}$ (resp. $\bar{u}$) of (1.21), the next estimate holds*

$$\underline{u} \ll \theta_{[\lambda,D,b]} \qquad (\text{resp. } \theta_{[\lambda,D,b]} \ll \bar{u}),$$

where $\theta_{[\lambda,D,b]}$ is the unique positive solution of (1.21).

Proof: Suppose $\underline{u} > 0$ is a strict subsolution of (1.21). Then, by Theorem 1.7(a), $\underline{u} \leq \theta_{[\lambda,D,b]}$. Therefore,

$$\underline{u} < \theta_{[\lambda,D,b]},$$

because $\underline{u}$ is not a solution of (1.21). Consequently,

$$w := \theta_{[\lambda,D,b]} - \underline{u} > 0. \tag{1.40}$$

On the other hand, by adapting the uniqueness argument of the proof of Theorem 1.7, it becomes apparent that

$$\begin{cases} (-d\Delta - \lambda + V)w = 0 & \text{in } D, \\ w \geq 0 & \text{on } \partial D, \end{cases} \tag{1.41}$$

where V is the potential defined by

$$V := -a \int_0^1 \frac{\partial f}{\partial u}(\cdot, t\theta_{[\lambda,D,b]} + (1-t)\underline{u})(t\theta_{[\lambda,D,b]} + (1-t)\underline{u})\, dt$$

$$- a \int_0^1 f(\cdot, t\theta_{[\lambda,D,b]} + (1-t)\underline{u})\, dt.$$

Subsequently, we distinguish two different situations. If $w|_{\partial D} > 0$, then $w > 0$ is a positive strict supersolution of $-d\Delta - \lambda + V$ in D under homogeneous Dirichlet boundary conditions. Consequently, by Theorem 1.1, $w \gg 0$ in D and therefore,

$$\underline{u} \ll \theta_{[\lambda,D,b]}.$$

If $w|_{\partial D} = 0$, then $w > 0$ provides us with a principal eigenfunction of

$$\lambda_1[-d\Delta - \lambda + V, D] = 0$$

and hence, $w \gg 0$. Indeed, since

$$\begin{cases} (-d\Delta - \lambda + V + 1)w = w > 0 & \text{in } D, \\ w = 0 & \text{on } \partial D, \end{cases}$$

and

$$\lambda_1[-d\Delta - \lambda + V + 1, D] = 1 > 0,$$

it follows from Theorem 1.1 that $w \gg 0$. This ends the proof. □

Finally, we conclude this section by providing an extremely useful property.

Lemma 1.9 *Suppose f satisfies* (Hf), *D is a subdomain of Ω of class $C^{2+\nu}$ such that $a < 0$ in D, $V \in L^\infty(D)$, and the problem*

$$\begin{cases} (-d\Delta + V)u = \lambda u + af(\cdot, u)u & in \ D, \\ u = 0 & on \ \partial D, \end{cases} \tag{1.42}$$

possesses a subsolution, $\underline{u}$, such that $\underline{u} = 0$ on ∂D and $\underline{u} \gg 0$. Then,

$$\lambda > \lambda_1[-d\Delta + V, D]. \tag{1.43}$$

Proof: Let $\varphi \gg 0$ be an eigenfunction associated with $\lambda_1[-d\Delta + V, D]$ and choose a sufficiently large constant $\kappa > 0$ such that $\kappa\varphi > \underline{u}$. Then,

$$\begin{aligned} (-d\Delta + V)(\kappa\varphi - \underline{u}) &= \kappa\lambda_1[-d\Delta + V, D]\varphi - (-d\Delta + V)\underline{u} \\ &\geq \kappa\lambda_1[-d\Delta + V, D]\varphi - \lambda\underline{u} - af(\cdot, \underline{u})\underline{u} \\ &> \kappa\lambda_1[-d\Delta + V, D]\varphi - \lambda\underline{u} \\ &= \lambda_1[-d\Delta + V, D](\kappa\varphi - \underline{u}) + (\lambda_1[-d\Delta + V, D] - \lambda)\underline{u}. \end{aligned}$$

Consequently,

$$(-d\Delta + V - \lambda_1[-d\Delta + V, D])(\kappa\varphi - \underline{u}) > (\lambda_1[-d\Delta + V, D] - \lambda)\underline{u}. \tag{1.44}$$

Suppose (1.43) fails. Then, it follows from (1.44) that $\kappa\varphi - \underline{u}$ is a positive strict supersolution of

$$\mathfrak{L} := -d\Delta + V - \lambda_1[-d\Delta + V, D].$$

Hence, according to Theorem 1.1, $\lambda_1[\mathfrak{L}, D] > 0$. But

$$\lambda_1[\mathfrak{L}, D] = \lambda_1[-d\Delta + V - \lambda_1[-d\Delta + V, D], D]$$

$$= \lambda_1[-d\Delta + V, D] - \lambda_1[-d\Delta + V, D] = 0,$$

which is a contradiction. This ends the proof. $\quad\square$

1.6   Logistic equation in population dynamics

Let us denote by $p(t)$ the number of individuals of a population at time $t \geq 0$ and suppose the instantaneous rate of variation of the population is proportional to the total population $p(t)$ through some constant $m \in \mathbb{R}$ measuring either the growth of the population, if $m > 0$, or its decline, if $m < 0$. Then, the equation governing the evolution of the population is

$$p'(t) = mp(t). \tag{1.45}$$

Thus, if the initial population equals $p_0 := p(0) > 0$, then

$$p(t) = e^{mt}p_0 \qquad \text{for all } t \geq 0.$$

Hence, the time the population needs to double itself is given by the value of T for which $e^{mT} = 2$. In other words,

$$T = \frac{\log 2}{m}.$$

According to the celebrated essay on the *Principle of Population* of the political economist and Church of England priest, Thomas R. Malthus [187],

"In the United States of America, where the means of subsistence have been more ample, the manners of the people more pure, and consequently the checks to early marriages fewer, than in any of the modern states of Europe, the population has been found to double itself in twenty-five years. This ratio of increase, though short of the utmost power of population, yet as the result of actual experience, we will take as our rule, and say, that population, when unchecked, goes on doubling itself every twenty-five years or increases in a geometrical ratio."

It becomes apparent that, around 1798, the neat birth rate of the population of the United States of America was

$$m = \frac{\log 2}{25}.$$

Actually, Th. R. Malthus [187] established a more general principle:

"Assuming then my postulata as granted, I say, that the power of population is indefinitely greater than the power in the earth to produce subsistence for man. Population, when unchecked, increases in a geometrical ratio. Subsistence increases only in an arithmetical ratio. A slight acquaintance with numbers will shew the immensity of the first power in comparison of the second."

It was 36 years later, in 1835, when the astronomer and statistician A. Quetelet [212] went back to the essay of Th. R. Malthus in his extremely pioneering *Essay on Social Physics*, to complete it with the following, rather independent, general principle:

"The resistance, or the sum of the obstacles to the development of a population, is like the square of the speed of variation of the population."

Nevertheless, neither the moral father of the population dynamics nor the father of the scientific sociology translated his laws into mathematical terms. Instead, it was the mathematician P. F. Verhulst [228], who, inspired by the these pioneering ideas, added (translated by J. L. Mawhin [192]):

"I have tried since a long time to determine, with the help of analysis, the credible law of the population; but I have abandoned this type of research, because the observation data are too scarce to allow the verification of the formulas. [...] However, as the methodology I have followed seems to lead necessarily to the knowledge of the true law, [...] I have thought that I have to comply the invitation of Mr. Quatelet of making it public."

Then, denoting by $p(t)$ the size at time t of the population, P. F. Verhulst translated mathematically the Malthus law as (1.45) for some constant m, and, crucially, observed that

"As the speed of the increase of the population is diminished by its very increase, we must subtract from mp an unknown function $\varphi(p)$ of p. [...] The simplest hypothesis [...] consists in taking $\varphi(p) = np^2$."

This led him to the famous differential equation

$$p'(t) = mp(t) - np^2(t), \qquad (1.46)$$

often known as the *logistic equation*, undoubtedly, the most paradigmatic one of population dynamics. But, after P. F. Verhulst died, his contributions remained in almost complete oblivion for almost 80 years until the biologists R. Pearl and L. L. Reed [207] rediscovered them in the same demographical context and compared them with a series of field data. Since then, (1.46) is also known as *Verhulst equation*, or *Verhulst-Pearl equation*.

Incidentally, perhaps by the lack of any mathematical background, the essay of Th. R. Malthus [187] had a very strong influence on the work of Ch. Darwin [69], who recognized that in October 1838 he had read it just for entertainment. As he was very well prepared to appreciate the struggles for life of animals and plants, he realized that the favorable variations should be maintained, while the unfavorable ones should be destroyed; the final result being the formation of new species. The deep influence of the celebrated essay by Th. R. Malthus [187] on Ch. Darwin [69] can be easily documented by simply reading the famous essay of Ch. Darwin [69] on *The Origin of the Species by Means of Natural Selection*, where the genius recognized that

"A struggle for existence inevitably follows from the high rate at which all organic beings tend to increase. Every being, which during its natural lifetime produces several eggs or seeds, must suffer destruction during some period of its life, and during some season or occasional year, otherwise, on the principle of geometrical increase, its numbers would quickly become so inordinately great that no country could support the product. Hence, as more individuals are produced than can possibly survive, there must in every case be a struggle for existence, either one individual with another of the same species, or with the individuals of distinct species, or with the physical conditions of life. It is the doctrine of Malthus applied with manifold force to the whole animal and vegetable kingdoms; for in this case there can be no artificial increase of food, and no prudential restraint from marriage. Although some species may be now increasing, more or less rapidly, in numbers, all cannot do so, for the world would not hold them."

Seventeen years after the relevance of the work of P. F. Verhulst [228] was recognized by R. Pearl and L. L. Reed [207], in 1937, the reaction diffusion equation

$$\frac{\partial p}{\partial t} - d\Delta p = mp - np^2 \tag{1.47}$$

with $d > 0$, was introduced by A. N. Kolmogorov, I. G. Petrovsky and N. S. Piskunov [124], and independently by R. A. Fisher [88], to study some problems of biological nature.

While in the non-spatial model (1.46), the evolution of the species does not depend on the spatial location of the individuals in the inhabiting region Ω in the spatial model (1.47) the individuals are assumed to disperse randomly in Ω, like the particles of an ideal gas governed by a Brownian motion. So, each individual moves around with no preference for a particular direction.

Rather astonishingly, random motion provokes a regular migration from highly populated areas to less populated areas in complete agreement with the heat transport law of the mathematician J. B. J. Fourier [89] and the mass transfer laws governing most of chemical diffusion processes of the physiologist A. Fick [87].

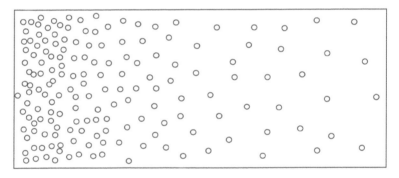

FIGURE 1.5: Brownian motion.

The influential essay of E. Shrödinger [223], *What is Life? The Physical Aspect of the Living Cell*, contains a very illuminating introduction to the role played by the spatial dispersion in population dynamics, as well some rather pioneering ideas which strongly influenced the evolution of biological sciences during the twentieth century.

1.7   Comments on Chapter 1

A function $u : \Omega \to \mathbb{R}$ is said to be (globally) Hölder continuous in Ω, with exponent $\nu \in (0, 1]$, if there exists a constant $C \geq 0$ such that

$$|u(x) - u(y)| \leq C|x - y|^{\nu} \qquad \text{for all} \ \ x, y \in \Omega.$$

Obviously, u is Hölder continuous with exponent $\nu = 1$ if, and only if, u is (globally) Lipschitz continuous. Throughout this book, for any given $\nu \in (0,1]$, we denote by $\mathcal{C}^{\nu}(\bar{\Omega})$ the Banach space of all continuous functions $u : \bar{\Omega} \to \mathbb{R}$ that are Hölder continuous in Ω with exponent ν, endowed with the norm

$$\|u\|_{\mathcal{C}^{\nu}(\bar{\Omega})} := \|u\|_{\mathcal{C}(\bar{\Omega})} + \sup_{\substack{x,y\in\bar{\Omega} \\ x\neq y}} \frac{|u(x) - u(y)|}{|x-y|^{\nu}}, \qquad u \in \mathcal{C}^{\nu}(\bar{\Omega}),$$

where

$$\|u\|_{\mathcal{C}(\bar{\Omega})} := \max_{x\in\bar{\Omega}} \|u(x)\|.$$

More generally, for every integer $k \geq 1$ and $\nu \in (0,1]$, we denote by $\mathcal{C}^{k+\nu}(\bar{\Omega})$ the Banach subspace of $\mathcal{C}^{k}(\bar{\Omega})$ consisting of all functions $u \in \mathcal{C}^{k}(\bar{\Omega})$ such that $D^{\alpha}u \in \mathcal{C}^{\nu}(\bar{\Omega})$ for all multi-index $\alpha = (\alpha_1, ..., \alpha_N) \in \mathbb{N}^N$ with

$$|\alpha| = \alpha_1 + \alpha_2 + \cdots + \alpha_N,$$

where we are denoting

$$D^{\alpha}u = \frac{\partial^{|\alpha|}u}{\partial^{\alpha_1}x_1 \cdots \partial^{\alpha_N}x_N}.$$

Naturally, the Banach space $\mathcal{C}^{k+\nu}(\bar{\Omega})$ is equipped with the norm

$$\|u\|_{\mathcal{C}^{k+\nu}(\bar{\Omega})} := \|u\|_{\mathcal{C}^{k}(\bar{\Omega})} + \sum_{|\alpha|=k} \sup_{\substack{x,y\in\bar{\Omega} \\ x\neq y}} \frac{|D^{\alpha}u(x) - D^{\alpha}u(y)|}{|x-y|^{\nu}}, \qquad u \in \mathcal{C}^{k+\nu}(\bar{\Omega}),$$

where

$$\|u\|_{\mathcal{C}^{k}(\bar{\Omega})} = \sum_{|\alpha|\leq k} \|D^{\alpha}u\|_{\mathcal{C}(\bar{\Omega})}.$$

Similarly, for a given function $f : \bar{\Omega} \times [0,\infty) \to \mathbb{R}$, two integers $k_1, k_2 \geq 0$ and $\nu_1, \nu_2 \in (0,1)$, it is said that $f \in \mathcal{C}^{k_1+\nu_1, k_2+\nu_2}(\bar{\Omega} \times [0,\infty))$ if $f(\cdot, u) \in \mathcal{C}^{k_1+\nu_1}(\bar{\Omega})$ for all $u \in [0,\infty)$ and $f(x, \cdot) \in \mathcal{C}^{k_2+\nu_2}[0, M]$ for all $M > 0$ and $x \in \bar{\Omega}$.

In the one-dimensional problem, $N = 1$, one does not need to use the spaces of Hölder continuous functions for the validity of the results. Indeed, in such case it suffices to impose

$$a \in \mathcal{C}(\bar{\Omega}), \qquad f \in \mathcal{C}^1(\bar{\Omega} \times [0,\infty)),$$

to get *classical* solutions of class $\mathcal{C}^2$ everywhere.

Most of the results of this chapter are valid for a general linear second order elliptic operator of the form

$$\mathfrak{L} = \operatorname{div}\left(A(x)\nabla\cdot\right) + \langle b, \nabla\cdot\rangle + c \tag{1.48}$$

where

$$A(x) = (a_{ij}(x))_{1\leq i,j\leq N}$$

is a symmetric matrix with

$$a_{ij} = a_{ji} \in \mathcal{C}^{1+\nu}(\bar{\Omega}), \quad 1 \le i, j \le N,$$

$$b = (b_j)_{1 \le j \le N} \quad \text{with} \quad b_j \in \mathcal{C}^{\nu}(\bar{\Omega}),$$

and $c \in \mathcal{C}^{\nu}(\bar{\Omega})$. Moreover, instead of Dirichlet boundary conditions, given two open and closed pieces of the boundary, Γ_0 and Γ_1, of class $\mathcal{C}^{2+\nu}$ with $\partial\Omega = \Gamma_0 \cup \Gamma_1$, one can work with a general boundary operator

$$\mathfrak{B} : \mathcal{C}^{2+\nu}(\bar{\Omega}) \to \mathcal{C}^{1+\nu}(\partial\Omega)$$

of mixed type

$$\mathfrak{B}\psi := \begin{cases} \psi & \text{on } \Gamma_0, \\ \partial_\nu\psi + \beta\psi & \text{on } \Gamma_1, \end{cases} \quad \psi \in \mathcal{C}^{2+\nu}(\bar{\Omega}), \quad (1.49)$$

where $\nu = An$ is the co-normal vector field and $\beta \in \mathcal{C}^{1+\nu}(\Gamma_1)$.

Actually, it is far from necessary to work within the setting of the Schauder theory neither. Indeed, one can also work in a general domain Ω of class $\mathcal{C}^2$ with a general second order elliptic operators of the form (1.48) such that

$$a_{ij} = a_{ji} \in W^{1,\infty}(\Omega), \quad b_j, c \in L^\infty(\Omega), \quad 1 \le i, j \le N,$$

as those introduced in Chapter 4 of [163]. Similarly, instead of Dirichlet boundary conditions, one can also consider a boundary operator $\mathfrak{B}$ like those introduced on page 92 of [163]. Naturally, under these general conditions the solutions will be strong, in the sense that they satisfy

$$u \in \bigcap_{p > N} W_0^{2,p}(\Omega),$$

with all the pertinent implications of this feature (see [163, Ch. 5]).

This chapter has been elaborated from Section 3 of [160].

Chapter 2

Classical diffusive logistic equation

This chapter studies the dynamics of

$$\begin{cases} \frac{\partial u}{\partial t} - d\Delta u = \lambda u + a(x)f(x,u)u & \text{in } D \times (0, \infty), \\ u = M & \text{on } \partial D \times (0, \infty), \\ u(\cdot, 0) = u_0 > 0 & \text{in } D, \end{cases} \qquad (2.1)$$

where $M \in [0, \infty)$ is a constant, D is a subdomain of Ω of class $\mathcal{C}^{2+\nu}$ such that $D \subset \Omega_-$, and f satisfies (Hf) and (Hg). In particular,

$$a(x) < 0 \qquad \text{for all } x \in D.$$

Under these conditions, (2.1) is a classical (diffusive) logistic problem, however $a(x)$ might vanish on some piece of ∂D. The problem (2.1) is said to be *unperturbed* if $M = 0$, and *perturbed* (from $M = 0$) if $M > 0$.

The main result of this chapter establishes that the dynamic of (2.1) is governed by its maximal non-negative steady state, i.e., by the maximal non-negative solution of the semilinear elliptic problem

$$\begin{cases} -d\Delta u = \lambda u + a(x)f(x,u)u & \text{in } D, \\ u = M & \text{on } \partial D. \end{cases} \qquad (2.2)$$

In particular, it will be shown that, for every $M > 0$, (2.2) possesses a unique positive solution, which is a global attractor for the solutions of (2.1). Throughout this book, to be consistent with the notation introduced in Theorem 1.7, this solution will be denoted by $\theta_{[\lambda, D, M]}$. One of the main results of this chapter establishes that

$$\lim_{M \downarrow 0} \theta_{[\lambda, D, M]} = \begin{cases} 0 & \text{if } \lambda \leq \lambda_1[-d\Delta, D], \\ \theta_{[\lambda, D]} & \text{if } \lambda > \lambda_1[-d\Delta, D], \end{cases}$$

where

$$\theta_{[\lambda, D]} := \theta_{[\lambda, D, 0]}$$

stands for the unique positive solution of

$$\begin{cases} -d\Delta u = \lambda u + a(x)f(x,u)u & \text{in } D, \\ u = 0 & \text{on } \partial D. \end{cases} \tag{2.3}$$

The problem (2.3) admits a (unique) positive solution if, and only if,

$$\lambda > \lambda_1[-d\Delta, D].$$

Moreover, the maximal non-negative solution of (2.3) is a global attractor for (2.1) if $M = 0$. As a byproduct, there is a continuous transition between the dynamics of (2.1) as $M > 0$ perturbs from $M = 0$.

The analysis of (2.1) in the special case when $\bar{D} \subset \Omega_-$, i.e., when

$$a(x) < 0 \qquad \text{for all } x \in \bar{D},$$

is substantially easier than the analysis of the general case when $D \subset \Omega_-$, because sufficiently large positive constants are supersolutions of (2.2) if $\bar{D} \subset \Omega_-$, while no positive constant can be a supersolution of (2.2) if $\lambda > 0$ and $a(x)$ vanishes somewhere on ∂D.

This chapter has been distributed in three sections. Section 2.1 studies the unperturbed problem, Section 2.2 studies the perturbed problem, and Section 2.3 establishes the structural stability of (2.1) as $M > 0$ perturbs from $M = 0$.

The following corollary of Lemma 1.6 is very useful.

Lemma 2.1 *Suppose $u \in \mathcal{C}^{2+\nu}(\bar{D})$, $u > 0$, is a solution of (2.2) for some $M \in [0, \infty)$. Then, $u \gg 0$ and*

$$\lambda \leq \lambda_1[-d\Delta - af(\cdot, u), D].$$

If, in addition, $M = 0$, then,

$$\lambda = \lambda_1[-d\Delta - af(\cdot, u), D]. \tag{2.4}$$

2.1 Unperturbed logistic problem

Throughout this section we assume that

$$M = 0.$$

The following result establishes the existence and the uniqueness of the positive solution of (2.3).

Theorem 2.2 *Suppose $M = 0$ and f satisfies* (Hf) *and* (Hg). *Then,* (2.3) *possesses a positive solution if, and only if,*

$$\lambda > \lambda_1[-d\Delta, D].$$

Moreover, it is unique if it exists, and if we denote it by $\theta_{[\lambda,D]}$, then $\theta_{[\lambda,D]} \gg 0$ and

$$\lim_{t\uparrow\infty} u_{[\lambda,D]}(\cdot, t; u_0) = \begin{cases} 0 & \text{if } \lambda \leq \lambda_1[-d\Delta, D], \\ \theta_{[\lambda,D]} & \text{if } \lambda > \lambda_1[-d\Delta, D], \end{cases} \quad \text{in } \mathcal{C}(\bar{D}),$$

where $u_{[\lambda,D]}$ stands for the unique solution of (2.1).

Proof: The existence and the uniqueness of $u_{[\lambda,D]}$ can be derived, e.g., from D. Daners and P. Koch-Medina [68]. According to Theorem 1.7, to prove this theorem it suffices to construct a positive supersolution of (2.3) for every $\lambda > \lambda_1[-d\Delta, D]$. By (Hg), in the special case when

$$\bar{D} \subset \Omega_-, \tag{2.5}$$

sufficiently large positive constants are positive supersolutions of (2.3). Consequently, for the rest of this proof we will assume that

$$D \subset \Omega_-$$

and fix $\lambda > \lambda_1[-d\Delta, D]$. As ∂D is of class $\mathcal{C}^{2+\nu}$, it possesses a finite number of components, say $n \geq 1$. Let $\Gamma_{D,j}$, $1 \leq j \leq n$, denote them, and, for sufficiently small $\varepsilon > 0$ and $j \in \{1, ..., n\}$, consider the open subset of D defined by

$$D_{\varepsilon,j} := \{x \in D \ : \ \text{dist}(x, \Gamma_{D,j}) < \varepsilon\}.$$

For sufficiently small $\varepsilon > 0$, we have that

$$\bar{D}_{\varepsilon,i} \cap \bar{D}_{\varepsilon,j} = \emptyset \quad \text{if } i \neq j, \qquad \bigcup_{j=1}^{n} D_{\varepsilon,j} \subset D.$$

Moreover, the open set $D_{\varepsilon,j}$ also is of class $\mathcal{C}^{2+\nu}$ for all $j \in \{1, ..., n\}$ and, since

$$\lim_{\varepsilon\downarrow 0} |D_{\varepsilon,j}| = 0, \qquad 1 \leq j \leq n,$$

we already know, by the analysis carried out at the beginning of Chapter 1, that

$$\lim_{\varepsilon\downarrow 0} \lambda_1[-d\Delta, D_{\varepsilon,j}] = \infty, \qquad 1 \leq j \leq n.$$

Thus, ε can be shortened, if necessary, so that

$$\min_{1\leq j\leq n} \lambda_1[-d\Delta, D_{\varepsilon,j}] > \lambda. \tag{2.6}$$

For every $1 \leq j \leq n$, let $\varphi_{\varepsilon,j} \gg 0$ be a principal eigenfunction associated to $\lambda_1[-d\Delta, D_{\varepsilon,j}]$. As $\varphi_{\varepsilon,j}(x) > 0$ for all $x \in D_{\varepsilon,j}$, it is apparent that

$$\min_{1 \leq j \leq n} \min_{D \cap \partial D_{\varepsilon/2,j}} \varphi_{\varepsilon,j} > 0. \tag{2.7}$$

Subsequently, we consider the auxiliary function Φ defined through

$$\Phi := \begin{cases} \varphi_{\varepsilon,j} & \text{in } \bar{D}_{\varepsilon/2,j}, \quad 1 \leq j \leq n, \\ \psi_{\varepsilon} & \text{in } D_{\text{int}} := D \setminus \bigcup_{j=1}^{n} \bar{D}_{\varepsilon/2,j}, \end{cases} \tag{2.8}$$

where ψ_{ε} is any $\mathcal{C}^{2+\nu}$–extension of the function

$$\varphi_{\varepsilon,1} \otimes \cdots \otimes \varphi_{\varepsilon,n}$$

to the open set

$$D_{\text{int}} = \{x \in D \ : \ \text{dist}\,(x, \partial D) > \varepsilon/2\} \Subset D$$

with the special requirement that

$$\inf_{D_{\text{int}}} \psi_{\varepsilon} > 0. \tag{2.9}$$

Such a function ψ_{ε} exists because of (2.7).

We claim that $\kappa\Phi$ provides us with a supersolution of (2.3) for sufficiently large $\kappa > 1$. Indeed, by construction,

$$\Phi = 0 \quad \text{on} \quad \partial D = \bigcup_{j=1}^{n} \Gamma_{D,j} \tag{2.10}$$

and hence, $\kappa\Phi = 0$ on ∂D for all $\kappa > 0$. Moreover, for every $j \in \{1, ..., n\}$ and $\kappa > 0$, we find from (2.6) that, in $D_{\varepsilon/2,j}$,

$$-d\Delta(\kappa\Phi) = -\kappa d\Delta\varphi_{\varepsilon,j} = \kappa\lambda_1[-d\Delta, D_{\varepsilon,j}]\varphi_{\varepsilon,j}$$
$$> \lambda\kappa\varphi_{\varepsilon,j} = \lambda\kappa\Phi \geq \lambda\kappa\Phi + af(\cdot, \kappa\Phi)\kappa\Phi,$$

because

$$af(\cdot, \kappa\Phi)\kappa\Phi \leq 0.$$

Finally, in D_{int}, we have that

$$-d\Delta(\kappa\Phi) = -\kappa d\Delta\psi_{\varepsilon} \geq \lambda\kappa\psi_{\varepsilon} + af(\cdot, \kappa\psi_{\varepsilon})\kappa\psi_{\varepsilon} = \lambda\kappa\Phi + af(\cdot, \kappa\Phi)\kappa\Phi$$

for sufficiently large $\kappa > 1$, because $a(x)$ is negative and bounded away from zero in D_{int} and, owing to (2.9),

$$\lim_{\kappa\uparrow\infty} f(\cdot, \kappa\psi_{\varepsilon}) = \infty \quad \text{uniformly in} \quad D_{\text{int}}.$$

Therefore, for sufficiently large $\kappa > 1$,

$$af(\cdot, \kappa\psi_\varepsilon) \leq \frac{-d\Delta\psi_\varepsilon}{\psi_\varepsilon} - \lambda \qquad \text{in } D_{\text{int}}.$$

By Theorem 1.7, the proof is complete. $\quad\square$

2.2 Solution set for the unperturbed problem

The next result establishes that the set of positive solutions (λ, u) of (2.3) consists of a differentiable curve emanating from $u = 0$ at $\lambda = \lambda_1[-d\Delta, D]$, where the former attractive character of $u = 0$ as a steady-state solution of (2.1) is lost.

Theorem 2.3 *Suppose $M = 0$ and f satisfies (Hf) and (Hg). Then, the solution operator*

$$\begin{array}{ccc} (\lambda_1[-d\Delta, D], \infty) & \xrightarrow{\theta} & \mathcal{C}(\bar{D}) \\ \lambda & \mapsto & \theta(\lambda) := \theta_{[\lambda, D]} \end{array} \tag{2.11}$$

is of class $\mathcal{C}^1$ and strongly increasing, in the sense that

$$\theta(\lambda) \gg \theta(\mu) \qquad \text{if } \lambda > \mu > \lambda_1[-d\Delta, D].$$

Moreover, $\theta(\lambda)$ bifurcates from $(\lambda, u) = (\lambda, 0)$ at $\lambda = \lambda_1[-d\Delta, D]$, i.e.,

$$\lim_{\lambda \downarrow \lambda_1[-d\Delta, D]} \theta(\lambda) = 0 \qquad \text{in } \mathcal{C}(\bar{D}). \tag{2.12}$$

Proof: Subsequently, we denote by $\mathcal{C}_0(\bar{D})$ the closed subspace of $\mathcal{C}(\bar{D})$ formed by all continuous functions u in $\bar{D}$ with $u = 0$ on ∂D. Then, the solutions of (2.3) are the zeroes of the nonlinear operator

$$\mathfrak{F} : \mathbb{R} \times \mathcal{C}_0(\bar{D}) \longrightarrow \mathcal{C}_0(\bar{D})$$

defined by

$$\mathfrak{F}(\lambda, u) := u - (-d\Delta)^{-1} [\lambda u + af(\cdot, u)u],$$

where $(-d\Delta)^{-1}$ stands for the resolvent operator of $-d\Delta$ in D under homogeneous Dirichlet boundary conditions. In other words, if $G(x, y)$ stands for the Green function of $-d\Delta$ in D, [110], whose existence is guaranteed by the Perron theorem, [208], then

$$\mathfrak{F}(\lambda, u) := u - \int_D G(\cdot, y)[\lambda u(y) + a(y)f(y, u(y))u(y)] \, dy$$

for all $u \in \mathcal{C}_0(\bar{D})$.

The operator $\mathfrak{F}$ is of class $\mathcal{C}^1$ and, by elliptic regularity, $\mathfrak{F}(\lambda, \cdot)$ is a nonlinear compact perturbation of the identity map for every $\lambda \in \mathbb{R}$. Moreover,

$$\mathfrak{F}(\lambda, 0) = 0 \qquad \text{for all } \lambda \in \mathbb{R}$$

and

$$D_u \mathfrak{F}(\lambda, 0)u = u - \lambda(-d\Delta)^{-1}u \qquad \text{for all } (\lambda, u) \in \mathbb{R} \times \mathcal{C}_0(\bar{D}).$$

Thus, $D_u \mathfrak{F}(\lambda, 0)$ is a Fredholm analytic pencil of index zero whose spectrum consists of the eigenvalues of $-d\Delta$ in D under homogeneous Dirichlet boundary conditions. In particular,

$$N[D_u \mathfrak{F}(\lambda_1[-d\Delta, D], 0)] = \text{span}[\varphi],$$

where $\varphi \gg 0$ is any principal eigenfunction of $\lambda_1[-d\Delta, D]$. We claim that

$$D_\lambda D_u \mathfrak{F}(\lambda_1[-d\Delta, D], 0)\varphi \notin R[D_u \mathfrak{F}(\lambda_1[-d\Delta, D], 0)]. \tag{2.13}$$

Consequently, the transversality condition of M. G. Crandall and P. H. Rabinowitz [59] holds. The proof of (2.13) proceeds by contradiction. Suppose

$$D_\lambda D_u \mathfrak{F}(\lambda_1[-d\Delta, D], 0)\varphi = -(-d\Delta)^{-1}\varphi \in R[D_u \mathfrak{F}(\lambda_1[-d\Delta, D], 0)].$$

Then, there exists $u \in \mathcal{C}_0(\bar{D})$ such that

$$u - \lambda_1[-d\Delta, D](-d\Delta)^{-1}u = -(-d\Delta)^{-1}\varphi.$$

By elliptic regularity, $u \in \mathcal{C}_0^{2+\nu}(\bar{D})$ and

$$(-d\Delta - \lambda_1[-d\Delta, D])u = -\varphi.$$

Multiplying this equation by φ, integrating in D and applying the formula of integration by parts gives

$$\int_D \varphi^2 = 0,$$

which is impossible. This contradiction shows (2.13). Therefore, by the theorem of M. G. Crandall and P. H. Rabinowitz [59], $(\lambda, u) = (\lambda_1[-d\Delta, D], 0)$ is a bifurcation point from $(\lambda, u) = (\lambda, 0)$ to a $\mathcal{C}^1$-curve of positive solutions of (2.3). By the uniqueness of the positive solution, already established by Theorem 2.2, (2.12) holds.

Now, let $(\lambda, u) = (\lambda_0, u_0)$ be a positive solution of (2.3). Then,

$$\mathfrak{F}(\lambda_0, u_0) = 0$$

and, according to Lemma 2.1, $u_0 \gg 0$ and

$$\lambda_0 = \lambda_1[-d\Delta - af(\cdot, u_0), D]. \tag{2.14}$$

Differentiating $\mathfrak{F}$ with respect to u,

$$D_u\mathfrak{F}(\lambda_0, u_0)u = u - (-d\Delta)^{-1}\left[\lambda_0 u + a\frac{\partial f}{\partial u}(\cdot, u_0)u_0 u + af(\cdot, u_0)u\right]$$

for all $u \in \mathcal{C}_0(\bar{D})$. Hence, $D_u\mathfrak{F}(\lambda_0, u_0)$ is a Fredholm operator of index zero. Moreover, as it is injective, it is a linear topological isomorphism. Indeed, if

$$u - (-d\Delta)^{-1}\left[\lambda_0 u + a\frac{\partial f}{\partial u}(\cdot, u_0)u_0 u + af(\cdot, u_0)u\right] = 0$$

for some $u \in \mathcal{C}_0(\bar{D})$, then, by elliptic regularity, $u \in \mathcal{C}_0^{2+\nu}(\bar{D})$ and

$$\mathfrak{L}(\lambda_0, u_0)u = 0 \qquad \text{in} \quad D, \tag{2.15}$$

where

$$\mathfrak{L}(\lambda_0, u_0) := -d\Delta - \lambda_0 - a\frac{\partial f}{\partial u}(\cdot, u_0)u_0 - af(\cdot, u_0). \tag{2.16}$$

On the other hand, since $D \subset \Omega_-$, we have $a(x) < 0$ for all $x \in D$. Thus, by the monotonicity of the principal eigenvalue with respect to the potential, it follows from (2.14) and (Hf) that

$$\lambda_1[\mathfrak{L}(\lambda_0, u_0), D] > \lambda_1[-d\Delta - \lambda_0 - af(\cdot, u_0), D] = 0.$$

Hence, (2.15) implies $u = 0$. Consequently, $D_u\mathfrak{F}(\lambda_0, u_0)$ is a linear topological isomorphism. Therefore, combining the uniqueness of the positive solution of (2.3), as a consequence from Theorem 2.2, with the implicit function theorem, establishes the regularity of the solution operator (2.11). Finally, differentiating the identity

$$\mathfrak{F}(\lambda, \theta(\lambda)) = 0, \qquad \lambda > \lambda_1[-d\Delta, D],$$

with respect to λ yields

$$D_\lambda\theta = (-d\Delta)^{-1}\left(\theta + \lambda D_\lambda\theta + a\frac{\partial f}{\partial u}(\cdot, \theta)\theta D_\lambda\theta + af(\cdot, \theta)D_\lambda\theta\right),$$

where $\theta = \theta(\lambda)$, or, equivalently,

$$\mathfrak{L}(\lambda, \theta(\lambda))D_\lambda\theta(\lambda) = \theta(\lambda), \qquad \lambda > \lambda_1[-d\Delta, D].$$

As $\theta(\lambda) \gg 0$ and

$$\lambda_1[\mathfrak{L}(\lambda, \theta(\lambda)), D] > 0 \qquad \text{for all } \lambda > \lambda_1[-d\Delta, D],$$

we find from Theorem 1.1 that

$$D_\lambda\theta(\lambda) = \left(-d\Delta - \lambda - a\frac{\partial f}{\partial u}(\cdot, \theta(\lambda))\theta(\lambda) - af(\cdot, \theta(\lambda))\right)^{-1}\theta(\lambda) \gg 0,$$

which concludes the proof. As $\theta_{[\mu,D]}$ is a strict subsolution of (2.3) if

$$\lambda_1[-d\Delta, D] < \mu < \lambda,$$

the estimate

$$\theta_{[\mu,D]} \ll \theta_{[\lambda,D]}$$

can be also derived from Lemma 1.8.   □

Figure 2.1 illustrates the results established by Theorems 2.2 and 2.3. For a given value of $x \in D$, it shows the curve

$$\lambda \mapsto \theta_{[\lambda,D]}(x).$$

It bifurcates from the horizontal axis, $u = 0$, at $\lambda = \lambda_1[-d\Delta, D]$ and it increases for all further values of λ. The direction of the arrows represents the flow of (2.1).

According to Theorem 2.2, $u_{[\lambda,D]}(x, t; u_0)$ decays to zero as $t \uparrow \infty$ if $\lambda \leq \lambda_1[-d\Delta, D]$, while it approximates $\theta_{[\lambda,D]}(x)$ for all $\lambda > \lambda_1[-d\Delta, D]$. Thus, the trivial equilibrium $u = 0$ of (2.1) is a global attractor if $\lambda \leq \lambda_1[-d\Delta, D]$, while it is unstable for $\lambda > \lambda_1[-d\Delta, D]$. The stability lost by $u = 0$ as λ

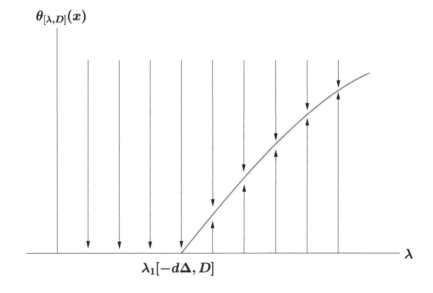

FIGURE 2.1: The dynamics of (2.1) for $M = 0$.

crosses the critical threshold $\lambda_1[-d\Delta, D]$ is gained by the positive solution $\theta_{[\lambda,D]}$ bifurcating from $u = 0$ at $\lambda_1[-d\Delta, D]$, in complete agreement with the *exchange stability principle* of M. G. Crandall and P. H. Rabinowitz [60].

2.3 Perturbed logistic problem

Throughout this section we assume that

$$M > 0.$$

In this case, the next result holds.

Theorem 2.4 *Suppose $M > 0$ and f satisfies* (Hf) *and* (Hg). *Then,* (2.2) *possesses a unique positive solution, $\theta_{[\lambda,D,M]}$, for each $\lambda \in \mathbb{R}$. Moreover,*

$$\lim_{t\uparrow\infty} u_{[\lambda,D,M]}(\cdot, t; u_0) = \theta_{[\lambda,D,M]} \quad in \quad \mathcal{C}(\bar{D}), \tag{2.17}$$

where $u_{[\lambda,D,M]}(x, t; u_0)$ stands for the solution of (2.1).
  If, in addition, $\underline{u} > 0$ (resp. $\bar{u} > 0$) is a strict subsolution (resp. supersolution) of (2.2), *then*

$$\underline{u} \ll \theta_{[\lambda,D,M]} \quad (resp. \quad \theta_{[\lambda,D,M]} \ll \bar{u}).$$

Consequently, the estimates

$$0 < M_1 \le M_2 < \infty, \quad -\infty < \lambda_1 \le \lambda_2 < \infty, \quad M_2 - M_1 + \lambda_2 - \lambda_1 > 0,$$

imply

$$\theta_{[\lambda_1,D,M_1]} \ll \theta_{[\lambda_2,D,M_2]}.$$

Moreover,

$$\theta_{[\lambda,D]} \ll \theta_{[\lambda,D,M]} \tag{2.18}$$

for all $\lambda > \lambda_1[-d\Delta, D]$ and $M > 0$.

Proof: By Theorem 1.7, it suffices to construct a positive supersolution of (2.2) for each $\lambda \in \mathbb{R}$. As in the proof of Theorem 2.2, sufficiently large positive constants provide us with such a supersolution if $\bar{D} \subset \Omega_-$. It remains to construct a supersolution in the general case when

$$D \subset \Omega_-.$$

The construction is a variant of the one already carried out in the proof of Theorem 2.2. Let $\Gamma_{D,j}$, $1 \le j \le n$, denote the components of ∂D and, for sufficiently small $\varepsilon > 0$ and each $j \in \{1, ..., n\}$, consider the open subset

$$D_{\varepsilon,j} := \{x \in \mathbb{R}^N : \text{dist}(x, \Gamma_{D,j}) < \varepsilon\}.$$

For sufficiently small $\varepsilon > 0$, we have that

$$\bar{D}_{\varepsilon,i} \cap \bar{D}_{\varepsilon,j} = \emptyset \quad \text{if} \quad i \neq j, \qquad \Gamma_{D,j} \subset D_{\varepsilon,j} \quad 1 \leq j \leq n,$$

and the open set $D_{\varepsilon,j}$ also is of class $\mathcal{C}^{2+\nu}$ for all $j \in \{1, ..., n\}$. Moreover,

$$\lim_{\varepsilon \downarrow 0} |D_{\varepsilon,j}| = 0, \qquad 1 \leq j \leq n,$$

implies that

$$\lim_{\varepsilon \downarrow 0} \lambda_1[-d\Delta, D_{\varepsilon,j}] = \infty, \qquad 1 \leq j \leq n.$$

Thus, as in the proof of Theorem 2.2, ε can be shortened to get (2.6).

For each $1 \leq j \leq n$, let $\varphi_{\varepsilon,j} \gg 0$ be a principal eigenfunction associated to $\lambda_1[-d\Delta, D_{\varepsilon,j}]$. As $\varphi_{\varepsilon,j}(x) > 0$ for all $x \in D_{\varepsilon,j}$, it is apparent that

$$\min_{1 \leq j \leq n} \min_{\partial D} \varphi_{\varepsilon,j} > 0 \quad \text{and} \quad \min_{1 \leq j \leq n} \min_{D \cap \partial D_{\varepsilon/2,j}} \varphi_{\varepsilon,j} > 0. \qquad (2.19)$$

Thus, there exists $\kappa_0 > 0$ such that

$$\kappa \Phi > M \quad \text{on} \quad \partial D \quad \text{for all} \quad \kappa \geq \kappa_0,$$

where Φ is given by (2.8). Reasoning as in the proof of Theorem 2.2, it is apparent that, for sufficiently large $\kappa > 1$, the function $\bar{u} := \kappa \Phi$ provides us with a strict supersolution of (2.2). This ends the proof. $\quad \square$

As a byproduct of Theorem 2.4, the point-wise limit

$$\theta_{[\lambda, D, \infty]} := \lim_{M \uparrow \infty} \theta_{[\lambda, D, M]} \quad \text{in} \ D \qquad (2.20)$$

is well defined, though it might equal ∞ in some subregion of D if no further restriction on f is imposed. The main goal of Chapter 3 is to establish how (2.20) equals the minimal positive solution of the singular problem

$$\begin{cases} -d\Delta u = \lambda u + a(x) f(x, u) u & \text{in} \ \ D, \\ u = \infty & \text{on} \ \ \partial D, \end{cases}$$

when, in addition, f satisfies Hypothesis (KO).

2.4   Structural stability as $M > 0$ perturbs from $M = 0$

The next result establishes the *structural stability* of (2.1) when $M > 0$ perturbs from $M = 0$.

Proposition 2.5 *Suppose f satisfies* (Hf) *and* (Hg). *Then,*

$$\lim_{M\downarrow 0} \theta_{[\lambda,D,M]} = \begin{cases} 0, & \text{if } \lambda \leq \lambda_1[-d\Delta, D], \\ \theta_{[\lambda,D]}, & \text{if } \lambda > \lambda_1[-d\Delta, D]. \end{cases}$$

Proof: According to Theorem 2.4, the point-wise limit

$$0 \leq \Theta_0 := \lim_{M\downarrow 0} \theta_{[\lambda,D,M]} \tag{2.21}$$

is well defined, because

$$0 \ll \theta_{[\lambda,D,M]} \ll \theta_{[\lambda,D,\tilde{M}]} \qquad \text{if } 0 < M < \tilde{M} \tag{2.22}$$

for all $\lambda \in \mathbb{R}$. We claim that Θ_0 provides us with a (classical) solution of (2.3). Indeed, by (2.22), for every $\tilde{M} > 0$, there exists a constant $C := C(\tilde{M}) > 0$ such that

$$\|\theta_{[\lambda,D,M]}\|_{\mathcal{C}(\bar{D})} \leq C \qquad \text{for all } M \in (0, \tilde{M}).$$

Thus, by the Schauder estimates (e.g., D. Gilbarg and N. S. Trudinger [103, Th. 6.6]), there exists a positive constant $C_1 > 0$ such that

$$\|\theta_{[\lambda,D,M]}\|_{\mathcal{C}^{2+\nu}(\bar{D})} \leq C_1 \qquad \text{for all } M \in (0, \tilde{M}).$$

As the injection

$$\mathcal{C}^{2+\nu}(\bar{D}) \hookrightarrow \mathcal{C}^2(\bar{D})$$

is compact (see J. López-Gómez [163, p. 197]), and the point-wise limit (2.21) is unique, necessarily $\Theta_0 \in \mathcal{C}^2(\bar{D})$ and

$$\lim_{M\downarrow 0} \|\theta_{[\lambda,D,M]} - \Theta_0\|_{\mathcal{C}^2(\bar{D})} = 0.$$

Consequently, $\Theta_0 \geq 0$ solves (2.3) in D. Actually, by elliptic regularity, $\Theta_0 \in \mathcal{C}^{2+\nu}(\bar{D})$.

Thanks to Theorem 2.2, $\Theta_0 = 0$ if $\lambda \leq \lambda_1[-d\Delta, D]$. Suppose

$$\lambda > \lambda_1[-d\Delta, D].$$

Then, letting $M \downarrow 0$ in (2.18) yields

$$0 \ll \theta_{[\lambda,D]} \leq \Theta_0$$

and therefore $\Theta_0 \gg 0$. By the uniqueness of the positive solution of (2.3), we conclude that $\Theta_0 = \theta_{[\lambda,D]}$. This ends the proof. $\square$

According to Proposition 2.5, the curve of maximal non-negative solutions of (2.3) perturbs into the curve of positive solutions of (2.2) as the parameter

$M > 0$ leaves the level $M = 0$. In Figure 2.2, fixing $M > 0$ and $x \in D$, we have represented the curve

$$\lambda \mapsto \theta_{[\lambda,D,M]}(x).$$

Thanks to Proposition 2.5, it approximates 0, as $M \downarrow 0$, for all $\lambda \leq \lambda_1[-d\Delta, D]$, while it approximates $\theta_{[\lambda,D]}(x) > 0$ for all $\lambda > \lambda_1[-d\Delta, D]$. The dashed curves in Figure 2.2 represent the non-negative solutions of (2.3). It should be compared with Figure 2.1.

The next result provides us with an alternative proof of Proposition 2.5 through the implicit function theorem.

Proposition 2.6 *Suppose f satisfies* (Hf) *and* (Hg), *and*

$$\lambda \in \mathbb{R} \setminus \{\lambda_1[-d\Delta, D]\}.$$

Then, the solution operator

$$
\begin{array}{ccc}
(0, \infty) & \xrightarrow{S} & \mathcal{C}(\bar{D}) \\
M & \mapsto & S(M) := \theta_{[\lambda,D,M]}
\end{array}
$$

is of class $\mathcal{C}^1$. Moreover,

$$\lim_{M \downarrow 0} S(M) = \left\{ \begin{array}{ll} 0, & \text{if } \lambda < \lambda_1[-d\Delta, D], \\ \theta_{[\lambda,D]}, & \text{if } \lambda > \lambda_1[-d\Delta, D]. \end{array} \right.$$

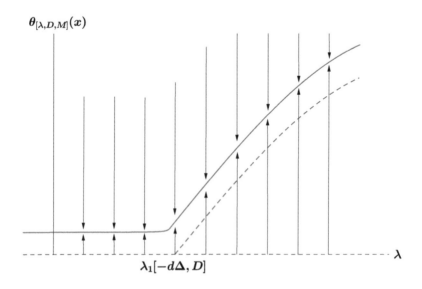

FIGURE 2.2: The dynamics of (2.1) in case $M > 0$.

Proof: The change of variable

$$u = v + M, \qquad M \in \mathbb{R},$$

transforms (2.3) into

$$\begin{cases} -d\Delta v = \lambda(v + M) + af(\cdot, v+M)(v+M) & \text{in } D, \\ v = 0 & \text{on } \partial D, \end{cases} \tag{2.23}$$

whose solutions are the zeroes of the nonlinear operator

$$\mathfrak{G} \; : \; \mathbb{R} \times \mathcal{C}_0(\bar{D}) \longrightarrow \mathcal{C}_0(\bar{D})$$

defined by

$$\begin{aligned} \mathfrak{G}(M, v) &:= v - (-d\Delta)^{-1} \left[\lambda(v + M) + af(\cdot, v + M)(v + M) \right] \\ &= \mathfrak{F}(\lambda, v + M) - M \end{aligned}$$

for all $(M, v) \in \mathbb{R} \times \mathcal{C}_0(\bar{D})$, where $\mathfrak{F}$ is the operator introduced in the proof of Theorem 2.3. The operator $\mathfrak{G}$ is of class $\mathcal{C}^1$ and, by elliptic regularity, $\mathfrak{G}(M, \cdot)$ is a nonlinear compact perturbation of the identity map for all $M \in \mathbb{R}$. Moreover,

$$\begin{aligned} \mathfrak{G}(0,0) &= 0 & \text{for all } \lambda \in \mathbb{R}, \\ \mathfrak{G}(0, \theta_{[\lambda, D]} - M) &= 0 & \text{for all } \lambda > \lambda_1[-d\Delta, D], \end{aligned}$$

$$D_v \mathfrak{G}(0,0)v = v - \lambda(-d\Delta)^{-1}v \quad \text{for all } (\lambda, v) \in \mathbb{R} \times \mathcal{C}_0(\bar{D}),$$

and

$$D_v \mathfrak{G}(0, \theta_{[\lambda, D]} - M)v = D_u \mathfrak{F}(\lambda, \theta_{[\lambda, D]})v, \tag{2.24}$$

for all $\lambda > \lambda_1[-d\Delta, D]$ and $v \in \mathcal{C}_0(\bar{D})$. Suppose

$$\lambda < \lambda_1[-d\Delta, D].$$

Then, $D_v \mathfrak{G}(0,0)$ is an isomorphism and, thanks to the implicit function theorem, there exist $\varepsilon > 0$, $0 < \delta \leq \varepsilon$, and a map of class $\mathcal{C}^1$,

$$\begin{array}{ccc} [-\varepsilon, \varepsilon] & \overset{v}{\longrightarrow} & \mathcal{C}_0(\bar{D}) \\ M & \mapsto & v(M) \end{array}$$

such that $v(0) = 0$,

$$\mathfrak{G}(M, v(M)) = 0 \qquad \text{for all } M \in [-\varepsilon, \varepsilon] \tag{2.25}$$

and

$$\left. \begin{array}{c} \mathfrak{G}(M, v) = 0 \\ |M| + \|v\|_{\mathcal{C}(\bar{D})} \leq \delta \end{array} \right\} \quad \Longrightarrow \quad v = v(M).$$

Differentiating (2.25) with respect to M at $(M, v) = (0, 0)$ yields

$$D_M \mathfrak{G}(0,0) + D_v \mathfrak{G}(0,0) D_M v(0) = 0,$$

and hence,

$$D_M v(0) - \lambda(-d\Delta)^{-1} D_M v(0) = \lambda(-d\Delta)^{-1} 1.$$

Thus, by elliptic regularity, $D_M v(0) \in \mathcal{C}^{2+\nu}(\bar{D})$ is the unique solution of

$$\begin{cases} (-d\Delta - \lambda) D_M v(0) = \lambda & \text{in } D, \\ D_M v(0) = 0 & \text{on } \partial D. \end{cases}$$

So,

$$D_M v(0) = \lambda(-d\Delta - \lambda)^{-1} 1.$$

Thanks to Theorem 1.1, $(-d\Delta - \lambda)^{-1}$ is well defined because $\lambda < \lambda_1[-d\Delta, D]$. Consequently, setting

$$u(M) := v(M) + M, \qquad M \in [-\varepsilon, \varepsilon], \tag{2.26}$$

it becomes apparent that

$$D_M u(0) = D_M v(0) + 1 = 1 + \lambda(-d\Delta - \lambda)^{-1} 1 \qquad \text{in } D$$

is the unique classical solution of

$$\begin{cases} (-d\Delta - \lambda) D_M u(0) = 0 & \text{in } D, \\ D_M u(0) = 1 & \text{on } \partial D. \end{cases}$$

Therefore, by Theorem 1.1,

$$D_M u(0) \gg 0 \qquad \text{in } D. \tag{2.27}$$

As $u(0) = 0$, (2.27) implies $u(M) > 0$ for sufficiently small $M > 0$. Thus, by the uniqueness of the positive solution of (2.3), already established by Theorem 2.4, we have

$$u(M) = \theta_{[\lambda, D, M]} = S(M)$$

for sufficiently small $M > 0$. Consequently, as $u(M)$ is of class $\mathcal{C}^1$ in M, also is $S(M)$ and the proof is complete in this case.

Now, suppose

$$\lambda > \lambda_1[-d\Delta, D].$$

In this case, in the proof of Theorem 2.3 we have already shown that $D_u \mathfrak{F}(\lambda, \theta_{[\lambda, D]})$ is an isomorphism. Thus, thanks to (2.24), $D_v \mathfrak{G}(M, \theta_{[\lambda, D]} - M)$ also is an isomorphism. Hence, by the implicit function theorem, there exist $\varepsilon > 0$, $0 < \delta \le \varepsilon$, and a map of class $\mathcal{C}^1$,

$$\begin{array}{ccc} [-\varepsilon, \varepsilon] & \xrightarrow{v} & \mathcal{C}_0(\bar{D}) \\ M & \mapsto & v(M) \end{array}$$

such that $v(0) = \theta_{[\lambda, D]}$,

$$\mathfrak{G}(M, v(M) - M) = 0 \qquad \text{for all } M \in [-\varepsilon, \varepsilon],$$

and

$$\left.\begin{array}{l} \mathfrak{G}(M, v - M) = 0 \\ |M| + \|v - \theta_{[\lambda,D]}\|_{\mathcal{C}(\bar{D})} \leq \delta \end{array}\right\} \quad \Longrightarrow \quad v = v(M).$$

As $v(0) = \theta_{[\lambda,D]} \gg 0$, we have that

$$u(M) = M + v(M) \gg 0$$

for sufficiently small $M > 0$. Therefore, by Theorem 2.4,

$$u(M) = \theta_{[\lambda,D,M]} = S(M).$$

This completes the proof. □

2.5 Comments on Chapter 2

The classical diffusive logistic equation arises in the very special case when $\bar{D} \subset \Omega_-$, i.e.,

$$a(x) < 0 \qquad \text{for all } x \in \bar{D}. \tag{2.28}$$

In such case, sufficiently large positive constants provide us with supersolutions of (2.2) for all $M \geq 0$ and consequently, Theorem 1.7 applies to the problems (2.2) and (2.3). If (2.28) fails, the large positive constants do not provide positive supersolutions of (2.2).

The construction of the supersolutions in these degenerated situations in order to prove Theorems 2.2 and 2.4, goes back to J. López-Gómez [141]. Originally, they were introduced to prove the next property

$$\lim_{\mu \to \infty} \lambda_1[-d\Delta + \mu V, D] = \lambda_1[-d\Delta, D_0] \tag{2.29}$$

for all potential $V > 0$ such that $D_0 = V^{-1}(0)$ is a nice region. The identity (2.29) has a number of rather striking applications in population dynamics. For instance, it allowed us to establish that the principle of competitive exclusion is false in the presence of refuges or protection zones because the species can segregate to each of them as the intensity of the aggressions from the competitors increases. Actually, we will use this property in Chapter 10.

The reader interested in this particular issue might wish to have a look at J. López-Gómez [143], J. López-Gómez and J. C. Sabina de Lis [180], S. Cano-Casanova and J. López-Gómez [36], S. Cano-Casanova, J. López-Gómez and M. Molina-Meyer [41], and J. López-Gómez and M. Molina-Meyer [168]. Originally, (2.29) was used the characterize the existence of principal eigenvalues for some general classes of linear weighted boundary value problems (see J. López-Gómez [141, 144, 163]).

The first time that these supersolutions were used to characterize the existence of positive solutions in a generalized logistic equation of degenerate type was in J. M. Fraile et al. [90]. Some degenerated logistic equations, in the sense that $a(x)$ is allowed to vanish within the support domain, had been previously studied by H. Brézis and L. Oswald [31] by means of some classical minimization techniques and by T. Ouyang [203] through global continuation, but J. M. Fraile et al. [90] wrote the first paper in which the method of subsolutions and supersolutions was incorporated to dealt with a very general class of degenerate logistic equations of diffusive type, and it was the first work where the problem of analyzing the dynamics of the associated parabolic problem was addressed successfully, as will become apparent in the next chapters.

This chapter has completed and refined the former results of Section 4 of [160]. As in Chapter 1, all the results of this chapter are valid for wide classes of linear second order elliptic operators like (1.48) under mixed boundary conditions.

Chapter 3

A priori bounds in Ω_-

This chapter refines the classical a priori bounds of J. B. Keller [123] and R. Osserman [202] to establish the existence of a minimal and a maximal positive solution for the singular boundary value problem

$$\begin{cases} -d\Delta u = \lambda u + a(x)f(x,u)u & \text{in} \quad D, \\ u = \infty & \text{on} \quad \partial D, \end{cases} \tag{3.1}$$

under condition (KO). As a byproduct, it will become apparent that

$$\theta_{[\lambda,D,\infty]} := \lim_{M\uparrow\infty} \theta_{[\lambda,D,M]} \tag{3.2}$$

provides us with the minimal positive solution of (3.1), where $\theta_{[\lambda,D,M]}$ stands for the unique positive solution of (2.2).

As in Chapter 2, throughout this chapter we assume that

$$D \subset \Omega_-.$$

The next result establishes the strong positivity of any positive solution of (3.1). By a solution of (3.1) we mean a function $u \in \mathcal{C}^{2+\nu}(D)$ such that

$$\lim_{\substack{x\in D \\ d(x)\downarrow 0}} u(x) = \infty, \quad \text{where} \quad d(x) := \text{dist}\,(x, \partial D) \quad \text{for all} \quad x \in \mathbb{R}^N.$$

These solutions are usually referred to as *large* or *explosive* solutions of

$$-d\Delta u = \lambda u + a(x)f(x,u)u.$$

Lemma 3.1 *Suppose f satisfies* (Hf) *and $u \in \mathcal{C}^{2+\nu}(D)$, $u > 0$, solves* (3.1). *Then, $u(x) > 0$ for all $x \in D$.*

Proof: For sufficiently large n, the function u provides us with a positive supersolution of $-d\Delta - \lambda - af(\cdot, u)$ in

$$D_n := \{x \in D \;:\; \text{dist}(x, \partial D) > 1/n\}$$

such that $u > 0$ on ∂D_n. Thus, thanks to Lemma 1.6, $u \gg 0$ in D_n, which concludes the proof. $\qquad \square$

Section 3.1 studies the problem (3.1) in the special, but pivotal case, when

$$D = B_R(x_0) := \left\{ x \in \mathbb{R}^N : |x - x_0| < R \right\}, \qquad \bar{D} \subset \Omega_-, \qquad (3.3)$$

for some $x_0 \in \Omega_-$. The main result establishes the existence of a minimal large solution of the problem in an arbitrary ball. Section 3.2 establishes the existence of a minimal positive solution of (3.1) in the general case when $D \subset \Omega_-$. Section 3.3 shows the existence of a maximal solution and characterizes it as the point-wise limit of the minimal solutions in smaller domains, D_n, approximating D from the interior as $n \uparrow \infty$. Finally, Section 3.4 gives a sufficient condition for f to satisfy the Keller–Osserman condition (KO).

3.1 Singular problem in a ball $B_R(x_0) \Subset \Omega_-$

Throughout this section we impose (3.3). So,

$$\max_{\bar{D}} a < 0.$$

Its main result can be summarized as follows.

Theorem 3.2 *Suppose* (3.3) *and assume* f *satisfies Hypothesis* (KO). *Then, the point-wise limit* (3.2) *is finite in* D *and it provides us with the minimal positive solution of* (3.1) *for all* $\lambda \in \mathbb{R}$.

Proof: As f satisfies (KO), it also satisfies (Hf) and (Hg). By (Hg), we have

$$
\begin{aligned}
-d\Delta\theta_{[\lambda,D,M]} &= \lambda\theta_{[\lambda,D,M]} + a(x)f(x,\theta_{[\lambda,D,M]})\theta_{[\lambda,D,M]} \\
&\leq \lambda\theta_{[\lambda,D,M]} + a(x)g(\theta_{[\lambda,D,M]})\theta_{[\lambda,D,M]} \\
&\leq \max\{1,\lambda\}\theta_{[\lambda,D,M]} + \left(\max_{\bar{D}} a\right) g(\theta_{[\lambda,D,M]})\theta_{[\lambda,D,M]} \\
&= \max\{1,\lambda\} \left(1 + \frac{\max_{\bar{D}} a}{\max\{1,\lambda\}} g(\theta_{[\lambda,D,M]})\right) \theta_{[\lambda,D,M]} \\
&= -\max\{1,\lambda\} \left(-\frac{\max_{\bar{D}} a}{\max\{1,\lambda\}} g(\theta_{[\lambda,D,M]}) - 1\right) \theta_{[\lambda,D,M]}.
\end{aligned}
$$

Hence, $\theta_{[\lambda,D,M]}$ is a positive subsolution of the problem

$$
\begin{cases}
-d\Delta u = -h(u) & \text{in } D = B_R(x_0), \\
u = M & \text{on } \partial D,
\end{cases}
\qquad (3.4)
$$

where

$$h(u) := \beta(\alpha g(u) - 1)u, \qquad u \geq 0, \tag{3.5}$$

with

$$\beta := \max\{1, \lambda\} > 0 \quad \text{and} \quad \alpha := -\max_{\bar{D}} a/\beta > 0.$$

By Theorem 2.4, (3.4) possesses a unique positive solution, Θ_M. Moreover,

$$\theta_{[\lambda, D, M]} \leq \Theta_M \quad \text{for all} \quad M > 0 \tag{3.6}$$

and

$$\Theta_{M_1} \ll \Theta_{M_2} \qquad \text{if} \quad 0 < M_1 < M_2.$$

Therefore, the point-wise limit

$$\Theta_\infty := \lim_{M \uparrow \infty} \Theta_M \tag{3.7}$$

is well defined in D. Thanks to (3.6), the point-wise limit (3.2) is finite in D provided Θ_∞ is finite in D.

As (3.4) is invariant by rotations, Θ_M must be radially symmetric for each $M > 0$, by uniqueness. Hence,

$$\Theta_M(x) = \Psi_M(r), \qquad r := |x - x_0|, \qquad x \in D = B_R(x_0),$$

where Ψ_M stands for the unique positive solution of

$$\begin{cases} d\left(\psi''(r) + \frac{N-1}{r}\psi'(r)\right) = h(\psi(r)), & 0 < r < R, \\ \psi'(0) = 0, \quad \psi(R) = M. \end{cases} \tag{3.8}$$

The existence of a solution of (3.8) can be inferred by taking into account that $\underline{\psi} := 0$ is a subsolution and that $\bar{\psi} := C$, with $C > 0$ constant, is a supersolution of (3.8) provided

$$\bar{\psi} = C \geq \max\{M, g^{-1}(1/\alpha)\} > 0 = \underline{\psi}.$$

Necessarily, it is unique, as in the contrary case the problem (3.4) should admit two radially symmetric solutions, which is impossible.

Subsequently, without loss of generality, we can assume that

$$M > u^* > g^{-1}(1/\alpha), \tag{3.9}$$

where $u^* = u^*(\alpha)$ is the constant arising in Hypothesis (KO). By (3.9), the constant

$$\underline{u} := g^{-1}(1/\alpha)$$

is a positive strict subsolution of (3.4), because $h(\underline{u}) = 0$ and $\underline{u} < M$ on ∂D, by (3.9). As every constant $\bar{u} = C > M$ provides us with a strict supersolution of (3.4), it follows from Theorem 2.4 that

$$g^{-1}(1/\alpha) < \Theta_M(x) = \Psi_M(r) \qquad \text{for all} \quad x \in \bar{D}. \tag{3.10}$$

Consequently, since $h(z) > 0$ for all $z > g^{-1}(1/\alpha)$,

$$h(\Psi_M(r)) > 0 \qquad \text{for all } r \in [0, R]. \tag{3.11}$$

On the other hand, since for every $z \geq g^{-1}(1/\alpha)$

$$h'(z) = \beta\left(\alpha g(z) + \alpha z g'(z) - 1\right) \geq \alpha\beta z g'(z) > 0,$$

it is apparent that

$$h \text{ is increasing in } [g^{-1}(1/\alpha), \infty).$$

Multiplying the differential equation of (3.8) by r^{N-1} and rearranging terms yields

$$d\left(r^{N-1}\Psi'_M(r)\right)' = r^{N-1}h(\Psi_M(r)), \qquad 0 < r < R. \tag{3.12}$$

Thus, integrating (3.12) in $(0, r)$, shows that

$$d\Psi'_M(r) = r^{1-N}\int_0^r s^{N-1}h(\Psi_M(s))\,ds, \qquad 0 < r < R. \tag{3.13}$$

Consequently, according to (3.11) and (3.13) we find that

$$\Psi'_M(r) > 0 \qquad \text{for all } r \in (0, R)$$

and hence, the map $r \mapsto \Psi_M(r)$ is increasing, as well as

$$r \mapsto h(\Psi_M(r)), \qquad 0 \leq r \leq R,$$

because h is increasing in $[g^{-1}(1/\alpha), \infty)$ and, thanks to (3.10),

$$\Psi_M(r) > g^{-1}(1/\alpha) \qquad \text{for all } r \in [0, R].$$

Therefore, it follows from (3.13) that

$$d\Psi'_M(r) \leq r^{1-N}h(\Psi_M(r))\int_0^r s^{N-1}\,ds = \frac{r}{N}h(\Psi_M(r)) \tag{3.14}$$

for all $r \in (0, R)$. Thus, going back to (3.8), we find from (3.14) that

$$h(\Psi_M(r)) = d\left(\Psi''_M(r) + \frac{N-1}{r}\Psi'_M(r)\right) \leq d\Psi''_M(r) + \frac{N-1}{N}h(\Psi_M(r)).$$

So,

$$d\Psi''_M(r) \geq \frac{h(\Psi_M(r))}{N}, \qquad 0 < r < R.$$

Similarly, since $\Psi'_M \geq 0$, (3.8) also implies that

$$d\Psi''_M(r) \leq h(\Psi_M(r)), \qquad 0 < r < R.$$

Consequently,

$$h(\Psi_M(r)) \geq d\Psi_M''(r) \geq \frac{h(\Psi_M(r))}{N}, \qquad 0 < r < R. \tag{3.15}$$

Multiplying (3.15) by $\Psi_M'(r)$ and integrating in $(0, r)$ gives

$$\int_0^r h(\Psi_M(s))\Psi_M'(s)\,ds \geq d\int_0^r \Psi_M''(s)\Psi_M'(s)\,ds \geq \frac{1}{N}\int_0^r h(\Psi_M(s))\Psi_M'(s)\,ds$$

for all $r \in [0, R)$. Thus, performing the change of variable

$$z = \Psi_M(s), \qquad 0 \leq s \leq r,$$

and taking into account that $\Psi_M'(0) = 0$, it becomes apparent that

$$2\int_{\Psi_M(0)}^{\Psi_M(r)} h(z)\,dz \geq d\left(\Psi_M'(r)\right)^2 \geq \frac{2}{N}\int_{\Psi_M(0)}^{\Psi_M(r)} h(z)\,dz \tag{3.16}$$

for all $r \in [0, R)$. Hence, taking the square root of the reciprocal of (3.16), we find that

$$\frac{1}{\sqrt{2}}\left(\int_{\Psi_M(0)}^{\Psi_M(r)} h\right)^{-\frac{1}{2}} \leq \frac{1}{\sqrt{d}\,\Psi_M'(r)} \leq \sqrt{\frac{N}{2}}\left(\int_{\Psi_M(0)}^{\Psi_M(r)} h\right)^{-\frac{1}{2}}$$

for all $r \in (0, R)$. So, multiplying these inequalities by $\Psi_M'(r)$, and integrating in (r, R) yields

$$\frac{1}{\sqrt{2}}\int_r^R \frac{\Psi_M'(s)}{\sqrt{\int_{\Psi_M(0)}^{\Psi_M(s)} h}}\,ds \leq \frac{R-r}{\sqrt{d}} \leq \sqrt{\frac{N}{2}}\int_r^R \frac{\Psi_M'(s)}{\sqrt{\int_{\Psi_M(0)}^{\Psi_M(s)} h}}\,ds$$

for all $r \in [0, R)$. Consequently, the change of variable

$$u = \Psi_M(s), \qquad r \leq s \leq R,$$

transforms the previous inequalities into

$$\frac{1}{\sqrt{2}}\int_{\Psi_M(r)}^M \frac{du}{\sqrt{\int_{\Psi_M(0)}^u h}} \leq \frac{R-r}{\sqrt{d}} \leq \sqrt{\frac{N}{2}}\int_{\Psi_M(r)}^M \frac{du}{\sqrt{\int_{\Psi_M(0)}^u h}} \tag{3.17}$$

for all $r \in [0, R)$.

Pick a $r \in [0, R)$. As Ψ_M is increasing, we have

$$\Psi_M(r) \geq \Psi_M(0) > g^{-1}(1/\alpha),$$

which implies

$$\int_{\Psi_M(r)}^u h \leq \int_{\Psi_M(0)}^u h \qquad \text{for all } u \geq \Psi_M(r),$$

because $h \geq 0$ in $[g^{-1}(1/\alpha), \infty)$. Consequently, the second inequality of (3.17) provides us with

$$0 < \frac{R-r}{\sqrt{d}} \leq \sqrt{\frac{N}{2}} \int_{\Psi_M(r)}^{M} \frac{du}{\sqrt{\int_{\Psi_M(0)}^{u} h}} \leq \sqrt{\frac{N}{2}} \int_{\Psi_M(r)}^{\infty} \frac{du}{\sqrt{\int_{\Psi_M(r)}^{u} h}} \, du$$

for all $r \in [0, R)$. In other words,

$$0 < \frac{R-r}{\sqrt{d}} \leq \sqrt{\frac{N}{2}} \, J(\Psi_M(r)), \qquad 0 \leq r < R, \tag{3.18}$$

where we are denoting

$$J(z) := \int_{z}^{\infty} \frac{du}{\sqrt{\int_{z}^{u} h(s) \, ds}}, \qquad z > g^{-1}(1/\alpha). \tag{3.19}$$

For any given $z > g^{-1}(1/\alpha)$, the successive changes of variable

$$u = z\theta, \qquad s = zt,$$

transform (3.19) into

$$J(z) = \int_{1}^{\infty} \frac{z \, d\theta}{\sqrt{\int_{z}^{z\theta} h(s) \, ds}} = \int_{1}^{\infty} \frac{z \, d\theta}{\sqrt{\int_{1}^{\theta} h(zt) z \, dt}} = \int_{1}^{\infty} \frac{d\theta}{\sqrt{\int_{1}^{\theta} \frac{h(zt)}{z} \, dt}}.$$

Hence, according to (3.5) and (1.8),

$$J(z) = \frac{1}{\sqrt{\beta}} \int_{1}^{\infty} \frac{d\theta}{\sqrt{\int_{1}^{\theta} (\alpha \, g(zt) - 1) \, t \, dt}} = \frac{I(z)}{\sqrt{\beta}},$$

and therefore, (3.18) can be equivalently written as

$$0 < \frac{R-r}{\sqrt{d}} \leq \sqrt{\frac{N}{2\beta}} \, I(\Psi_M(r)), \qquad 0 \leq r < R. \tag{3.20}$$

Fix $r \in [0, R)$ and consider the values $\Psi_M(r)$ for $M > u^*$. As the map

$$M \mapsto \Psi_M(r)$$

is increasing, the limit

$$\Psi_\infty(r) := \lim_{M \uparrow \infty} \Psi_M(r)$$

is well defined. Moreover, by construction,

$$\Theta_\infty(x) = \Psi_\infty(|x - x_0|) \qquad \text{for all} \quad x \in D = B_R(x_0),$$

where Θ_∞ is the point-wise limit (3.7).

Suppose $\Psi_\infty(r) = \infty$. Then, letting $M \uparrow \infty$ in (3.20), it becomes apparent from the Keller–Osserman condition (1.9) that

$$0 < \frac{R - r}{\sqrt{d}} \leq 0,$$

which is impossible. Therefore,

$$\Psi_\infty(r) < \infty \qquad \text{for all } r \in [0, R).$$

Hence, $\Theta_\infty < \infty$ in D, as required.

On the other hand, since

$$\Psi_\infty(0) = \lim_{M \uparrow \infty} \Psi_M(0),$$

by continuous dependence with respect to the initial values, $\Psi_\infty(r)$ must be the unique solution of the Cauchy problem

$$\begin{cases} d\left(\psi''(r) + \frac{N-1}{r}\psi'(r)\right) = h(\psi(r)), & 0 < r < R, \\ \psi(0) = \Psi_\infty(0), \quad \psi'(0) = 0. \end{cases}$$

Multiplying the differential equation by r^{N-1}, rearranging terms and integrating in $(0, r)$ yields

$$d\Psi'_\infty(r) = r^{1-N} \int_0^r s^{N-1} h(\Psi_\infty(s)) \, ds > 0, \qquad 0 < r < R.$$

Thus, $r \mapsto \Psi_\infty(r)$, $0 \leq r < R$, is increasing. So, the limit

$$\Psi_\infty(R) := \lim_{r \uparrow R} \Psi_\infty(r)$$

is well defined. By continuous dependence, $\Psi_M(R) = M$, $M > u^*$ should be bounded if $\Psi_\infty(R) < \infty$, which is impossible. Consequently,

$$\lim_{r \uparrow R} \Psi_\infty(r) = \infty$$

and Θ_∞ provides us with a radially symmetric positive solution of

$$\begin{cases} -d\Delta u = -h(u) & \text{in } D = B_R(x_0), \\ u = \infty & \text{on } \partial D. \end{cases} \tag{3.21}$$

Moreover, since Θ_∞ is finite in D, due to (3.6) and (3.7), the function $\theta_{[\lambda, D, \infty]}$ defined by (3.2) is finite in D. Fix $\varepsilon \in (0, R/2)$. Then, thanks to (3.6),

$$\theta_{[\lambda, D, M]} \leq \Theta_\infty \quad \text{in } B_{R-\varepsilon}(x_0)$$

for all $M > 0$. Hence, by the Schauder interior estimates, there is a constant $C = C(\varepsilon) > 0$ such that

$$\|\theta_{[\lambda,D,M]}\|_{C^{2+\nu}(\bar{B}_{R-2\varepsilon}(x_0))} \leq C(\varepsilon) \qquad \text{for all} \quad M > 0.$$

Thus, since the injection

$$C^{2+\nu}(\bar{B}_{R-2\varepsilon}(x_0)) \hookrightarrow C^2(\bar{B}_{R-2\varepsilon}(x_0))$$

is compact and the limit (3.2) is uniquely determined, we find that

$$\lim_{M\uparrow\infty} \|\theta_{[\lambda,D,M]} - \theta_{[\lambda,D,\infty]}\|_{C^2(\bar{B}_{R-2\varepsilon}(x_0))} = 0.$$

Consequently, $\theta_{[\lambda,D,\infty]}$ solves

$$-d\Delta u = \lambda u + a(x)f(x,u)u \tag{3.22}$$

in $\bar{B}_{R-2\varepsilon}(x_0)$. By elliptic regularity, we actually have that

$$\theta_{[\lambda,D,\infty]} \in C^{2+\nu}(\bar{B}_{R-2\varepsilon}(x_0)).$$

As this holds for sufficiently small $\varepsilon > 0$, $\theta_{[\lambda,D,\infty]}$ must solve the singular boundary value problem (3.1).

Lastly, we will prove that $\theta_{[\lambda,D,\infty]}$ is the minimal positive solution of (3.1). In particular, Θ_∞ must be the minimal positive solution of (3.21). Indeed, let L be any positive solution of (3.1). Then, for every $M > 0$, there exists a constant $C > M$ and a sufficiently small $\varepsilon > 0$ such that

$$\theta_{[\lambda,D,M]} \leq C \leq L \qquad \text{in the region} \quad R - \varepsilon < |x - x_0| < R. \tag{3.23}$$

By Theorem 2.4, it follows from (3.23) that also

$$\theta_{[\lambda,D,M]} \leq \theta_{[\lambda,D,C]} \leq L \qquad \text{in} \quad B_{R-\varepsilon}(x_0)$$

for sufficiently large $n \geq 1$. Consequently,

$$\theta_{[\lambda,D,M]} \leq L \qquad \text{in} \quad B_R(x_0) \quad \text{for all} \quad M > 0. \tag{3.24}$$

Therefore, letting $M \uparrow \infty$ in (3.24) yields

$$\theta_{[\lambda,D,\infty]} \leq L \qquad \text{in} \quad B_R(x_0),$$

which shows the minimality of $\theta_{[\lambda,D,\infty]}$ and ends the proof. $\qquad \square$

3.2   Singular problem in a general $D \subset \Omega_-$

As a consequence of Theorem 3.2 the next result holds.

Proposition 3.3 *Suppose* $D \subset \Omega_-$ *is a subdomain of class* $\mathcal{C}^{2+\nu}$ *and* f *satisfies Hypothesis* (KO). *Then, for every* $\lambda \in \mathbb{R}$, *the point-wise limit*

$$\theta_{[\lambda,D,\infty]} := \lim_{M \uparrow \infty} \theta_{[\lambda,D,M]} \quad in \;\; D \tag{3.25}$$

is finite and it provides us with the minimal positive solution of the singular problem (3.1). *Moreover, for any positive solution* $u \in \mathcal{C}^2(D) \cap \mathcal{C}(\bar{D})$ *of* (3.22) *in* D, *necessarily*

$$u \le \theta_{[\lambda,D,\infty]} \quad in \;\; D. \tag{3.26}$$

Proof: Let $x_0 \in D$ be and $R > 0$ such that $\bar{B}_R(x_0) \subset D$, and set

$$\gamma_M := \max_{\partial B_R(x_0)} \theta_{[\lambda,D,M]}, \quad M > 0.$$

According to Theorem 2.4,

$$\theta_{[\lambda,D,M]} \ll \theta_{[\lambda,B_R(x_0),\gamma_M+1]} \quad in \;\; B_R(x_0)$$

for all $M > 0$. Thus, due to Theorem 3.2, we have that

$$\theta_{[\lambda,D,M]} \ll \theta_{[\lambda,B_R(x_0),\infty]} < \infty \quad in \;\; B_R(x_0).$$

Hence, letting $M \uparrow \infty$, we find that

$$\theta_{[\lambda,D,\infty]} \le \theta_{[\lambda,B_R(x_0),\infty]} < \infty \quad in \;\; B_R(x_0).$$

Consequently, since x_0 and $R > 0$ are arbitrary, $\theta_{[\lambda,D,\infty]}$ must be finite in D as soon as $\bar{B}_R(x_0) \subset D$. Moreover, for each compact subset $K \subset D$ there exists a constant $C(K) > 0$ such that

$$\theta_{[\lambda,D,M]} \le C(K) \quad in \;\; K \;\; for \; all \;\; M > 0. \tag{3.27}$$

Subsequently, for every $n \in \mathbb{N}$, we consider the open subset

$$D_n := \{x \in D \; : \; \mathrm{dist}\,(x, \partial D) > 1/n\}. \tag{3.28}$$

By construction,

$$\bar{D}_n \subset D_{n+1} \subset D \subset \Omega_-, \quad \max_{\bar{D}_n} a < 0, \quad D = \bigcup_{n \ge m} D_n, \tag{3.29}$$

for all $m \geq 1$. Moreover, there exists $n_0 \in \mathbb{N}$ such that D_n is a subdomain of D of class $\mathcal{C}^{2+\nu}$ for all $n \geq n_0$. Pick $n \geq n_0$. As $K := \bar{D}_{n+1}$ is a compact subset of D, by (3.27), there is a constant $C = C(n) > 0$ such that

$$\theta_{[\lambda,D,M]} \leq C(n) \quad \text{in} \quad \bar{D}_{n+1} \quad \text{for all} \quad M > 0.$$

Thus, since $\bar{D}_n \subset D_{n+1}$, by the Schauder interior estimates, there exists a constant $C_1 = C_1(n) > 0$ such that

$$\|\theta_{[\lambda,D,M]}\|_{\mathcal{C}^{2+\nu}(\bar{D}_n)} \leq C_1(n) \quad \text{for all} \quad M > 0.$$

Consequently, as the injection

$$\mathcal{C}^{2+\nu}(\bar{D}_n) \hookrightarrow \mathcal{C}^2(\bar{D}_n)$$

is compact and the point-wise limit (3.25) is unique, it becomes apparent that

$$\lim_{M \uparrow \infty} \|\theta_{[\lambda,D,M]} - \theta_{[\lambda,D,\infty]}\|_{\mathcal{C}^2(\bar{D}_n)} = 0 \quad \text{for all} \quad n \geq n_0.$$

Therefore, by (3.29), $\theta_{[\lambda,D,\infty]}$ solves (3.22) in D. Actually, due to (3.25), $\theta_{[\lambda,D,\infty]}$ solves the singular problem (3.1). Moreover, by elliptic regularity,

$$\theta_{[\lambda,D,\infty]} \in \mathcal{C}^{2+\nu}(D).$$

Next, we will prove that $\theta_{[\lambda,D,\infty]}$ is the minimal positive solution of (3.1). Indeed, let L be any positive solution of (3.1). Then, for every $M > 0$, there exists a constant $C > M$ such that, for sufficiently large $n \in \mathbb{N}$,

$$\theta_{[\lambda,D,M]} \leq C \leq L \quad \text{in} \quad D \setminus D_n.$$

Thanks to Theorem 2.4, this estimate implies

$$\theta_{[\lambda,D,M]} \leq \theta_{[\lambda,D_n,C]} \leq L \quad \text{in} \quad D_n$$

and hence,

$$\theta_{[\lambda,D,M]} \leq L \quad \text{in} \quad D \quad \text{for all} \quad M > 0.$$

Therefore, letting $M \uparrow \infty$ yields

$$\theta_{[\lambda,D,\infty]} \leq L,$$

which shows the minimality of $\theta_{[\lambda,D,\infty]}$.

To prove (3.26), let $u \in \mathcal{C}^2(D) \cap \mathcal{C}(\bar{D})$ be a positive solution of (3.22) and denote

$$\gamma := \max_{\partial D} u.$$

Then, for every $M > \gamma$, we have

$$u \leq \gamma < M = \theta_{[\lambda,D,M]} \quad \text{on} \quad \partial D,$$

and hence, due to Theorem 2.4,

$$u \ll \theta_{[\lambda,D,M]} \leq \theta_{[\lambda,D,\infty]} \quad \text{in} \quad D.$$

The proof is complete. □

3.3 Existence of minimal and maximal solutions

Proposition 3.3 can be sharpened to obtain the existence of minimal and maximal solutions.

Theorem 3.4 *Suppose* $D \subset \Omega_-$ *is a subdomain of class* $\mathcal{C}^{2+\nu}$ *and* f *satisfies Hypothesis* (KO). *Then,* (3.1) *possesses minimal and maximal positive solutions, denoted by* $L_{[\lambda,D]}^{\min}$ *and* $L_{[\lambda,D]}^{\max}$, *respectively, in the sense that any other positive solution* L *of* (3.1) *satisfies*

$$L_{[\lambda,D]}^{\min} \leq L \leq L_{[\lambda,D]}^{\max}.$$

Moreover,

$$L_{[\lambda,D]}^{\min} = \theta_{[\lambda,D,\infty]} := \lim_{M\uparrow\infty} \theta_{[\lambda,D,M]} \tag{3.30}$$

and

$$L_{[\lambda,D]}^{\max} = \lim_{n\uparrow\infty} L_{[\lambda,D_n]}^{\min}, \tag{3.31}$$

where D_n, *for sufficiently large* n, *are those defined by* (3.28). *See Figure 3.1.*

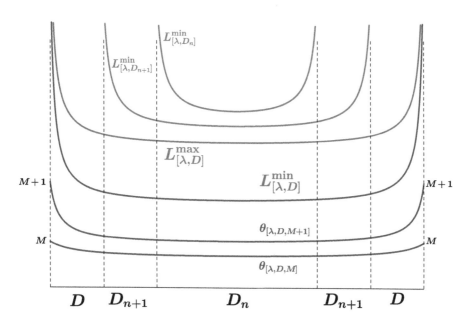

FIGURE 3.1: Scheme of the construction of $L_{[\lambda,D]}^{\min}$ and $L_{[\lambda,D]}^{\max}$.

Proof: The fact that (3.30) is the minimal positive solution of (3.1) has been already established by Proposition 3.3. It remains to prove that (3.31) is the maximal one.

Subsequently, for any $M > 0$ and sufficiently large n, say $n \geq n_0$, we set

$$\gamma := \max_{\bar{D}_n} \theta_{[\lambda, D_{n+1}, M]}.$$

Then, by Theorem 2.4 and Proposition 3.3,

$$\theta_{[\lambda, D_{n+1}, M]} \ll \theta_{[\lambda, D_n, \gamma+1]} \leq \theta_{[\lambda, D_n, \infty]} \quad \text{in} \quad D_n, \qquad n \geq n_0,$$

for all $M > 0$. Thus, letting $M \uparrow \infty$ yields

$$\theta_{[\lambda, D_{n+1}, \infty]} \leq \theta_{[\lambda, D_n, \infty]} \quad \text{in} \quad D_n, \qquad n \geq n_0.$$

According to Proposition 3.3, this estimate can be equivalently written as

$$L_{[\lambda, D_{n+1}]}^{\min} \leq L_{[\lambda, D_n]}^{\min} \quad \text{in} \quad D_n \qquad \text{for all } n \geq n_0. \tag{3.32}$$

Therefore, by (3.29), the point-wise limit

$$L := \lim_{n \to \infty} L_{[\lambda, D_n]}^{\min} \quad \text{in} \quad D \tag{3.33}$$

is well defined.

On the other hand, as $\bar{D}_n \subset D_{n+1}$ for all $n \geq n_0$, thanks to the Schauder interior estimates, it follows from (3.32) that, for every $n \geq n_0$, there exists a constant $C(n) > 0$ such that

$$\|L_{[\lambda, D_m]}^{\min}\|_{\mathcal{C}^{2+\nu}(\bar{D}_n)} \leq C(n) \qquad \text{for all} \quad m \geq n+1.$$

Consequently, as the injection

$$\mathcal{C}^{2+\nu}(\bar{D}_n) \hookrightarrow \mathcal{C}^2(\bar{D}_n)$$

is compact and the point-wise limit (3.33) unique, we find that

$$\lim_{n \to \infty} \|L_{[\lambda, D_n]}^{\min} - L\|_{\mathcal{C}^2(K)} = 0$$

for all compact subset $K \subset D$. Consequently, $L \in \mathcal{C}^2(D)$ must be a solution of (3.22) in D. By elliptic regularity, $L \in \mathcal{C}^{2+\nu}(D)$.

Let $\tilde{L}$ be an arbitrary positive solution of (3.1) and set

$$\gamma_n := \max_{\bar{D}_n} \tilde{L}, \qquad n \geq n_0.$$

Then, $\tilde{L}|_{D_n}$ provides us with a subsolution of the problem

$$\begin{cases} -d\Delta u = \lambda u + a(x) f(x, u) u & \text{in} \quad D_n, \\ u = \gamma_n & \text{on} \quad \partial D_n, \end{cases}$$

for all $n \geq n_0$. Hence, it follows from Theorem 2.4 and Proposition 3.3 that

$$\tilde{L} \leq \theta_{[\lambda, D_n, \gamma_n]} \leq L^{\min}_{[\lambda, D_n]} \quad \text{in} \quad D_n \quad \text{for all} \quad n \geq n_0.$$

Consequently, letting $n \to \infty$, we obtain from (3.33) that

$$\tilde{L} \leq L \quad \text{in} \quad D.$$

Therefore, L must be the maximal positive solution of (3.1). □

Under the general assumptions of this section, we conjecture that

$$L^{\min}_{[\lambda, D]} = L^{\max}_{[\lambda, D]}. \tag{3.34}$$

Part II of this book is devoted to the proof of (3.34), which entails the uniqueness of the solution of (3.1) in some special cases. The proof in the general case remains an open problem.

3.4   Some sufficient conditions for (KO)

Suppose f satisfies (Hf) and (Hg). As, according to (Hg), $g(u)$ is increasing, the integral function $I(u)$ defined by (1.8) is decreasing. Thus, if there exists $u^* > g^{-1}(1/\alpha)$ such that $I(u^*) < \infty$, the limit

$$I_\infty := \lim_{u \uparrow \infty} I(u)$$

is well defined and it satisfies $I_\infty \in [0, \infty)$. Consequently, if $I(u^*) < \infty$ for some $u^* > g^{-1}(1/\alpha)$, the condition (KO) can be equivalently expressed as

$$I_\infty = 0. \tag{3.35}$$

The next result provides us with a simple condition ensuring (3.35).

Lemma 3.5 *Suppose f satisfies (Hf) and (Hg), $I(u^*) < \infty$ for some $u^* > g^{-1}(1/\alpha)$ and there exists $\tilde{u} > u^*$ such that*

$$Q(u) := \min_{t \geq 1} \frac{g(ut)}{g(\tilde{u}t)}, \qquad u \geq \tilde{u}, \tag{3.36}$$

satisfies

$$\lim_{u \uparrow \infty} Q(u) = \infty. \tag{3.37}$$

Then, (3.35) (and hence, (KO)) holds.

Proof: For every $t \geq 1$ and $u \geq \tilde{u}$, we have that

$$(\alpha g(ut) - 1)t = \left(\alpha g(\tilde{u}t)\frac{g(ut)}{g(\tilde{u}t)} - 1\right)t = \frac{g(ut)}{g(\tilde{u}t)}\left(\alpha g(\tilde{u}t) - \frac{g(\tilde{u}t)}{g(ut)}\right)t.$$

According to (Ag),

$$\frac{g(\tilde{u}t)}{g(ut)} \leq 1 \quad \text{for all } t \geq 1 \text{ and } u \geq \tilde{u}.$$

Thus, since $\tilde{u} > u^* > g^{-1}(1/\alpha)$,

$$\alpha g(\tilde{u}t) - \frac{g(\tilde{u}t)}{g(ut)} \geq \alpha g(\tilde{u}t) - 1 > 0 \qquad \text{for all } t \geq 1.$$

Hence, for every $t \geq 1$ and $u \geq \tilde{u}$,

$$(\alpha g(ut) - 1)t \geq Q(u)(\alpha g(\tilde{u}t) - 1)t$$

and therefore,

$$I(u) \leq \frac{1}{\sqrt{Q(u)}} I(\tilde{u}) \qquad \text{for all } u \geq \tilde{u}. \tag{3.38}$$

As

$$I(\tilde{u}) < I(u^*) < \infty,$$

letting $u \uparrow \infty$ in (3.38) yields (3.35). □

The next result provides us with a very simple sufficient condition for (3.37).

Lemma 3.6 *Suppose f satisfies* (Hf) *and* (Hg)*, $I(u^*) < \infty$ for some $u^* > g^{-1}(1/\alpha)$ and there exists $s_0 > 0$ such that*

$$G(s) := \frac{g'(s)}{g(s)}s, \qquad s \geq s_0, \tag{3.39}$$

is non-decreasing in $[s_0, \infty)$. Then, there exists $\tilde{u} > u^*$ such that the quotient function $Q(u)$ defined through (3.36) *satisfies* (3.37)*.*

Proof: The result holds with the choice

$$\tilde{u} := \max\{u^*, s_0\} + 1.$$

Indeed, for every $u \geq \tilde{u}$, set

$$\xi(t) := \frac{g(ut)}{g(\tilde{u}t)}, \qquad t \in [1, \infty).$$

Since $g \in \mathcal{C}^1[0, \infty)$ and $g(z) > 0$ for all $z > 0$, $\xi \in \mathcal{C}^1[1, \infty)$ and

$$\xi'(t) = [g(\tilde{u}t)]^{-2}\left[g'(ut)g(\tilde{u}t)u - g'(\tilde{u}t)g(ut)\tilde{u}\right]$$

for all $t \geq 1$. Moreover, since

$$ut \geq \tilde{u}t \geq \tilde{u} > s_0,$$

and G is non-decreasing, we find that

$$G(ut) \geq G(\tilde{u}t) \qquad \text{for all } t \geq 1.$$

Equivalently,

$$\frac{g'(ut)}{g(ut)} ut \geq \frac{g'(\tilde{u}t)}{g(\tilde{u}t)} \tilde{u}t$$

and hence,

$$\xi'(t) \geq 0 \qquad \text{for all } t \geq 1.$$

In particular,

$$\xi(t) \geq \xi(1) \qquad \text{for all } t \geq 1,$$

and so,

$$Q(u) = \min_{t \geq 1} \xi(t) = \xi(1) = \frac{g(u)}{g(\tilde{u})}.$$

Therefore, by (Hg),

$$\lim_{u \uparrow \infty} Q(u) = \infty.$$

The proof is complete. □

In the special case when $g(s) = s^p$, $s \geq 0$, for some $p > 0$, we have that

$$G(s) := \frac{g'(s)}{g(s)} s = p > 0$$

for all $s \geq 0$. Thus, G is increasing. Therefore, owing to (1.12), we find from Lemmas 3.5 and 3.6 that condition (KO) holds.

Obviously, the function G is non-decreasing if

$$\frac{g'}{g} = (\log g)'$$

is non-decreasing, i.e., if $g(s)$ is *logarithmically convex* for $s \geq s_0$. When $g \in C^2[0, \infty)$, this occurs if

$$\frac{g''g - (g')^2}{g^2} \geq 0 \qquad \text{in } [s_0, \infty). \tag{3.40}$$

As $g'(s) > 0$ for all $s > 0$, (3.40) can be equivalently written in the form

$$\frac{gg''}{(g')^2} \geq 1 \qquad \text{in } [s_0, \infty). \tag{3.41}$$

Logarithmically convex functions will arise again throughout Part II in connection with the problem of the uniqueness for the singular problem (3.1).

Although the function $g(u) = e^u$ satisfies (3.41), the function $g(u) = u^p$, $u \geq 0$, $p > 0$, cannot satisfy it.

3.5   Comments on Chapter 3

The first results concerning the existence of large solutions of

$$d\Delta u = h(u) \tag{3.42}$$

in a bounded domain of $\mathbb{R}^N$ with $N \geq 1$ go back L. Bieberbach [28] and H. Rademacher [216], who considered the equation

$$d\Delta u = e^u$$

in two and three dimensions, respectively. Later, these results were generalized by J. B. Keller [123] and, independently, by R. Osserman [202] to any spatial dimension and any increasing function $h(u)$ satisfying

$$\int_0^\infty \frac{dx}{\sqrt{\int_0^x h}} < \infty, \tag{3.43}$$

which explains why these types of conditions are referred to in the specialized literature as *Keller–Osserman conditions*.

Essentially, in these classical papers it was established that if $h \in \mathcal{C}^1[0, \infty)$ satisfies $h(u) > 0$ and $h'(u) \geq 0$ for all $u \geq 0$, then (3.42) possesses a radially symmetric solution globally defined in $\mathbb{R}^N$ if, and only if,

$$\int_0^\infty \frac{dx}{\sqrt{\int_0^x h}} = \infty.$$

As a consequence, it was inferred that if a simply-connected surface $\mathcal{S}$ has a Riemannian metric with negative Gauss curvature everywhere, then $\mathcal{S}$ is conformally equivalent to the interior of the unit circle (see J. B. Keller [123] and R. Osserman [202]).

In the context of population dynamics, when dealing with the diffusive logistic equation, the nonlinearity $h(u)$ is given by

$$h(u) = Ag(u)u - \lambda u \tag{3.44}$$

for some positive constants A, $\lambda > 0$, and a certain function $g(u)$ satisfying (Hg). So, the nonlinearity $h(u)$ changes sign at

$$u = g^{-1}(\lambda/A) > 0.$$

As a result, the classical assumptions of J. B. Keller [123] and R. Osserman [202] are not satisfied.

The most pioneering results for changing sign nonlinearities were those of A. C. Lazer and P. J. McKenna [129, 130]. For the choice (3.44) it was assumed in [130] that $h \in \mathcal{C}^1[g^{-1}(\lambda/A), \infty)$, with $h' \geq 0$ and $h'(u)$ non-decreasing for sufficiently large u and that, in addition,

$$\liminf_{u \uparrow \infty} \frac{h'(u)}{\sqrt{\int_{g^{-1}(\lambda/A)}^u h}} > 0. \tag{3.45}$$

Instead of these conditions, J. López-Gómez [147] only imposed the existence of some $u^* > g^{-1}(\lambda/A)$ for which (1.8) and (1.9) hold, which is much weaker than (3.45) for the special, but very important, choice

$$g(u) = u^p, \qquad u \geq 0.$$

Indeed, we already know that (KO) holds for all $p > 0$, while a direct calculation shows that (3.45) fails if $0 < p < 2$.

The generalized Keller–Osserman condition (KO), which is optimal in the context of this book, was originally introduced under the form

$$\mathcal{I}(u) := \int_u^\infty \frac{dx}{\sqrt{\int_u^x h}} < \infty, \qquad \lim_{u \uparrow \infty} \mathcal{I}(u) = 0, \tag{3.46}$$

for some $u^* > g^{-1}(\lambda/A)$ and all $u \geq u^*$ (see [147]). Although at first glance (3.46) might look different from (KO), they are equivalent. Indeed, the successive changes of variable $x = u\theta$ and $z = ut$ show that

$$\mathcal{I}(u) = \int_1^\infty \frac{u \, d\theta}{\sqrt{\int_u^{u\theta} h(z) \, dz}} = \int_1^\infty \frac{u \, d\theta}{\sqrt{\int_1^\theta h(ut) u \, dt}}$$

$$= \int_1^\infty \frac{d\theta}{\sqrt{\int_1^\theta (Ag(ut) - \lambda) t \, dt}} = \frac{I(u)}{\sqrt{\lambda}},$$

where $I(u)$ is given by (1.8) with $\alpha = A/\lambda$. Consequently, (KO) is indeed equivalent to (3.46).

In radical contrast with (3.43), for the choice (3.44) one has that

$$\mathcal{I}(g^{-1}(\lambda/A)) := \int_{g^{-1}(\lambda/A)}^\infty \frac{dx}{\sqrt{\int_{g^{-1}(\lambda/A)}^x h}} = \infty. \tag{3.47}$$

Consequently, the classical theory of J. B. Keller [123] and R. Osserman [202] cannot be applied *mutatis mutandis* to deal with changing sign nonlinearities. In order to prove (3.47), one should take into account that

$$h(z) = h'(g^{-1}(\lambda/A))(z - g^{-1}(\lambda/A)) + o(z - g^{-1}(\lambda/A)) \quad \text{as } z \downarrow g^{-1}(\lambda/A),$$

and that

$$h'(z) = Ag'(z)z + Ag(z) - \lambda, \qquad z \geq 0,$$

which implies

$$C := h'(g^{-1}(\lambda/A)) = Ag'(g^{-1}(\lambda/A))g^{-1}(\lambda/A) > 0.$$

Thus, the integrand of (3.47) is given by

$$\frac{1}{\sqrt{\int_{g^{-1}(\lambda/A)}^{x} h}} \sim \sqrt{\frac{2}{C}} \frac{1}{x - g^{-1}(\lambda/A)} \qquad \text{as } x \downarrow g^{-1}(\lambda/A),$$

which indeed entails (3.47).

It should be noted that all the results of this chapter still remain valid if we substitute the differential operator $-d\Delta$ by $-d\Delta + V$ with $V \in L^\infty(D)$.

Chapter 4

Generalized diffusive logistic equation

This chapter characterizes the existence of positive solutions of

$$\begin{cases} -d\Delta u = \lambda u + a(x)f(x,u)u & \text{in} \quad \Omega, \\ u = 0 & \text{on} \quad \partial\Omega, \end{cases} \tag{4.1}$$

when $a(x)$ and $f(x,u)$ satisfy Hypotheses (Ha) and (Hf)–(Hg), respectively, as well as the existence of positive solutions for the following families of singular boundary value problems

$$\begin{cases} -d\Delta u = \lambda u + a(x)f(x,u)u & \text{in} \quad \Omega_j, \\ u = 0 & \text{on} \quad \partial\Omega_j \cap \partial\Omega, \\ u = \infty & \text{on} \quad \partial\Omega_j \setminus \partial\Omega, \end{cases} \tag{4.2}$$

where f satisfies (KO) and

$$\Omega_j := \Omega \setminus \left(\bar{\Omega}_{0,1} \cup \cdots \cup \bar{\Omega}_{0,j} \right), \qquad 1 \leq j \leq q_0. \tag{4.3}$$

It should be remembered that

$$\Omega_{0,j} = \bigcup_{i=1}^{m_j} \Omega_{0,j}^i, \qquad 1 \leq j \leq q_0,$$

where, according to the notations of Hypothesis (Ha), $\Omega_{0,j}^i$, $1 \leq j \leq q_0$, $1 \leq i \leq m_j$, are the components of

$$\Omega_0 = \text{int } a^{-1}(0).$$

Those components were labeled to satisfy (1.3), i.e.,

$$\sigma_j := \lambda_1[-\Delta, \Omega_{0,j}^i], \qquad 1 \leq i \leq m_j, \; 1 \leq j \leq q_0,$$
$$\sigma_j < \sigma_{j+1}, \qquad\qquad 1 \leq j \leq q_0 - 1. \qquad (4.4)$$

By construction,

$$\Omega_{q_0} = \Omega \setminus \bar{\Omega}_0 = \Omega_-. \qquad (4.5)$$

Throughout the rest of this book, we will adopt the notation

$$\sigma_{q_0+1} := \infty.$$

By the results of Chapter 3, f should satisfy (KO) so that (4.2) can admit a positive solution. These solutions will regulate the dynamics of the parabolic model (1.1) according to the different ranges of values of the parameter $\lambda \in \mathbb{R}$, which explains our interest in analyzing (4.2). Precisely, the solutions of (4.1) regulate the dynamics of (1.1) for $\lambda < d\sigma_1$, while the solutions of (4.2) regulate the dynamics of (1.1) for $d\sigma_j \leq \lambda < d\sigma_{j+1}$, if $1 \leq j \leq q_0 - 1$, and for $\lambda \geq d\sigma_{q_0}$ if $j = q_0$. The solutions of (4.2) will be referred to as the *large solutions of order j of*

$$-d\Delta u = \lambda u + a(x)f(x,u)u$$

in Ω_j.

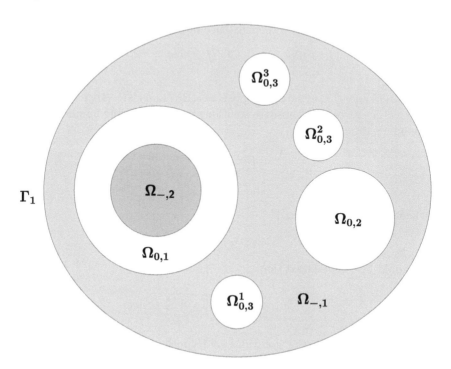

FIGURE 4.1: An intricate nodal configuration for $a(x)$.

Although in the special case when $a(x)$ has the nodal behavior described by Figure 1.1, we see that

$$\Omega_1 = \Omega \setminus \bar{\Omega}_{0,1} = \Omega_- \cup \bar{\Omega}_{0,2}, \qquad \Omega_2 = \Omega_-,$$

and hence, Ω_1 and Ω_2 are connected, in general, some or several of the Ω_j's, $1 \leq j \leq q_0 - 1$, might not be connected. Indeed, suppose $a(x)$ has the nodal configuration sketched in Figure 4.1. In this example, Ω_- has two components, named $\Omega_{-,1}$ and $\Omega_{-,2}$; Ω_0 consists of five components, named $\Omega_{0,1}$, $\Omega_{0,2}$, $\Omega_{0,3}^1$, $\Omega_{0,3}^2$ and $\Omega_{0,3}^3$, and $\partial\Omega$ consists of two components, Γ_1 and Γ_2. According to Hypothesis (Ha), we are assuming that

$$\sigma_1 = \lambda_1[-\Delta, \Omega_{0,1}] < \sigma_2 = \lambda_1[-\Delta, \Omega_{0,2}] < \sigma_3 = \lambda_1[-\Delta, \Omega_{0,3}^i], \quad 1 \in \{1,2,3\}.$$

In this example, $\Omega_1 := \Omega \setminus \bar{\Omega}_{0,1}$ possesses two components.

The organization of this chapter is the following. Section 4.1 characterizes the existence of positive solutions of (4.1) and Section 4.2 studies the inhomogeneous counterpart of (4.1). Section 4.3 studies a hierarchic chain of classical inhomogeneous problems closely related to (4.2) and Section 4.4 uses this analysis to establish the existence of minimal and maximal large solutions supported in Ω_j for each $1 \leq j \leq q_0$. Sections 4.5 and 4.6 ascertain the point-wise behavior of the positive solution of (4.1) as $\lambda \uparrow d\sigma_0$ by means of two complementary technical devices. Section 4.7 characterizes the limiting behavior of the minimal large solution of (4.2), $1 \leq j \leq q_0$, as $\lambda \uparrow d\sigma_{j+1}$. Finally, Section 4.8 studies the limiting behavior of the large positive solutions of (4.2) as $\lambda \downarrow -\infty$.

4.1 Classical positive solutions in Ω

The next result characterizes the existence of positive solutions for (4.1). These solutions regulate the dynamics of (1.1) within the range

$$d\sigma_0 < \lambda < d\sigma_1.$$

It should be remembered that $\sigma_0 := \lambda_1[-\Delta, \Omega]$.

Theorem 4.1 *Suppose f satisfies (Hf) and (Hg). Then, (4.1) possesses a positive solution if, and only if,*

$$d\sigma_0 < \lambda < d\sigma_1 \qquad \left(\Leftrightarrow \frac{\lambda}{\sigma_1} < d < \frac{\lambda}{\sigma_0}\right). \tag{4.6}$$

Moreover, it is unique if it exists and if we denote it by $\theta_{[\lambda,\Omega]}$ then the solution operator

$$
\begin{array}{ccc}
(d\sigma_0, d\sigma_1) & \xrightarrow{\theta} & C(\bar{\Omega}) \\
\lambda & \mapsto & \theta(\lambda) := \theta_{[\lambda,\Omega]}
\end{array}
\tag{4.7}
$$

is point-wise increasing and of class $\mathcal{C}^1$. *Furthermore,*

$$\lim_{\lambda\downarrow d\sigma_0} \theta_{[\lambda,\Omega]} = 0 \qquad in \quad \mathcal{C}(\bar{\Omega}), \tag{4.8}$$

and

$$\lim_{\lambda\uparrow d\sigma_1} \theta_{[\lambda,\Omega]} = \infty \quad uniformly \ in \ compact \ subsets \ of \quad \Omega_{0,1}. \tag{4.9}$$

Thus, $\theta_{[\lambda,\Omega]}$ *bifurcates from* $u = 0$ *at* $\lambda = d\sigma_0$, *and blows up to infinity in* $\Omega_{0,1}$ *as* $\lambda \uparrow d\sigma_1$.

Figure 4.2 shows the map $\lambda \mapsto \theta_{[\lambda,\Omega]}(x)$ for an arbitrary $x \in \Omega_{0,1}$. According to Theorem 4.1, it bifurcates from 0 at $\lambda = d\sigma_0$, increases and blows up as $\lambda \uparrow d\sigma_1$. The dashed line emphasizes the loss of the stability of $(\lambda, 0)$ as λ crosses $d\sigma_0$.

Proof of Theorem 4.1: Suppose f satisfies (Hf) and (Hg), and u is a positive solution of (4.1). Then, by Lemma 1.6, $u \gg 0$ and

$$\lambda = \lambda_1[-d\Delta - af(\cdot, u), \Omega]. \tag{4.10}$$

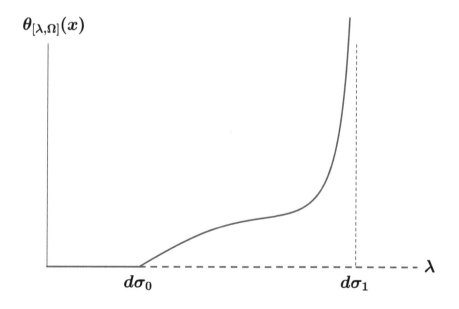

FIGURE 4.2: The map $\lambda \mapsto \theta_{[\lambda,\Omega]}(x)$ for $x \in \Omega_{0,1}$.

As $af(\cdot, u) < 0$ in Ω, by the monotonicity of the principal eigenvalue with respect to the potential, (4.10) implies

$$\lambda = \lambda_1[-d\Delta - af(\cdot, u), \Omega] > \lambda_1[-d\Delta, \Omega] = d\sigma_0.$$

Similarly, by the monotonicity with respect to the domain, we find from (4.10) that

$$\lambda < \lambda_1[-d\Delta - af(\cdot, u), \Omega_{0,1}^1] = \lambda_1[-d\Delta, \Omega_{0,1}^1] = d\sigma_1$$

because $a = 0$ in $\Omega_{0,1}^1$. Therefore, (4.6) is necessary for the existence of a positive solution of (4.1).

Now, we will prove that (4.6) is also sufficient for the existence of a positive solution. Suppose (4.6). According to Theorem 1.7, it suffices to construct a positive supersolution of (4.1). It should be noted that, thanks again to Theorem 1.7, (4.1) cannot admit a positive supersolution for $\lambda \geq d\sigma_1$. To construct the supersolution we proceed as follows. For each $1 \leq j \leq q_0$ and sufficiently small $\delta > 0$ consider the open δ–neighborhoods

$$\Omega_{\delta,j}^i := \left\{ x \in \Omega \; : \; \text{dist}\,(x, \Omega_{0,j}^i) < \delta \right\}, \qquad 1 \leq j \leq q_0, \quad 1 \leq i \leq m_j,$$

which have been represented in Figure 4.3 in the special case when $a(x)$ has the nodal configuration sketched in Figure 1.1. In Figure 4.3, $\Omega_{\delta,1}$ consists of $\Omega_{0,1} \cup \gamma_1$ and the set of points $x \in \Omega_-$ such that $\text{dist}\,(x, \gamma_1) < \delta$, and $\Omega_{\delta,2}$ consists of $\bar{\Omega}_{0,2}$ and the set of points $x \in \Omega_-$ with $\text{dist}\,(x, \gamma_2) < \delta$.

By the continuous dependence of the principal eigenvalues with respect to the domain,

$$\lim_{\delta \downarrow 0} \lambda_1[-\Delta, \Omega_{\delta,j}^i] = \lambda_1[-\Delta, \Omega_{0,j}^i] = \sigma_j, \qquad 1 \leq j \leq q_0, \quad 1 \leq i \leq m_j.$$

Thus, by their monotonicity properties, it becomes apparent from (4.4) that there is $\delta_0 > 0$ such that, for every $\delta \in (0, \delta_0)$,

$$\begin{cases} d\sigma_0 < \lambda < \lambda_1[-d\Delta, \Omega_{\delta,1}^i] < d\sigma_1, & 1 \leq i \leq m_1, \\ d\sigma_{j-1} < \lambda_1[-d\Delta, \Omega_{\delta,j}^i] < d\sigma_j, & 2 \leq j \leq q_0, \; 1 \leq i \leq m_j. \end{cases} \qquad (4.11)$$

By Hypothesis (Ha), $\partial\Omega \cap \partial\Omega_-$ consists of a finite number of components of $\partial\Omega$ that simultaneously are components of $\partial\Omega_-$, say Γ_j, $1 \leq j \leq n$, if it is non-empty. In the situation described by Figure 1.1, $\partial\Omega \cap \partial\Omega_- = \emptyset$, while in the context described by Figure 4.1, Γ_1 is the unique component of $\partial\Omega \cap \partial\Omega_-$.

Suppose $\partial\Omega \cap \partial\Omega_-$ is non-empty and consider the open neighborhoods of Γ_j defined by

$$\mathcal{N}_{-,j,\delta} := \{x \in \Omega_- \; : \; \text{dist}\,(x, \Gamma_j) < \delta\}, \qquad 1 \leq j \leq n.$$

For sufficiently small $\delta > 0$, $\mathcal{N}_{-,j,\delta}$ is a $\mathcal{C}^{2+\nu}$–subdomain of Ω_- such that

$$\lim_{\delta \downarrow 0} |\mathcal{N}_{-,j,\delta}| = 0 \qquad \text{for all } 1 \leq j \leq n.$$

Thus, δ_0 can be shortened, if necessary, so that

$$\min_{1 \leq j \leq n} \lambda_1[-d\Delta, \mathcal{N}_{-,j,\delta}] > \lambda \qquad \text{for all } 0 < \delta < \delta_0. \tag{4.12}$$

Pick $\lambda \in (d\sigma_0, d\sigma_1)$, $\delta \in (0, \delta_0)$, and, for every $1 \leq j \leq q_0$ and $1 \leq i \leq m_j$, let $\varphi^i_{\delta,j} \gg 0$ be a principal eigenfunction of $\lambda_1[-d\Delta, \Omega^i_{\delta,j}]$; it is unique up to a multiplicative constant. Similarly, for every $1 \leq k \leq n$, let $\psi_{-,k,\delta}$ be a principal eigenfunction of $\lambda_1[-d\Delta, \mathcal{N}_{-,k,\delta}]$. Then, consider the function Φ defined through

$$\Phi := \begin{cases} \varphi^i_{\delta,j} & \text{in } \bar{\Omega}^i_{\delta/2,j}, \quad 1 \leq j \leq q_0, \ 1 \leq i \leq m_j, \\ \psi_{-,k,\delta} & \text{in } \bar{\mathcal{N}}_{-,k,\delta/2}, \quad 1 \leq k \leq n, \\ \xi_- & \text{in } K_\delta := \{x \in \Omega_- \ : \ \text{dist}(x, \partial\Omega_-) \geq \delta/2\}, \end{cases} \tag{4.13}$$

where ξ_- is any smooth extension, positive and bounded away from zero, of the function

$$\bigotimes_{j=1}^{q_0} \bigotimes_{i=1}^{m_j} \varphi^i_{\delta,j} \bigotimes_{k=1}^{n} \psi_{-,k,\delta}$$

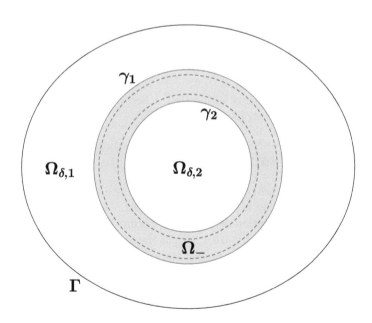

FIGURE 4.3: The δ-neighborhoods $\Omega_{\delta,1}$ and $\Omega_{\delta,2}$.

to the compact set K_δ. Note that ξ_- exists because, for every $1 \le j \le q_0$ and $1 \le i \le m_j$, the function $\varphi_{\delta,j}^i$ is positive and bounded away from zero on $\Omega_- \cap \partial\Omega_{\delta/2,j}^i$, and, similarly, for every $1 \le k \le n$, the function $\psi_{-,k,\delta}$ is positive and bounded away from zero on $\Omega_- \cap \partial\mathcal{N}_{-,k,\delta/2}$. Figure 4.4 shows a genuine profile of Φ when $a(x)$ has the nodal configuration of Figure 1.1.

We claim that the function

$$\bar{u} := \kappa\Phi \qquad (4.14)$$

is a supersolution of (4.1) for sufficiently large $\kappa > 1$. Indeed, by construction,

$$\kappa\Phi = 0 \qquad \text{on} \quad \partial\Omega,$$

because, thanks to (Ha), any component of $\partial\Omega$ must be either a component of $\partial\Omega_-$, say Γ_k, where $\Phi = \psi_{-,k,\delta} = 0$, or a component of $\partial\Omega_0$ and, in this case it is a component of $\partial\Omega_{\delta,j}^i$ for some $1 \le j \le q_0$ and $1 \le i \le m_j$, where also $\Phi = \varphi_{\delta,j}^i = 0$.

Moreover, thanks to (4.11), for every $1 \le j \le q_0$ and $1 \le i \le m_j$, in $\Omega_{\delta/2,j}^i$ we have that,

$$-d\Delta(\kappa\Phi) = -\kappa d\Delta\varphi_{\delta,j}^i = \kappa\lambda_1[-d\Delta, \Omega_{\delta,j}^i]\varphi_{\delta,j}^i > \kappa\lambda\varphi_{\delta,j}^i$$
$$\ge \lambda\kappa\varphi_{\delta,j}^i + af(\cdot, \kappa\varphi_{\delta,j}^i)\kappa\varphi_{\delta,j}^i = \lambda\kappa\Phi + af(\cdot, \kappa\Phi)\kappa\Phi$$

for all $\kappa > 0$. Similarly, thanks to (4.12), for every $1 \le k \le n$, in $\mathcal{N}_{-,k,\delta/2}$ we have that

$$-d\Delta(\kappa\Phi) = -\kappa d\Delta\psi_{-,k,\delta} = \kappa\lambda_1[-d\Delta, \mathcal{N}_{-,k,\delta}]\psi_{-,k,\delta} > \kappa\lambda\psi_{-,k,\delta}$$
$$\ge \lambda\kappa\psi_{-,k,\delta} + af(\cdot, \kappa\psi_{-,k,\delta})\kappa\psi_{-,k,\delta} = \lambda\kappa\Phi + af(\cdot, \kappa\Phi)\kappa\Phi.$$

Finally, note that

$$-d\Delta(\kappa\Phi) \ge \lambda\kappa\Phi + af(\cdot, \kappa\Phi)\kappa\Phi \qquad \text{in} \ K_\delta$$

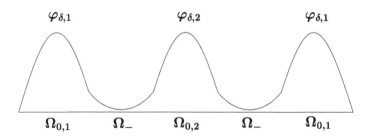

FIGURE 4.4: The profile of the *supersolution element* Φ.

if and only if

$$\frac{-d\Delta\xi_-}{\xi_-} \geq \lambda + af(\cdot, \kappa\xi_-) \qquad \text{in } K_\delta. \tag{4.15}$$

As $-a(x)$ and $\xi_-(x)$ are positive and bounded away from zero in $K_\delta \Subset \Omega_-$, by Hypothesis (Hg), (4.15) holds for sufficiently large $\kappa > 1$. Therefore, (4.14) indeed provides us with a supersolution of (4.1) for sufficiently large κ. Consequently, by Theorem 1.7, (4.1) possesses a positive solution if, and only if, (4.6) holds. Moreover, it is strongly positive and unique if it exists. Throughout the rest of this book, we will denote it by

$$\theta(\lambda) = \theta_{[\lambda,\Omega]} = \theta_{[\lambda,\Omega,0]}.$$

The proof of Theorem 2.3 can be adapted *mutatis mutandis* to show (4.8) and to establish the monotonicity of the solution operator (4.7). Actually, since $\theta_{[\mu,\Omega]}$ is a strict positive supersolution of (4.1) for all $\mu \in (\lambda, d\sigma_1)$, it follows from Lemma 1.8 that

$$\theta_{[\lambda,\Omega]} \ll \theta_{[\mu,\Omega]} \qquad \text{if } d\sigma_0 < \lambda < \mu < d\sigma_1.$$

So, the technical details of these proofs are omitted here.

To complete the proof of the theorem it remains to show (4.9). Fix $\tilde{\lambda} \in (d\sigma_0, d\sigma_1)$ and $1 \leq i \leq m_1$, and let $\eta > 0$ be such that

$$\theta_{[\tilde{\lambda},\Omega]} > \eta\,\varphi_{0,1}^i \qquad \text{in } \Omega_{0,1}^i, \tag{4.16}$$

where $\varphi_{0,1}^i \gg 0$ is a principal eigenfunction associated with

$$\sigma_1 = \lambda_1[-\Delta, \Omega_{0,1}^i].$$

Differentiating with respect to λ the realization of (4.1) at $\theta(\lambda)$ and rearranging terms yields

$$\begin{cases} \mathfrak{L}(\lambda, \theta(\lambda))D_\lambda\theta(\lambda) = \theta(\lambda) & \text{in } \Omega, \\ D_\lambda\theta(\lambda) = 0 & \text{on } \partial\Omega, \end{cases} \tag{4.17}$$

where

$$\mathfrak{L}(\lambda, \theta(\lambda)) := -d\Delta - a\frac{\partial f}{\partial u}(\cdot, \theta(\lambda))\theta(\lambda) - af(\cdot, \theta(\lambda)) - \lambda. \tag{4.18}$$

Indeed, (4.17) follows straight away from the differentiability of the solution operator $\lambda \mapsto \theta(\lambda) = \theta_{[\lambda,\Omega]}$, by adapting the proof of Theorem 2.3. Since

$$-a\frac{\partial f}{\partial u}(\cdot, \theta(\lambda))\theta(\lambda) > 0 \qquad \text{in } \Omega, \qquad \lambda = \lambda_1[-d\Delta - af(\cdot, \theta(\lambda)), \Omega],$$

from the monotonicity of the principal eigenvalue with respect to the potential, we can infer that

$$\lambda_1[\mathfrak{L}(\lambda, \theta(\lambda)), \Omega] > \lambda_1[-d\Delta - af(\cdot, \theta(\lambda)) - \lambda, \Omega] = 0.$$

Therefore, by Theorem 1.1, $\mathfrak{L}(\lambda, \theta(\lambda))$ satisfies the strong maximum principle in Ω under Dirichlet boundary conditions. In particular, since $\theta(\lambda) > 0$ in Ω, it follows from (4.17) that

$$D_\lambda \theta(\lambda) = \mathfrak{L}^{-1}(\lambda, \theta(\lambda))\theta(\lambda) \gg 0 \tag{4.19}$$

and, consequently, the map $\lambda \mapsto \theta(\lambda)$ is *strongly increasing*.

On the other hand, by (4.16), we find that, for every $\lambda \in (\tilde{\lambda}, d\sigma_1)$,

$$\theta_{[\lambda,\Omega]} \gg \theta_{[\tilde{\lambda},\Omega]} > \eta\,\varphi_{0,1}^i \quad \text{in} \quad \Omega_{0,1}^i.$$

Hence, owing to (4.17),

$$\begin{cases} (-d\Delta - \lambda)\, D_\lambda\theta(\lambda) = \theta(\lambda) > \eta\,\varphi_{0,1}^i & \text{in} \quad \Omega_{0,1}^i, \\ D_\lambda\theta(\lambda) \geq 0 & \text{on} \quad \partial\Omega_{0,1}^i, \end{cases} \tag{4.20}$$

because $a = 0$ in $\Omega_{0,1}^i$ and $D_\lambda\theta(\lambda) \geq 0$ in $\bar{\Omega}$. Moreover,

$$\lambda_1[-d\Delta - \lambda, \Omega_{0,1}^i] = d\sigma_1 - \lambda > 0$$

for all $\lambda \in (\tilde{\lambda}, d\sigma_1)$. Therefore, thanks again to Theorem 1.1, (4.20) implies

$$D_\lambda\theta(\lambda) \gg \Psi_\lambda \quad \text{in} \quad \Omega_{0,1}^i, \tag{4.21}$$

where Ψ_λ is the unique solution of

$$\begin{cases} (-d\Delta - \lambda)\,\Psi_\lambda = \eta\,\varphi_{0,1}^i & \text{in} \quad \Omega_{0,1}^i, \\ \Psi_\lambda = 0 & \text{on} \quad \partial\Omega_{0,1}^i. \end{cases}$$

A direct calculation shows that

$$\Psi_\lambda = \frac{\eta}{d\sigma_1 - \lambda}\,\varphi_{0,1}^i \qquad \text{for all} \ \ \lambda \in (\tilde{\lambda}, d\sigma_1)$$

and hence, it follows from (4.21) that

$$\theta(\lambda) > \theta(\tilde{\lambda}) + \eta\log\frac{\tilde{\lambda} - d\sigma_1}{\lambda - d\sigma_1}\varphi_{0,1}^i \quad \text{in} \quad \Omega_{0,1}^i.$$

Consequently,

$$\lim_{\lambda \uparrow d\sigma_1} \theta(\lambda) = \infty \quad \text{uniformly in compact subsets of} \ \ \Omega_{0,1}^i.$$

As this occurs for all $i \in \{1, ..., m_1\}$, (4.9) holds, which ends the proof. □

Remark 4.2 When either $\partial\Omega \cap \partial\Omega_- = \emptyset$, or $\partial\Omega \cap \partial\Omega_- \neq \emptyset$ but $a(x) > 0$ for all $x \in \partial\Omega \cap \partial\Omega_-$, the construction of the supersolution is easier, as in such case in the construction of Φ one does not need to invoke the eigenfunctions $\psi_{-,k,\delta}$, $1 \leq k \leq n$.

4.2   Associated inhomogeneous problems

The main result of this section is the next one.

Theorem 4.3 *Let $M > 0$ and suppose a and f satisfy (Ha) and (Hf)–(Hg), respectively. Then, the problem*

$$\begin{cases} -d\Delta u = \lambda u + a(x)f(x, u)u & in \ \ \Omega, \\ u = M & on \ \ \partial\Omega, \end{cases} \tag{4.22}$$

possesses a positive solution if, and only if,

$$\lambda < d\sigma_1 \qquad \left(\Leftrightarrow \frac{\lambda}{\sigma_1} < d \right). \tag{4.23}$$

Moreover, it is unique if it exists, and if we denote it by $\theta_{[\lambda,\Omega,M]}$, then

(a) *For every $\lambda < d\sigma_1$ and any positive strict subsolution (resp. supersolution) $\underline{u}$ (resp. $\bar{u}$) of (4.22),*

$$\underline{u} \ll \theta_{[\lambda,D,M]} \qquad (resp. \ \ \bar{u} \gg \theta_{[\lambda,D,M]}).$$

(b) $0 \ll \theta_{[\lambda,\Omega,M_1]} \ll \theta_{[\mu,\Omega,M_2]}$ *in Ω if*

$$-\infty < \lambda \le \mu < d\sigma_1, \quad 0 < M_1 \le M_2, \quad \mu - \lambda + M_2 - M_1 > 0.$$

(c) $\lim_{\lambda\uparrow d\sigma_1} \theta_{[\lambda,\Omega,M]} = \infty$ *uniformly in compact subsets of $\Omega_{0,1}$.*

(d) *Naturally,*

$$\lim_{M\downarrow 0} \theta_{[\lambda,\Omega,M]} = \begin{cases} 0, & if \ \lambda \le d\sigma_0, \\ \theta_{[\lambda,\Omega]}, & if \ d\sigma_0 < \lambda < d\sigma_1, \end{cases}$$

where $\theta_{[\lambda,\Omega]}$ is the unique positive solution of (4.1).

(e) *Furthermore,*

$$\lim_{\lambda\downarrow-\infty} \theta_{[\lambda,\Omega,M]} = 0 \quad uniformly \ in \ compact \ subsets \ of \ \Omega. \tag{4.24}$$

Proof: Since the solutions of (4.22) are strict supersolutions of (4.1), by Theorems 1.7 and 4.1, the problem (4.22) cannot admit a positive solution for $\lambda \ge d\sigma_1$. Thus, $\lambda < d\sigma_1$ is necessary for its existence. Suppose $\lambda < d\sigma_1$. Then, again by Theorem 1.7, to establish the existence and the uniqueness of a positive solution of (4.22), it suffices to construct a positive supersolution. As the supersolution already constructed in the proof of Theorem 4.1, $\bar{u} = \kappa\Phi$, with Φ given by (4.13), for sufficiently large $\kappa > 1$, satisfies $\bar{u} = 0$ on $\partial\Omega$, we should modify Φ on a neighborhood of $\partial\Omega$ in order to get $\Phi > 0$ on $\partial\Omega$. This can be easily accomplished by taking the *extended δ–neighborhoods*

$$\tilde{\Omega}^i_{\delta,j} := \left\{ x \in \mathbb{R}^N \; : \; \text{dist}\,(x, \Omega^i_{0,j}) < \delta \right\}, \qquad 1 \le j \le q_0, \quad 1 \le i \le m_j,$$

and

$$\tilde{\mathcal{N}}_{-,k,\delta} := \left\{ x \in \mathbb{R}^N \; : \; \text{dist}\,(x, \Gamma_j) < \delta \right\}, \qquad 1 \le k \le n,$$

for sufficiently small $\delta > 0$, instead of $\Omega^i_{\delta,j}$ and $\mathcal{N}_{-,k,\delta}$, respectively. As the components of $\partial\Omega$ must lie in the interior of these open δ–neighborhoods, and, consequently, the principal eigenfunctions associated with $\lambda_1[-d\Delta, \tilde{\Omega}^i_{\delta,j}]$ and $\lambda_1[-d\Delta, \tilde{\mathcal{N}}_{-,k,\delta}]$ are positive on them, the corresponding Φ must be positive on $\partial\Omega$ and hence, for sufficiently large $\kappa > 1$, it becomes apparent that $\kappa\Phi > M$ on $\partial\Omega$, which ends the proof of the existence and the uniqueness.

Part (a) is a straightforward consequence of Lemma 1.8, and Part (b) is a corollary of Part (a). As an easy consequence, $\theta_{[\lambda,\Omega]} \ll \theta_{[\lambda,\Omega,M]}$ for all $M > 0$. Therefore, Part (c) follows from (4.9). The proof of Proposition 2.5 can be adapted *mutatis mutandis* to show Part (d). It remains to prove Part (e).

Suppose $\lambda < 0$. Then, since $a \le 0$ and $f \ge 0$, we have that

$$-d\Delta\theta_{[\lambda,\Omega,M]} \le 0 \qquad \text{in } \Omega$$

and hence, $\theta_{[\lambda,\Omega,M]}$ is subharmonic in Ω. Thus, by the maximum principle,

$$M = \max_{\partial\Omega} \theta_{[\lambda,\Omega,M]} = \max_{\Omega} \theta_{[\lambda,\Omega,M]}$$

and so, $\theta_{[\lambda,\Omega,M]}$ provides us with a positive strict subsolution of the linear problem

$$\begin{cases} -d\Delta u = \lambda u & \text{in } \Omega, \\ u = M & \text{on } \partial\Omega. \end{cases} \tag{4.25}$$

Since

$$\lambda_1[-d\Delta - \lambda] = d\sigma_0 - \lambda > 0,$$

it follows from Theorem 1.1 that $\theta_{[\lambda,\Omega,M]} \ll u_\lambda$, where $u_\lambda \gg 0$ stands for the unique solution of (4.25). Consequently, it suffices to show that

$$\lim_{\lambda\downarrow-\infty} u_\lambda = 0 \quad \text{uniformly in compact subsets of } \Omega. \tag{4.26}$$

As the change of variable $u = v + M$ transforms (4.25) in

$$\begin{cases} \left(\frac{d}{\lambda}\Delta + 1\right) v = -M & \text{in } \Omega, \\ v = 0 & \text{on } \partial\Omega, \end{cases}$$

(4.26) follows from the fact that the unique solution of

$$\begin{cases} (-\varepsilon\Delta + 1) w = -M & \text{in } \Omega, \\ w = 0 & \text{on } \partial\Omega, \end{cases} \tag{4.27}$$

denoted by w_ε, satisfies

$$\lim_{\varepsilon \downarrow 0} w_\varepsilon = -M \quad \text{uniformly in compact subsets of } \Omega. \tag{4.28}$$

Indeed, let $K \subset \Omega$ be compact and

$$\delta := \operatorname{dist}(K, \partial\Omega) > 0.$$

Then, setting $R := \delta/2$, for every $x \in K$ we have that $B_R(x) \subset \Omega$. Let $\varphi \gg 0$ be the principal eigenfunction associated to $-\Delta$ in $B_R := B_R(0)$, subject to homogeneous Dirichlet boundary conditions, normalized so that $\varphi(0) = \|\varphi\|_{\mathcal{C}(\bar{B}_R)} = 1$. Then, for every $x \in K$, $\varphi_x := \varphi(\cdot - x)$ provides us with the principal eigenfunction of $-\Delta$ in $B_R(x)$ normalized so that

$$\varphi_x(x) = \|\varphi_x\|_{\mathcal{C}(\bar{B}_R(x))} = \varphi(0) = 1.$$

Note that

$$\lambda_1[-\Delta, B_R(x)] = \lambda_1[-\Delta, B_R] \qquad \text{for all } x \in K.$$

We claim that, for every $x \in K$,

$$\overline{w} := \frac{-M}{\varepsilon\lambda_1[-\Delta, B_R] + 1}\,\varphi_x$$

is a supersolution of

$$\begin{cases} (-\varepsilon\Delta + 1)\,w = -M & \text{in } B_R(x), \\ w = 0 & \text{on } \partial B_R(x). \end{cases} \tag{4.29}$$

Indeed, by construction,

$$(-\varepsilon\Delta + 1)\overline{w} = -M \quad \text{in } B_R(x),$$

and $\overline{w} = 0 > -M$ on $\partial B_R(x)$. Moreover, $\underline{w} := -M$ is a subsolution of (4.29) for all $x \in K$. Therefore, since

$$\lambda_1[-\varepsilon\Delta + 1, B_R(x)] = \varepsilon\lambda_1[-\Delta, B_R] + 1 > 0$$

for all $x \in K$ and $\varepsilon > 0$, it becomes apparent (e.g., from Theorem 1.1) that

$$\underline{w} = -M \le w_\varepsilon \le \overline{w} = \frac{-M}{\varepsilon\lambda_1[-\Delta, B_R] + 1}\,\varphi_x \quad \text{in } B_R(x)$$

for all $x \in K$. Consequently,

$$-M \le w_\varepsilon(x) \le \frac{-M}{\varepsilon\lambda_1[-\Delta, B_R] + 1} \qquad \text{for all } x \in K.$$

As these global estimates, for every compact subset $K \subset \Omega$, provide us with (4.28), the proof is complete. $\quad\square$

As a byproduct of Theorem 4.3, the next result holds.

Corollary 4.4 *Suppose $M > 0$, a satisfies* (Ha), *and f satisfies* (Hf) *and* (Hg). *Let Γ_j, $1 \leq j \leq m$, be the components of $\partial\Omega$ and consider any nonempty proper subset $J \subset \{1, ..., m\}$. Then, the problem*

$$
\begin{cases}
-d\Delta u = \lambda u + a(x)f(x, u)u & in \quad \Omega, \\
u = M & on \quad \cup_{j \in J} \Gamma_j, \\
u = 0 & on \quad \cup_{j \notin J} \Gamma_j,
\end{cases}
\tag{4.30}
$$

possesses a positive solution if, and only if, $\lambda < d\sigma_1$. Moreover, it is unique if it exists and if we denote it by $\theta_{[\lambda,\Omega,M,J]}$, then:

(a) *For every $\lambda < d\sigma_1$ and any positive strict subsolution (resp. supersolution) $\underline{u}$ (resp. $\overline{u}$) of* (4.30),

$$
\underline{u} \ll \theta_{[\lambda,\Omega,M,J]} \qquad (resp. \quad \overline{u} \gg \theta_{[\lambda,\Omega,M,J]}).
$$

(b) $0 \ll \theta_{[\lambda,\Omega,M_1,J]} \ll \theta_{[\mu,\Omega,M_2,J]}$ *in Ω if*

$$
-\infty < \lambda \leq \mu < d\sigma_1, \quad 0 < M_1 \leq M_2, \quad \mu - \lambda + M_2 - M_1 > 0.
$$

(c) $\lim_{\lambda \uparrow d\sigma_1} \theta_{[\lambda,\Omega,M,J]} = \infty$ *uniformly in compact subsets of $\Omega_{0,1}$.*

(d) *Naturally,*

$$
\lim_{M \downarrow 0} \theta_{[\lambda,\Omega,M,J]} = \begin{cases} 0, & if \ \lambda \leq d\sigma_0, \\ \theta_{[\lambda,\Omega]}, & if \ d\sigma_0 < \lambda < d\sigma_1, \end{cases}
$$

where $\theta_{[\lambda,\Omega]}$ is the unique positive solution of (4.1).

(e) *Furthermore,*

$$
\lim_{\lambda \downarrow -\infty} \theta_{[\lambda,\Omega,M,J]} = 0 \quad uniformly \ in \ compact \ subsets \ of \ \Omega.
$$

Proof: Suppose (4.30) has a positive solution, θ, and $\lambda > d\sigma_0$. Then, $\overline{u} := \theta$ provides us with a positive supersolution of (4.1) and hence, thanks to Theorem 1.7, (4.1) possesses a positive solution. Thus, by Theorem 4.1, $\lambda < d\sigma_1$. Therefore, $\lambda < d\sigma_1$ is necessary for the existence of a positive solution of (4.30). Suppose $\lambda < d\sigma_1$. Then, the unique positive solution of (4.22), $\theta_{[\lambda,\Omega,M]}$, given by Theorem 4.3, provides us with a positive supersolution of (4.30). Consequently, by Theorem 1.7, (4.30) possesses a unique positive solution, denoted by $\theta_{[\lambda,\Omega,M,J]}$.

Part (a) follows from Lemma 1.8 and Part (b) is a byproduct of Part (a). When $\lambda \uparrow d\sigma_1$, (4.6) holds and hence, by Theorem 4.1, (4.1) has a unique positive solution, $\theta_{[\lambda,\Omega]}$, which is a strict subsolution of (4.30). Thus, by Part (a), we find that

$$
\theta_{[\lambda,\Omega]} \ll \theta_{[\lambda,\Omega,M,J]} \qquad in \ \Omega. \tag{4.31}
$$

Part (c) is a direct consequence from (4.9) and (4.31). As J is a proper subset of $\{1, ..., m\}$, $\theta_{[\lambda,\Omega,M,J]}$ is a strict subsolution of (4.22) and hence, due to Theorem 4.3(a), we obtain that

$$\theta_{[\lambda,\Omega,M,J]} \ll \theta_{[\lambda,\Omega,M]} \quad \text{in } \Omega,$$

which, combined with (4.31), yields

$$\theta_{[\lambda,\Omega]} \ll \theta_{[\lambda,\Omega,M,J]} \ll \theta_{[\lambda,\Omega,M]} \quad \text{in } \Omega. \tag{4.32}$$

Therefore, Part (d) is a consequence from (4.32) and Theorem 4.3(d). Similarly, Part (e) follows from (4.32) and Theorem 4.3(e). □

4.3   Hierarchic chain of inhomogeneous problems

The main goal of this section is analyzing the following family of semilinear elliptic boundary value problems

$$\begin{cases} -d\Delta u = \lambda u + a(x)f(x,u)u & \text{in } \Omega_j, \\ u = 0 & \text{on } \partial\Omega_j \cap \partial\Omega, \\ u = M & \text{on } \partial\Omega_j \setminus \partial\Omega, \end{cases} \tag{4.33}$$

where $1 \leq j \leq q_0$ and $M > 0$. The solutions of (4.2) will be constructed from these solutions by letting $M \uparrow \infty$. The main result of this section can be stated as follows.

Theorem 4.5 *Suppose $M > 0$, $a(x)$ satisfies (Ha), f satisfies (Hf)–(Hg), and $1 \leq j \leq q_0$. Then, (4.33) possesses a positive solution, $\Theta_{[\lambda,\Omega_j,M]}$, if, and only if,*

$$\lambda < d\sigma_{j+1}. \tag{4.34}$$

Moreover, it is unique if it exists and

(a) *For every $\lambda < d\sigma_{j+1}$ and any positive strict subsolution (resp. supersolution) $\underline{u}$ (resp. $\bar{u}$) of (4.33),*

$$\underline{u} \ll \Theta_{[\lambda,\Omega_j,M]} \quad (\textit{resp. } \bar{u} \gg \Theta_{[\lambda,\Omega_j,M]}).$$

(b) $0 \ll \Theta_{[\lambda,\Omega_j,M_1]} \ll \Theta_{[\mu,\Omega_j,M_2]}$ *in Ω_j if*

$$-\infty < \lambda \leq \mu < d\sigma_{j+1}, \quad 0 < M_1 \leq M_2, \quad \mu - \lambda + M_2 - M_1 > 0.$$

(c) $\lim_{\lambda \uparrow d\sigma_{j+1}} \Theta_{[\lambda,\Omega_j,M]} = \infty$ *uniformly in compact subsets of $\Omega_{0,j+1}$.*

By (4.5), $\Omega_j = \Omega_-$ if $j = q_0$. In such case, since $\sigma_{q_0+1} := \infty$, according to Theorem 4.5, (4.33) possesses a unique positive solution for each $\lambda \in \mathbb{R}$, $\Theta_{[\lambda,\Omega_-,M]}$.

Proof: Suppose $j \le q_0 - 1$. Then, the lower order refuges of Ω_j, ordered by the size of the principal eigenvalue of $-\Delta$, are $\Omega_{0,j+1}^i$, $1 \le i \le m_{j+1}$. As

$$\lambda_1[-d\Delta, \Omega_{0,j+1}^i] = d\sigma_{j+1}, \qquad 1 \le i \le m_j,$$

Parts (a), (b) and (c) are immediate consequences from Corollary 4.4.

Suppose $j = q_0$. Then, the construction of a supersolution of (4.33) can be easily accomplished by adapting the proofs of Theorems 4.1 and 4.3. Indeed, let γ_j, $1 \le j \le n$, be the components of $\partial\Omega_-$ and consider the open neighborhoods of γ_j, $1 \le j \le n$, defined by

$$\mathcal{N}_{-,j,\delta} := \{x \in \Omega_- \; : \; \mathrm{dist}\,(x, \gamma_j) < \delta\}, \qquad 1 \le j \le n.$$

For sufficiently small $\delta > 0$, $\mathcal{N}_{-,j,\delta}$ is a $\mathcal{C}^{2+\nu}$–subdomain of Ω_- such that

$$\lim_{\delta \downarrow 0} |\mathcal{N}_{-,j,\delta}| = 0 \qquad \text{for all } 1 \le j \le n.$$

Thus, δ_0 can be shortened, if necessary, so that

$$\min_{1 \le j \le n} \lambda_1[-d\Delta, \mathcal{N}_{-,j,\delta}] > \lambda \qquad \text{for all } 0 < \delta < \delta_0.$$

Now, for every $1 \le j \le n$, let $\psi_{-,j,\delta}$ be a principal eigenfunction associated to $\lambda_1[-d\Delta, \mathcal{N}_{-,j,\delta}]$ and consider

$$\Phi := \begin{cases} \psi_{-,j,\delta} & \text{in } \bar{\mathcal{N}}_{-,j,\delta/2}, \quad 1 \le j \le n, \\[2mm] \xi_- & \text{in } K_\delta := \{x \in \Omega_- \; : \; \mathrm{dist}\,(x, \partial\Omega_-) \ge \delta/2\}, \end{cases}$$

where ξ_- is any smooth extension, positive and bounded away from zero, of the function

$$\bigotimes_{j=1}^n \psi_{-,j,\delta}$$

to the compact set K_δ. Note that ξ_- exists because, for every $1 \le j \le n$, $\psi_{-,j,\delta}$ is positive and bounded away from zero on $\Omega_- \cap \partial\mathcal{N}_{-,j,\delta/2}$. Arguing as in the proofs of Theorems 4.1 and 4.3, it is easily seen that, under these conditions, for sufficiently large $\kappa > 1$, the function $\bar{u} := \kappa\Phi$ provides us with a supersolution of (4.33) for $j = q_0$. Therefore, when $j = q_0$, Theorem 1.7 and Lemma 1.8 end the proof. $\square$

By letting $M \uparrow \infty$, the problems (4.33) provide us with the minimal positive solutions of (4.2). To construct the maximal ones we must shorten the Ω_j's in order to construct the minimal large solutions supported in these shortened open subsets. As in Theorem 3.4, the maximal solution of (4.2) is the limit of these minimal large solutions.

It should be noted that, thanks to (Ha), for every $1 \le j \le q_0$, the components of Ω_j either are components of $\partial\Omega_-$ or are common components of $\partial\Omega$ and $\partial\Omega_{0,k}^i$ for some $j + 1 \le k \le q_0$ and $1 \le i \le m_k$. Consequently, $\partial\Omega_j \setminus \partial\Omega \subset \partial\Omega_-$. Moreover, although Ω_j is not necessarily connected, it consists of a finite number of (disjoint) components of class $\mathcal{C}^{2+\nu}$.

Subsequently, for sufficiently large $n \in \mathbb{N}$, say $n \ge n_0$, we also consider the open subsets of Ω_j defined through

$$\Omega_{j,n} := \{x \in \Omega_j \ : \ \text{dist}\,(x, \partial\Omega_j \setminus \partial\Omega) > 1/n\}. \tag{4.35}$$

Also, we suppose that n_0 has been chosen sufficiently large so that $\Omega_{j,n}$ is of class $\mathcal{C}^{2+\nu}$ for all $n \ge n_0$. By construction,

$$\bar{\Omega}_{j,n} \subset \Omega_{j,n+1} \subset \Omega_j, \qquad \Omega_j = \bigcup_{n \ge n_0} \Omega_{j,n}.$$

Lastly, we also set

$$\Omega_{-,n} := \Omega_{q_0,n}, \qquad n \ge n_0.$$

The next result is a direct consequence of Theorem 4.5.

Theorem 4.6 *Suppose $M > 0$, $a(x)$ satisfies (Ha), f satisfies (Hf)–(Hg), $n \ge n_0$, and $1 \le j \le q_0$. Then,*

$$\begin{cases} -d\Delta u = \lambda u + a(x)f(x, u)u & in \quad \Omega_{j,n}, \\ u = 0 & on \quad \partial\Omega_{j,n} \cap \partial\Omega, \\ u = M & on \quad \partial\Omega_{j,n} \setminus \partial\Omega, \end{cases} \tag{4.36}$$

possesses a positive solution if, and only if,

$$\lambda < d\sigma_{j+1}, \tag{4.37}$$

$\Theta_{[\lambda,\Omega_{j,n},M]}$. *Moreover, it is unique if it exists and*

(a) *For every $\lambda < d\sigma_{j+1}$ and any positive strict subsolution (resp. supersolution) $\underline{u}$ (resp. $\bar{u}$) of (4.36),*

$$\underline{u} \ll \Theta_{[\lambda,\Omega_{j,n},M]} \qquad (resp. \ \ \bar{u} \gg \Theta_{[\lambda,\Omega_{j,n},M]}).$$

(b) $0 \ll \Theta_{[\lambda,\Omega_{j,n},M_1]} \ll \Theta_{[\mu,\Omega_{j,n},M_2]}$ *in Ω_j if*

$$-\infty < \lambda \le \mu < d\sigma_{j+1}, \quad 0 < M_1 \le M_2, \quad \mu - \lambda + M_2 - M_1 > 0.$$

(c) *If, in addition, $j \le q_0 - 1$, then*

$$\lim_{\lambda \uparrow d\sigma_{j+1}} \Theta_{[\lambda,\Omega_{j,n},M]} = \infty \qquad uniformly \ in \ compact \ subsets \ of \ \Omega_{0,j+1}.$$

4.4 Large solutions of arbitrary order $j \in \{1, ..., q_0\}$

Throughout this section, the notation introduced in the previous ones will be kept. Its main result can be stated as follows.

Theorem 4.7 *Suppose $a(x)$ satisfies* (Ha), *f satisfies* (KO), *and $1 \leq j \leq q_0$. Then,* (4.2) *possesses a positive solution if and only if $\lambda < d\sigma_{j+1}$. Moreover, in such case, the point-wise limit*

$$\overset{\min}{L_{[\lambda,\Omega_j]}} := \lim_{M\uparrow\infty} \Theta_{[\lambda,\Omega_j,M]} \qquad (4.38)$$

provides us with the minimal positive large solution of (4.2). *Similarly, the point-wise limit*

$$\overset{\max}{L_{[\lambda,\Omega_j]}} := \lim_{n\uparrow\infty} \overset{\min}{L_{[\lambda,\Omega_{j,n}]}} \qquad (4.39)$$

is the maximal positive large solution of (4.2).
    Furthermore, in case $1 \leq j \leq q_0 - 1$,

$$\lim_{\lambda\uparrow d\sigma_{j+1}} \overset{\min}{L_{[\lambda,\Omega_j]}} = \infty \quad \text{uniformly in compact subsets of } \Omega_{0,j+1}. \qquad (4.40)$$

Proof: Suppose $1 \leq j \leq q_0 - 1$ and $\lambda < d\sigma_{j+1}$, or $j = q_0$ and $\lambda \in \mathbb{R}$. Then, by Theorem 4.5, $\Theta_{[\lambda,\Omega_j,M]}$ is well defined. Moreover, $M \mapsto \Theta_{[\lambda,\Omega_j,M]}$ is increasing and hence, the point-wise limit (4.38) is well defined in Ω_j. To show that it is the minimal positive large solution of (4.2), we proceed as follows. First, we will prove that it is finite in Ω_-. Indeed, set

$$b_M := \max_{\bar{\Omega}_-} \Theta_{[\lambda,\Omega_j,M]} \geq M.$$

Since $\Omega_- \subset \Omega_j$, $\Theta_{[\lambda,\Omega_j,M]}$ is a subsolution of

$$\begin{cases} -d\Delta u = \lambda u + a(x)f(x,u)u & \text{in } \Omega_-, \\ u = b_M & \text{on } \partial\Omega_-. \end{cases} \qquad (4.41)$$

By Theorem 2.4, (4.41) has a unique positive solution, $\theta_{[\lambda,\Omega_-,b_M]}$, and

$$\Theta_{[\lambda,\Omega_j,M]} \leq \theta_{[\lambda,\Omega_-,b_M]} \quad \text{in } \Omega_-. \qquad (4.42)$$

According to Theorem 3.4, we already know that

$$\overset{\min}{L_{[\lambda,\Omega_-]}} := \lim_{b\uparrow\infty} \theta_{[\lambda,\Omega_-,b]}$$

is the minimal positive large solution of

$$-d\Delta u = \lambda u + a(x)f(x,u)u \qquad (4.43)$$

in Ω_-. Thus, letting $M \uparrow \infty$ in (4.41) yields

$$L^{\min}_{[\lambda,\Omega_j]} \leq L^{\min}_{[\lambda,\Omega_-]} \quad \text{in } \Omega_-,$$

and hence,

$$L^{\min}_{[\lambda,\Omega_j]} < \infty \qquad \text{in } \Omega_-, \tag{4.44}$$

as claimed above. Note that (4.44) establishes that $L^{\min}_{[\lambda,\Omega_j]} < \infty$ in Ω_j if $j = q_0$.
Suppose $j \leq q_0 - 1$ and $\lambda < d\sigma_{j+1}$, fix $n \geq n_0$ and set

$$b_n := \max_{\partial\Omega_{j,n}\setminus\partial\Omega} L^{\min}_{[\lambda,\Omega_j]}.$$

Then, since $\partial\Omega_{j,n} \setminus \partial\Omega \subset \Omega_-$, we find from (4.44) that $b_n < \infty$. Moreover, thanks to (4.38),

$$b_n \geq \max_{\partial\Omega_{j,n}\setminus\partial\Omega} \Theta_{[\lambda,\Omega_j,M]} \qquad \text{for all } M > 0$$

and so, $\Theta_{[\lambda,\Omega_j,M]}$ is a subsolution of

$$\begin{cases} -d\Delta u = \lambda u + a(x)f(x,u)u & \text{in } \Omega_{j,n}, \\ u = 0 & \text{on } \partial\Omega_{j,n} \cap \partial\Omega, \\ u = b_n & \text{on } \partial\Omega_{j,n} \setminus \partial\Omega, \end{cases} \tag{4.45}$$

for all $M > 0$. Since $\lambda < d\sigma_{j+1}$, by Theorem 4.6, (4.45) has a unique positive solution, $\Theta_{[\lambda,\Omega_{j,n},b_n]}$, and

$$\Theta_{[\lambda,\Omega_j,M]} \leq \Theta_{[\lambda,\Omega_{j,n},b_n]} \quad \text{in } \Omega_{j,n} \tag{4.46}$$

for all $M > 0$. Consequently, letting $M \uparrow \infty$ in (4.46) yields

$$L^{\min}_{[\lambda,\Omega_j]} \leq \Theta_{[\lambda,\Omega_{j,n},b_n]} \quad \text{in } \Omega_{j,n} \quad \text{for all } n \geq n_0.$$

Therefore, $L^{\min}_{[\lambda,\Omega_j]} < \infty$ in Ω_j, as claimed above.

Furthermore, owing to (4.46), one can easily adapt the proof of Proposition 3.3 to prove that $L^{\min}_{[\lambda,\Omega_j]} \in \mathcal{C}^{2+\nu}(\Omega_j)$ provides us with a positive solution of (4.2). Actually, $L^{\min}_{[\lambda,\Omega_j]}$ is the minimal positive solution. Indeed, let L be any positive solution of (4.2). Then, for every $M > 0$, there exist a constant $C > 0$ and an integer $n \in \mathbb{N}$ such that

$$\Theta_{[\lambda,\Omega_j,M]} \leq C \leq L \quad \text{in } \Omega_j \setminus \bar{\Omega}_{j,n}.$$

By Theorem 4.6, this estimate implies

$$\Theta_{[\lambda,\Omega_j,M]} \leq \Theta_{[\lambda,\Omega_{j,n},C]} \leq L \quad \text{in } \Omega_{j,n},$$

and consequently,

$$\Theta_{[\lambda,\Omega_j,M]} \leq L \quad \text{in } \Omega_j$$

for all $M > 0$. Therefore, letting $M \uparrow \infty$ yields

$$L_{[\lambda,\Omega_j]}^{\min} \leq L,$$

which establishes the minimality of $L_{[\lambda,\Omega_j]}^{\min}$.

Next, we will show that the point-wise limit (4.39) is well defined and that it provides us with the maximal positive solution of (4.2). As in the proof of Theorem 3.4, for every $M > 0$ and $n \geq n_0$, we set

$$b := \max_{\bar{\Omega}_{j,n}} \Theta_{[\lambda,\Omega_{j,n+1},M]}.$$

Then, owing to Theorem 4.6, it follows from (4.38) that

$$\Theta_{[\lambda,\Omega_{j,n+1},M]} \ll \Theta_{[\lambda,\Omega_{j,n},b+1]} \ll L_{[\lambda,\Omega_{j,n}]}^{\min} \quad \text{in} \quad \Omega_{j,n}$$

for all $n \geq n_0$ and $M > 0$. Thus, letting $M \uparrow \infty$ yields

$$L_{[\lambda,\Omega_{j,n+1}]}^{\min} \leq L_{[\lambda,\Omega_{j,n}]}^{\min} \quad \text{in} \quad \Omega_{j,n} \quad \forall\, n \geq n_0. \tag{4.47}$$

Hence, the point-wise limit (4.39) is well defined in Ω_j. Moreover, since $\bar{\Omega}_{j,n} \subset \Omega_{j,n+1}$ for all $n \geq n_0$, by the Schauder interior estimates, it follows from (4.47) that, for every $n \geq n_0$, there exists a constant $C(n) > 0$ such that

$$\|L_{[\lambda,\Omega_{j,m}]}^{\min}\|_{\mathcal{C}^{2+\nu}(\bar{\Omega}_{j,n})} \leq C(n) \quad \text{for all} \quad m \geq n+1.$$

Consequently, since the injection

$$\mathcal{C}^{2+\nu}(\bar{\Omega}_{j,n}) \hookrightarrow \mathcal{C}^2(\bar{\Omega}_{j,n})$$

is compact and the point-wise limit (4.39) unique, we find that

$$\lim_{n \to \infty} \|L_{[\lambda,\Omega_{j,n}]}^{\min} - L_{[\lambda,\Omega_j]}^{\max}\|_{\mathcal{C}^2(K)} = 0$$

for all compact subsets $K \subset \Omega_j$. Therefore, $L_{[\lambda,\Omega_j]}^{\max} \in \mathcal{C}^2(\Omega_j)$ solves (4.43) in Ω_j. By elliptic regularity, $L_{[\lambda,\Omega_j]}^{\max} \in \mathcal{C}^{2+\nu}(\Omega_j)$.

Let L be an arbitrary positive solution of (4.2) and set

$$b_n := \max_{\bar{\Omega}_{j,n}} L, \qquad n \geq n_0.$$

Then, $L|_{\Omega_{j,n}}$ is a subsolution of

$$\begin{cases} -d\Delta u = \lambda u + a(x)f(x,u)u & \text{in} \quad \Omega_{j,n}, \\ u = b_n & \text{on} \quad \partial\Omega_{j,n}, \end{cases}$$

for all $n \geq n_0$. Hence, by Theorem 4.6 and (4.38), we find that

$$L \leq \Theta_{[\lambda,\Omega_{j,n},b_n]} \leq L_{[\lambda,\Omega_{j,n}]}^{\min} \quad \text{in} \quad \Omega_{j,n} \quad \forall\, n \geq n_0.$$

Consequently, letting $n \uparrow \infty$ yields

$$L \leq L^{\max}_{[\lambda,\Omega_j]} \qquad \text{in } \Omega_j.$$

Therefore, $L^{\max}_{[\lambda,\Omega_j]}$ indeed is the maximal positive solution of (4.2).

To conclude the proof of the existence, it remains to show that $\lambda < d\sigma_{j+1}$ is necessary for the existence of a positive solution of (4.2) if $1 \leq j \leq q_0 - 1$. Suppose $1 \leq j \leq q_0 - 1$ and let L be a positive solution of (4.2) for some $\lambda \in \mathbb{R}$. Then, for every $n \geq n_0$, $L|_{\Omega_{j,n}}$ provides us with a positive strict supersolution of $-d\Delta - af(\cdot, L) - \lambda$ in $\Omega_{j,n}$ under homogeneous Dirichlet boundary conditions and hence, by Theorem 1.1,

$$\lambda_1[-d\Delta - af(\cdot, L) - \lambda, \Omega_{j,n}] > 0.$$

Therefore, by the monotonicity of the principal eigenvalue with respect to the domain,

$$\lambda < \lambda_1[-\Delta - af(\cdot, L), \Omega_{j,n}] < \lambda_1[-d\Delta - af(\cdot, L), \Omega^1_{0,j+1}] = d\sigma_{j+1},$$

because $a = 0$ in $\Omega^1_{0,j+1}$. Consequently, (4.2) admits a positive solution if, and only if, $\lambda < d\sigma_{j+1}$.

To complete the proof, it remains to show (4.40). Suppose $j \leq q_0 - 1$. Then, due to Theorem 4.1, for every $\lambda \in (d\lambda_1[-\Delta, \Omega_j], d\sigma_{j+1})$, the problem

$$\begin{cases} -d\Delta u = \lambda u + af(\cdot, u)u & \text{in } \Omega_j, \\ u = 0 & \text{on } \partial\Omega_j, \end{cases}$$

has a unique positive solution, $\theta_{[\lambda,\Omega_j]}$. As $\theta_{[\lambda,\Omega_j]}$ provides us with a positive strict subsolution of (4.33) for all $M > 0$, by Theorem 4.5, it is apparent that

$$\theta_{[\lambda,\Omega_j]} \ll \Theta_{[\lambda,\Omega_j,M]} \ll L^{\min}_{[\lambda,\Omega_j]} \qquad \text{in } \Omega_j.$$

Therefore, (4.40) is a direct consequence from (4.9).   □

4.5   Limiting behavior of the positive solution as $\lambda \uparrow d\sigma_1$

The next result ascertains the point-wise behavior of the unique positive solution, $\theta_{[\lambda,\Omega]}$, of (4.1) as $\lambda \uparrow d\sigma_1$. It sharpens (4.9) substantially.

Theorem 4.8 *Suppose $a(x)$ satisfies* (Ha) *and f satisfies* (KO). *Then,*

$$\lim_{\lambda \uparrow d\sigma_1} \theta_{[\lambda,\Omega]} = \begin{cases} \infty & \text{in } \bar{\Omega}_{0,1} \setminus \partial\Omega, \\ L^{\min}_{[d\sigma_1,\Omega_1]} & \text{in } \Omega_1 = \Omega \setminus \bar{\Omega}_{0,1}. \end{cases} \tag{4.48}$$

Moreover, $\theta_{[\lambda,\Omega]}$ approximates ∞ uniformly in compact subsets of $\bar{\Omega}_{0,1} \setminus \partial\Omega$.

Proof: Thanks to Theorem 4.5 with $j = 1$, $\Theta_{[d\sigma_1,\Omega_1,M]}$ is well defined for all $M > 0$. Set

$$M_\lambda := \max_{\partial\Omega_1} \theta_{[\lambda,\Omega]}, \qquad \lambda \in (d\sigma_0, d\sigma_1).$$

Then, according to Theorems 4.1, 4.5 and 4.7,

$$\theta_{[\lambda,\Omega]} \leq \Theta_{[d\sigma_1,\Omega_1,M_\lambda]} \leq \lim_{M\uparrow\infty} \Theta_{[d\sigma_1,\Omega_1,M]} = L_{[d\sigma_1,\Omega_1]}^{\min} \qquad \text{in } \Omega_1. \qquad (4.49)$$

On the other hand, since $\lambda \mapsto \theta_{[\lambda,\Omega]}$ is point-wise increasing, the limit

$$L := \lim_{\lambda\uparrow d\sigma_1} \theta_{[\lambda,\Omega]} \qquad \text{in } \Omega \qquad (4.50)$$

is well defined and, thanks to (4.9),

$$L = \infty \qquad \text{in } \Omega_{0,1}.$$

Moreover, letting $\lambda \uparrow d\sigma_1$ in (4.49) yields

$$L \leq L_{[d\sigma_1,\Omega_1]}^{\min} \qquad \text{in } \Omega_1 \qquad (4.51)$$

and therefore, L is finite in Ω_1. Let $\Omega_{1,n}$, $n \geq n_0$, be the open subsets of Ω_1 defined by (4.35). Due to (4.49), for every $n \geq n_0$, there exists a constant $C_n > 0$ such that

$$\theta_{[\lambda,\Omega]} \leq C_n \qquad \text{in } \bar\Omega_{1,n+1}$$

for all $\lambda \in (d\sigma_0, d\sigma_1)$. Thus, by the Schauder interior estimates, there exists a constant $\tilde{C}_n > 0$ such that

$$\|\theta_{[\lambda,\Omega]}\|_{C^{2+\nu}(\bar\Omega_{1,n})} \leq \tilde{C}_n \qquad \text{for all } \lambda \in (d\sigma_0, d\sigma_1).$$

Hence, as the injection

$$C^{2+\nu}(\bar\Omega_{1,n}) \hookrightarrow C^2(\bar\Omega_{1,n})$$

is compact and the point-wise limit (4.50) unique, necessarily

$$\lim_{\lambda\uparrow d\sigma_1} \|\theta_{[\lambda,\Omega_1]} - L\|_{C^2(\bar\Omega_{1,n})} = 0$$

for all $n \geq n_0$. Consequently, $L \in C^2(\Omega_1)$ and it solves (4.43) in Ω_1. By elliptic regularity, we actually have that $L \in C^{2+\nu}(\Omega_1)$.

According to (4.51), to prove the identity

$$L = L_{[d\sigma_1,\Omega_1]}^{\min} \qquad (4.52)$$

it suffices to show that $L = \infty$ on $\partial\Omega_1 \setminus \partial\Omega$. This property follows easily from

$$\lim_{\lambda\uparrow d\sigma_1} \min_{\partial\Omega_1\setminus\partial\Omega} \theta_{[\lambda,\Omega]} = \infty. \qquad (4.53)$$

Indeed, suppose (4.53) has been proven. Then, setting

$$m_\lambda := \min_{\partial\Omega_1\setminus\partial\Omega} \theta_{[\lambda,\Omega]}, \qquad \lambda \in (d\sigma_0, d\sigma_1),$$

it follows from Theorems 4.1 and 4.6 that

$$\theta_{[\lambda,\Omega]} \geq \Theta_{[\lambda,\Omega_1,m_\lambda]} \geq \Theta_{[d\sigma_0,\Omega_1,m_\lambda]} \qquad \text{in} \quad \Omega_1$$

for all $\lambda \in (d\sigma_0, d\sigma_1)$. Thus, letting $\lambda \uparrow d\sigma_1$, we find from (4.38), (4.50) and (4.53) that

$$L \geq L_{[d\sigma_0,\Omega_1]}^{\min} \qquad \text{in} \quad \Omega_1.$$

Therefore, L must solve (4.33) with $j = 1$ and $\lambda = d\sigma_1$. Consequently, (4.53) implies (4.52). Note that (4.9) and (4.53) imply

$$\lim_{\lambda\uparrow d\sigma_1} \theta_{[\lambda,\Omega]} = \infty \qquad \text{in} \quad \bar{\Omega}_{0,1} \setminus \partial\Omega.$$

Thus, $1/\theta_{[\lambda,\Omega]}$, $\lambda \in (d\sigma_0, d\sigma_1)$, provides us with a decreasing family of continuous functions point-wise converging to zero in $\bar{\Omega}_{0,1} \setminus \partial\Omega$. Therefore, by Dini's theorem, it approximates zero uniformly in compact subsets of $\bar{\Omega}_{0,1} \setminus \partial\Omega$. As a byproduct, $\theta_{[\lambda,\Omega]}$ approximates ∞ uniformly in compact subsets of $\bar{\Omega}_{0,1} \setminus \partial\Omega$ as $\lambda \uparrow d\sigma_1$. Consequently, to conclude the proof of the theorem, it suffices to prove (4.53). The proof will proceed by contradiction. Suppose (4.53) is not true. As $\partial\Omega_1 \setminus \partial\Omega$ consists of the components of $\partial\Omega_{0,1}^i$, $1 \leq i \leq m_1$, which are not components of $\partial\Omega$, there exist some of these components, e.g., γ, such that

$$b := \min_\gamma L = \min_{\partial\Omega_1\setminus\partial\Omega} L \in (0, \infty). \tag{4.54}$$

As γ is of class $\mathcal{C}^{2+\nu}$, there exist $R > 0$ and a map $Y : \gamma \to \Omega_{0,1}$ such that, for every $x \in \gamma$,

$$B_R(Y(x)) \subset \Omega_{0,1}, \quad \bar{B}_R(Y(x)) \cap \partial\Omega = \varnothing, \quad \bar{B}_R(Y(x)) \cap \gamma = \{x\}.$$

It should be noted that this can be done because, thanks to Theorem 1.9 of [163], the open set Ω_1 satisfies the uniform exterior sphere property in the strong sense on γ, because γ is of class $\mathcal{C}^2$.

By (4.50) and (4.54), for each $\lambda \in (d\sigma_0, d\sigma_1)$, there exists $x_\lambda \in \gamma$ such that

$$\theta_{[\lambda,\Omega]}(x_\lambda) = \min_\gamma \theta_{[\lambda,\Omega]} \leq \min_\gamma L = b.$$

Figure 4.5 sketches the construction of $B_R(Y(x_\lambda))$ for $a(x)$ with the nodal configuration of Figure 1.1. Note that, in that case, $\gamma = \gamma_1$.

By construction,

$$\text{dist}(x_\lambda, Y(x_\lambda)) = R \qquad \text{for all} \quad \lambda \in (d\sigma_0, d\sigma_1).$$

Moreover, the manifold γ_R defined by

$$\gamma_R := \{\, y \in \Omega_{0,1} \ : \ \mathrm{dist}\,(y,\gamma) = 2R \,\}$$

is a compact subset of $\Omega_{0,1}$ and hence, thanks to (4.9),

$$\lim_{\lambda \uparrow d\sigma_1} \theta_{[\lambda,\Omega]} = \infty \quad \text{uniformly in } \gamma_R. \tag{4.55}$$

Setting

$$\Omega_{0,1,R} := \{\, y \in \Omega_{0,1} \ : \ \mathrm{dist}\,(y,\gamma) < 2R \,\},$$

it turns out that $\partial\Omega_{0,1,R}$ consists of two components: γ and γ_R. Let $\tilde{x}_\lambda \in \bar{\Omega}_{0,1,R}$ be such that

$$\theta_{[\lambda,\Omega]}(\tilde{x}_\lambda) = \min_{\bar{\Omega}_{0,1,R}} \theta_{[\lambda,\Omega]} \le \min_\gamma \theta_{[\lambda,\Omega]} = \theta_{[\lambda,\Omega]}(x_\lambda) \le \min_\gamma L = b.$$

We claim that $\tilde{x}_\lambda = x_\lambda$ for λ sufficiently close to $d\sigma_1$. Indeed, suppose $\tilde{x}_\lambda \in \Omega_{0,1,R}$. Then,

$$\nabla\theta_{[\lambda,\Omega]}(\tilde{x}_\lambda) = 0, \qquad \Delta\theta_{[\lambda,\Omega]}(\tilde{x}_\lambda) \ge 0.$$

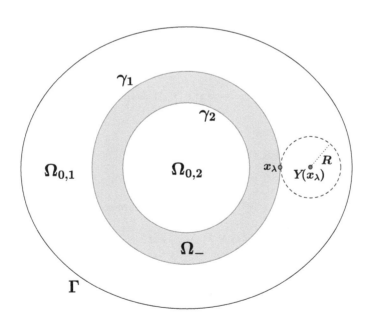

FIGURE 4.5: The ball $B_R(Y(x_\lambda))$.

But, since $\Omega_{0,1,R} \subset \Omega_{0,1}$ and $a = 0$ in $\Omega_{0,1}$, we have that

$$-d\Delta\theta_{[\lambda,\Omega]}(\tilde{x}_\lambda) = \lambda\,\theta_{[\lambda,\Omega]}(\tilde{x}_\lambda) > 0$$

for all $\lambda \in (d\sigma_0, d\sigma_1)$, because $\lambda > 0$ in such range, which is impossible. Thus,

$$\tilde{x}_\lambda \in \partial\Omega_{0,1,R} = \gamma \cup \gamma_R \quad \text{for all } \lambda \in (d\sigma_0, d\sigma_1).$$

Therefore, by (4.55), there exists $\hat{\sigma}_1 \in (\sigma_0, \sigma_1)$ such that

$$\min_{\bar{\Omega}_{0,1,R}} \theta_{[\lambda,\Omega]} = \min_\gamma \theta_{[\lambda,\Omega]} = \theta_{[\lambda,\Omega]}(x_\lambda) \leq b \quad \text{for each } \lambda \in [d\hat{\sigma}_1, d\sigma_1).$$

Consequently, $\tilde{x}_\lambda = x_\lambda$ and

$$\theta_{[\lambda,\Omega]}(x) \geq \theta_{[\lambda,\Omega]}(x_\lambda) \qquad \text{for all} \quad x \in \bar{B}_R(Y(x_\lambda)). \tag{4.56}$$

Subsequently, for every $\alpha > 0$ and $\lambda \in [d\hat{\sigma}_1, d\sigma_1)$, we will consider the *barrier function* ψ_λ defined by

$$\psi_\lambda(x) := e^{-\alpha|x-Y(x_\lambda)|^2} - e^{-\alpha R^2}, \qquad x \in \bar{B}_R(Y(x_\lambda)).$$

A direct calculation shows that, for every $x \in B_R(Y(x_\lambda))$,

$$(-d\Delta - \lambda)\psi_\lambda(x) = \left(2\alpha N d - 4\alpha^2 d|x - Y(x_\lambda)|^2 - \lambda\right)e^{-\alpha|x-Y(x_\lambda)|^2} + \lambda\,e^{-\alpha R^2}$$

and hence, for every

$$x \in A_R := B_R(Y(x_\lambda)) \setminus \bar{B}_{R/2}(Y(x_\lambda)),$$

we have that

$$(-d\Delta - \lambda)\psi_\lambda(x) \leq \left(2\alpha N d - 4\alpha^2 d|x - Y(x_\lambda)|^2 - \lambda\right)e^{-\alpha(R/2)^2} + \lambda\,e^{-\alpha R^2}$$
$$\leq \left(2\alpha N d - 4\alpha^2 d|x - Y(x_\lambda)|^2\right)e^{-\alpha(R/2)^2}$$

for sufficiently large $\alpha > 0$. Consequently, there exist $\alpha > 0$ and $\omega > 0$ such that

$$(-d\Delta - \lambda)\,\psi_\lambda \leq -\omega \quad \text{in} \quad A_R \tag{4.57}$$

for all $\lambda \in [d\hat{\sigma}_1, d\sigma_1)$. Throughout the rest of the proof, we will assume that α has been chosen to satisfy (4.57).

As $\bar{B}_{R/2}(Y(x_\lambda))$ is a compact subset of $\Omega_{0,1}$, (4.9) guarantees that

$$\lim_{\lambda \uparrow d\sigma_1} \min_{\bar{B}_{R/2}(Y(x_\lambda))} \theta_{[\lambda,\Omega]} = \infty.$$

Thus, setting

$$c_\lambda := \frac{\min_{\bar{B}_{R/2}(Y(x_\lambda))} \theta_{[\lambda,\Omega]} - \theta_{[\lambda,\Omega]}(x_\lambda)}{e^{-\alpha R^2/4} - e^{-\alpha R^2}},$$

we have that

$$\lim_{\lambda \uparrow d\sigma_1} c_\lambda = \infty, \tag{4.58}$$

because

$$\theta_{[\lambda,\Omega]}(x_\lambda) \le b \quad \text{for all} \quad \lambda \in [d\hat\sigma_1, d\sigma_1).$$

Moreover, by the definition of c_λ, for every $\lambda \in [d\hat\sigma_1, d\sigma_1)$, we have that

$$\theta_{[\lambda,\Omega]}(x) \ge \theta_{[\lambda,\Omega]}(x_\lambda) + c_\lambda \left(e^{-\alpha R^2/4} - e^{-\alpha R^2} \right) \quad \forall \ x \in \bar B_{R/2}(Y(x_\lambda)). \tag{4.59}$$

Subsequently, for every $\lambda \in [d\hat\sigma_1, d\sigma_1)$, we consider the auxiliary function

$$v_\lambda := \theta_{[\lambda,\Omega]} - \theta_{[\lambda,\Omega]}(x_\lambda) - c_\lambda \psi_\lambda \quad \text{in} \quad A_R.$$

According to (4.59),

$$v_\lambda \ge 0 \quad \text{on} \quad \partial B_{R/2}(Y(x_\lambda)).$$

Moreover, as $\psi_\lambda = 0$ on $\partial B_R(Y(x_\lambda))$, it follows from (4.56) that

$$v_\lambda = \theta_{[\lambda,\Omega]} - \theta_{[\lambda,\Omega]}(x_\lambda) \ge 0 \quad \text{on} \quad \partial B_R(Y(x_\lambda)).$$

Consequently,

$$v_\lambda \ge 0 \quad \text{on} \quad \partial A_R \quad \text{for all} \quad \lambda \in [d\hat\sigma_1, d\sigma_1). \tag{4.60}$$

On the other side, by (4.57), in the annular region A_R we have that

$$(-d\Delta - \lambda) v_\lambda = \lambda \, \theta_{[\lambda,\Omega]}(x_\lambda) - c_\lambda (-d\Delta - \lambda) \psi_\lambda \ge \lambda \theta_{[\lambda,\Omega]}(x_\lambda) + \omega c_\lambda.$$

Hence, thanks to (4.58), it becomes apparent that

$$(-d\Delta - \lambda) v_\lambda > 0 \quad \text{in} \quad A_R \tag{4.61}$$

for $\lambda < d\sigma_1$ sufficiently close to $d\sigma_1$. Therefore, since

$$\lambda < d\sigma_1 = \lambda_1[-d\Delta, \Omega_{0,1}^i] < \lambda_1[-d\Delta, A_R], \qquad 1 \le i \le m_1,$$

owing to Theorem 1.1, we can infer from (4.60) and (4.61) that

$$v_\lambda(x) > 0 \quad \text{if} \quad R/2 < |x - Y(x_\lambda)| < R.$$

Consequently,

$$\theta_{[\lambda,\Omega]}(x) \ge \theta_{[\lambda,\Omega]}(x_\lambda) + c_\lambda \psi_\lambda(x) \quad \text{for all} \quad x \in A_R \tag{4.62}$$

if $\lambda < d\sigma_1$ is sufficiently close to $d\sigma_1$.

Subsequently, we set

$$n_\lambda := \frac{Y(x_\lambda) - x_\lambda}{R}.$$

By definition,

$$\frac{\partial \theta_{[\lambda,\Omega]}}{\partial n_\lambda}(x_\lambda) = \lim_{t \downarrow 0} \frac{\theta_{[\lambda,\Omega]}(x_\lambda + t\, n_\lambda) - \theta_{[\lambda,\Omega]}(x_\lambda)}{t}.$$

Moreover, by (4.62), we find that, for every $t \in (0, R/2)$,

$$\frac{\theta_{[\lambda,\Omega]}(x_\lambda + t\, n_\lambda) - \theta_{[\lambda,\Omega]}(x_\lambda)}{t} \geq \frac{c_\lambda \psi_\lambda(x_\lambda + t\, n_\lambda)}{t}$$

$$= \frac{c_\lambda \left(e^{-\alpha |x_\lambda + t\, n_\lambda - Y(x_\lambda)|^2} - e^{-\alpha R^2} \right)}{t}$$

$$= \frac{c_\lambda \left(e^{-\alpha |t\, n_\lambda - R\, n_\lambda|^2} - e^{-\alpha R^2} \right)}{t}$$

$$= \frac{c_\lambda \left(e^{-\alpha (R-t)^2} - e^{-\alpha R^2} \right)}{t}.$$

Hence, since

$$\lim_{t \downarrow 0} \frac{e^{-\alpha(R-t)^2} - e^{-\alpha R^2}}{t} = 2\alpha R e^{-\alpha R^2},$$

we obtain that

$$\frac{\partial \theta_{[\lambda,\Omega]}}{\partial n_\lambda}(x_\lambda) \geq 2\alpha R e^{-\alpha R^2} c_\lambda.$$

Therefore, thanks to (4.58), we find that

$$\lim_{\lambda \uparrow d\sigma_1} \frac{\partial \theta_{[\lambda,\Omega]}}{\partial n_\lambda}(x_\lambda) = \infty. \tag{4.63}$$

Now, for each $\lambda < d\sigma_1$, $\lambda \sim d\sigma_1$, consider the boundary value problem

$$\begin{cases} -d\Delta u = \lambda u + a f(\cdot, u) u & \text{in} \quad \Omega_1 = \Omega \setminus \bar{\Omega}_{0,1}, \\ u = \theta_{[\lambda,\Omega]}(x_\lambda) & \text{on} \quad \gamma, \\ u = 0 & \text{on} \quad \partial\Omega_1 \setminus \gamma. \end{cases} \tag{4.64}$$

As $\lambda < d\sigma_2$, applying Corollary 4.4 with $\Omega = \Omega_1$, it is apparent that (4.64) possesses a unique positive solution, denoted by

$$\vartheta_\lambda := \theta_{[\lambda,\Omega_1,\theta_{[\lambda,\Omega]}(x_\lambda),\gamma]}.$$

Moreover, since $\theta_{[\lambda,\Omega]}|_{\Omega_1}$ provides us with a positive supersolution of (4.64), we find that

$$\vartheta_\lambda \leq \theta_{[\lambda,\Omega]} \quad \text{in} \quad \bar{\Omega}_1.$$

Therefore, since

$$\vartheta_\lambda(x_\lambda) = \theta_{[\lambda,\Omega]}(x_\lambda),$$

we find that

$$\frac{\partial \vartheta_\lambda}{\partial n_\lambda}(x_\lambda) \geq \frac{\partial \theta_{[\lambda,\Omega]}}{\partial n_\lambda}(x_\lambda).$$

Consequently, by (4.63), we also have that

$$\lim_{\lambda \uparrow d\sigma_1} \frac{\partial \vartheta_\lambda}{\partial n_\lambda}(x_\lambda) = \infty,$$

which is impossible, because ϑ_λ must approximate in $\mathcal{C}^1(\bar{D})$, as $\lambda \uparrow d\sigma_1$, the unique positive solution

$$\vartheta_{d\sigma_1} := \theta_{[d\sigma_1,\Omega_1,b,\gamma]}$$

of the problem

$$\begin{cases} -d\Delta u = d\sigma_1 u + af(\cdot,u)u & \text{in} \quad \Omega_1, \\ u = b & \text{on} \quad \gamma, \\ u = 0 & \text{on} \quad \partial\Omega_1 \setminus \gamma. \end{cases}$$

This contradiction shows (4.53) and completes the proof. $\quad\square$

4.6 Direct proof of Theorem 4.8 when $a \in \mathcal{C}^1(\Omega)$

Throughout this section, as in the proof of Theorem 4.1, we will denote

$$\Omega_{\delta,1}^i := \left\{ x \in \Omega \ : \ \text{dist}\,(x,\Omega_{0,1}^i) < \delta \right\}, \qquad 1 \leq i \leq m_1,$$

for sufficiently small $\delta > 0$. The main goal of this section is to give a direct proof of Theorem 4.8 when

$$\lim_{\delta \downarrow 0} \max_{1 \leq i \leq m_1} \frac{\sup_{\Omega_{\delta,1}^i \setminus \bar{\Omega}_{0,1}^i} |a(x)|}{\delta} = 0. \tag{4.65}$$

This condition holds if, for instance, $a \in \mathcal{C}^1$ in a neighborhood of $\partial\Omega_{0,1} \setminus \partial\Omega$, because, in such case,

$$a(x) = 0 \quad \text{and} \quad \nabla a(x) = 0 \qquad \text{for all } x \in \partial\Omega_{0,1}^i \setminus \partial\Omega, \qquad 1 \leq i \leq m_1.$$

Also, as in the proof of Theorem 4.1, for sufficiently small $\delta \geq 0$, we denote by $\varphi_{\delta,1}^i \gg 0$ the principal eigenfunction associated to $\lambda_1[-\Delta,\Omega_{\delta,1}^i]$, normalized so that

$$\|\varphi_{\delta,1}^i\|_{\mathcal{C}(\bar{\Omega}_{\delta,1}^i)} = 1, \qquad 1 \leq i \leq m_1.$$

According to J. López-Gómez and J. C. Sabina de Lis [181, Th. 3.2], for every $1 \leq i \leq m_1$, we have that

$$\lambda_1[-\Delta,\Omega_{\delta,1}^i] = \lambda_1[-\Delta,\Omega_{0,1}^i] + \mu_i\delta + O(\delta^2) \qquad \text{as } \delta \downarrow 0 \tag{4.66}$$

where

$$\mu_i := -\int_{\partial\Omega^i_{0,1}\setminus\partial\Omega} \left(\frac{\partial\varphi^i_{0,1}}{\partial n_i}\right)^2 dS < 0,$$

because $\Omega^i_{\delta,1}$ is a holomorphic perturbation of $\Omega^i_{0,1}$; n_i stands for the outward unit normal to $\Omega^i_{0,1}$ along its boundary.

Subsequently, we consider the function $u_\delta \in \mathcal{C}(\bar\Omega)$ defined by

$$u_\delta(x) := \begin{cases} C\varphi^i_{\delta,1}(x) & \text{if } x \in \Omega^i_{\delta,1}, \quad 1 \le i \le m_1, \\[2mm] 0 & \text{if } x \in \Omega \setminus \bigcup_{i=1}^{m_1}\Omega^i_{\delta,1}, \end{cases} \tag{4.67}$$

where $C(\delta) > 0$ is a constant to be chosen later so that u_δ becomes a weak subsolution of (4.1). Pick λ satisfying

$$\lambda_1[-d\Delta,\Omega^i_{\delta,1}] < \lambda_1[-d\Delta,\Omega^i_{\frac{\delta}{2},1}] < \lambda < \lambda_1[-d\Delta,\Omega^i_{0,1}] = d\sigma_1, \tag{4.68}$$

for sufficiently small $\delta > 0$ and all $1 \le i \le m_1$. According to a classical result of H. Berestycki and P. L. Lions [25], it is easily seen that u_δ is a weak subsolution of (4.1) if, and only if, for every $1 \le i \le m_1$,

$$-a(x)f(x,C\varphi^i_{\delta,1}(x)) \le \lambda - \lambda_1[-d\Delta,\Omega^i_{\delta,1}] \qquad \text{for all } x \in \Omega^i_{\delta,1},$$

because $u_\delta(x) = 0$ and, so, $f(x,u_\delta(x)) = f(x,0) = 0$ for all $x \in \Omega \setminus \cup_{i=1}^{m_1}\Omega^i_{\delta,1}$ and, according to (4.68), $\lambda > \lambda_1[-d\Delta,\Omega^i_{\delta,1}]$ for all $1 \le i \le m_1$. Thus, by (4.68) and (Ha), a sufficient condition is the following

$$-a(x)f(x,C\varphi^i_{\delta,1}(x)) \le \lambda_1[-d\Delta,\Omega^i_{\frac{\delta}{2},1}] - \lambda_1[-d\Delta,\Omega^i_{\delta,1}] \quad \forall\, x \in \Omega^i_{\delta,1}\setminus\bar\Omega^i_{0,1}.$$

The appropriate choice of C requires analyzing the decay rate as $\delta \downarrow 0$ of the several quantities arising in these estimates. First, note that, thanks to (4.66), we have that

$$\lambda_1[-d\Delta,\Omega^i_{\frac{\delta}{2},1}] - \lambda_1[-d\Delta,\Omega^i_{\delta,1}] = -d\frac{\mu_i}{2}\delta + O(\delta^2) \qquad \text{as } \delta \downarrow 0. \tag{4.69}$$

Moreover, by (Hf) and (Hg), the function

$$F(\xi) := \max_{x\in\bar\Omega} f(x,\xi), \qquad \xi \ge 0,$$

is non-decreasing and it satisfies

$$\lim_{\xi\uparrow\infty} F(\xi) = \infty. \tag{4.70}$$

Also, since the eigenfunctions $\varphi^i_{\delta,1}$, $1 \le i \le m_1$, have linear decay towards zero on $\partial\Omega^i_{\delta,1}$, reasoning as in J. López-Gómez and J. C. Sabina de Lis [181], it becomes apparent that there are two positive constants $C_2 > C_1 > 0$, independent of $\delta \sim 0$, such that, for any compact subset $K_i \subset \bar{\Omega}^i_{0,1} \setminus \partial\Omega$,

$$C_1\delta \leq \inf_{K_i} \varphi^i_{\delta,1}, \qquad \sup_{\Omega^i_{\delta,1}\setminus\bar{\Omega}^i_{0,1}} \varphi^i_{\delta,1} \leq C_2\,\delta, \qquad 1 \leq i \leq m_1. \qquad (4.71)$$

By the previous construction, it becomes apparent from (4.69) that u_δ is a weak subsolution of (4.1) provided

$$F(CC_2\delta) \leq \frac{-\frac{d\mu_i}{2}\delta + O(\delta^2)}{\sup_{\Omega^i_{\delta,1}\setminus\bar{\Omega}^i_{0,1}} a}.$$

Consequently, the optimal choice of the value of C is the following

$$C = C(\delta) := \max_{1\leq i\leq m_1} \frac{F^{-1}\left(\frac{-\frac{d\mu_i}{2}\delta + O(\delta^2)}{\sup_{\Omega^i_{\delta,1}\setminus\bar{\Omega}^i_{0,1}} a}\right)}{C_2\delta}.$$

Note that, thanks to (4.65) and (4.70),

$$\lim_{\delta\downarrow 0}(\delta C(\delta)) = \infty. \qquad (4.72)$$

Moreover, owing to (4.71), we find that

$$u_\delta(x) \geq C_1\delta C(\delta) \quad \text{for all } x \in \bigcup_{i=1}^{m_1} K_i$$

and therefore, thanks to (4.72),

$$\lim_{\delta\downarrow 0} u_\delta = \infty \quad \text{uniformly in } \bigcup_{i=1}^{m_1} K_i. \qquad (4.73)$$

As, according to the proof of Theorem 4.1, the problem (4.1) possesses arbitrarily large supersolutions in the interior of the cone of positive functions in $C^1_0(\bar{\Omega})$ if $\lambda < d\sigma_1$, by the weak counterpart of Theorem 1.2, we find that

$$u_\delta \leq \theta_{[\lambda,\Omega]} \leq \kappa\Phi \qquad \text{in } \bar{\Omega}$$

and consequently, by (4.73),

$$\lim_{\lambda\uparrow d\sigma_1} \theta_{[\lambda,\Omega]} = \infty \qquad \text{uniformly in } \bigcup_{i=1}^{m_1} K_i. \qquad (4.74)$$

As a byproduct,

$$\lim_{\lambda\uparrow d\sigma_1} \theta_{[\lambda,\Omega]}(x) = \infty \qquad \text{for all } x \in \bigcup_{i=1}^{m_1} \bar{\Omega}^i_{0,1} \setminus \partial\Omega = \bar{\Omega}_{0,1} \setminus \partial\Omega.$$

Finally, the proof of the first part of Theorem 4.8 shows (4.48).

4.7 Limiting behavior of the large solutions of order $1 \leq j \leq q_0 - 1$ as $\lambda \uparrow d\sigma_{j+1}$

The main result of this section can be stated as follows.

Theorem 4.9 *Suppose $a(x)$ and f satisfy (Ha) and (KO), respectively. Then, for every $1 \leq j \leq q_0 - 1$,*

$$\lim_{\lambda \uparrow d\sigma_{j+1}} L^{\min}_{[\lambda,\Omega_j]} = \begin{cases} \infty & in \ \bar{\Omega}_{0,j+1} \setminus \partial\Omega, \\ L^{\min}_{[d\sigma_{j+1},\Omega_{j+1}]} & in \ \Omega_{j+1} = \Omega_j \setminus \bar{\Omega}_{0,j+1}. \end{cases} \tag{4.75}$$

Proof: Fix $1 \leq j \leq q_0 - 1$. By Theorem 4.5, for every $M > 0$ and $\lambda, \mu \in \mathbb{R}$ with $\lambda < \mu < d\sigma_{j+1}$, we have that

$$\Theta_{[\lambda,\Omega_j,M]} \ll \Theta_{[\mu,\Omega_j,M]} \qquad in \ \Omega_j$$

and hence, letting $M \uparrow \infty$, we find from Theorem 4.7 that

$$L^{\min}_{[\lambda,\Omega_j]} \leq L^{\min}_{[\mu,\Omega_j]} \qquad in \ \ \Omega_j.$$

Consequently, the point-wise limit

$$L := \lim_{\lambda \uparrow d\sigma_{j+1}} L^{\min}_{[\lambda,\Omega_j]} \qquad in \ \ \Omega_j \tag{4.76}$$

is well defined. By construction, we find from Theorem 4.5 that

$$\theta_{[\lambda,\Omega_j]} \leq \Theta_{[\lambda,\Omega_j,M]} \leq L^{\min}_{[\lambda,\Omega_j]} \qquad for \ all \ M > 0. \tag{4.77}$$

As the lower order refuges of Ω_j, according to the size of the principal eigenvalues of $-\Delta$, are the components of $\Omega_{0,j+1}$, Theorem 4.8 guarantees that

$$\lim_{\lambda \uparrow d\sigma_{j+1}} \theta_{[\lambda,\Omega_j]} = \infty \qquad in \ \bar{\Omega}_{0,j+1} \setminus \partial\Omega. \tag{4.78}$$

Thus, letting $\lambda \uparrow d\sigma_{j+1}$ in (4.77), it follows from (4.76) and (4.78) that

$$L = \infty \qquad in \ \ \bar{\Omega}_{0,j+1} \setminus \partial\Omega. \tag{4.79}$$

Actually, according to (4.53), it becomes apparent that

$$\lim_{\lambda \uparrow d\sigma_{j+1}} m_\lambda = \infty, \qquad m_\lambda := \min_{\partial\Omega_{0,j+1} \setminus \partial\Omega} L^{\min}_{[\lambda,\Omega_j]}. \tag{4.80}$$

Subsequently, for every $M > 0$ and $\lambda < d\sigma_{j+1}$, we set

$$b_M := \max_{\partial\Omega_-} \Theta_{[\lambda,\Omega_j,M]}.$$

Then, by Theorems 4.5 and 4.7,

$$\Theta_{[\lambda,\Omega_j,M]} \ll \theta_{[d\sigma_{j+1},\Omega_-,b_M+1]} \ll \theta_{[d\sigma_{j+1},\Omega_-,\infty]} = L_{[d\sigma_{j+1},\Omega_-]}^{\min} \qquad \text{in} \quad \Omega_-$$

and hence, letting $M \uparrow \infty$, we find from Theorem 4.7 that

$$L_{[\lambda,\Omega_j]}^{\min} \le L_{[d\sigma_{j+1},\Omega_-]}^{\min} \quad \text{in } \Omega_- \text{ for all } \lambda < d\sigma_{j+1}.$$

Therefore, letting $\lambda \uparrow d\sigma_{j+1}$, (4.76) implies that

$$L \le L_{[d\sigma_{j+1},\Omega_-]}^{\min} \qquad \text{in } \Omega_-. \tag{4.81}$$

In particular, L is finite in Ω_-.

Next, we will consider, for sufficiently large $n \in \mathbb{N}$, say $n \ge n_0$, the open sets

$$\Omega_{-,n} := \{\, x \in \Omega_- \ : \ \text{dist}\,(x,\partial\Omega_-) > 1/n \,\};$$

n_0 should be chosen so that $\Omega_{-,n}$ is of class $\mathcal{C}^{2+\nu}$ for all $n \ge n_0$. By construction, it follows from (4.81) that, for every $n \ge n_0$, there exists a constant $C_n > 0$ such that

$$L_{[\lambda,\Omega_j]}^{\min} \le C_n \qquad \text{in } \bar\Omega_{-,n+1}$$

for all $\lambda < d\sigma_{j+1}$. Thus, by the Schauder interior estimates, there exists a constant $\tilde C_n > 0$ such that

$$\|L_{[\lambda,\Omega_j]}^{\min}\|_{\mathcal{C}^{2+\nu}(\bar\Omega_{-,n})} \le \tilde C_n \qquad \text{for all} \quad \lambda < d\sigma_{j+1}.$$

Hence, as the injection

$$\mathcal{C}^{2+\nu}(\bar\Omega_{-,n}) \hookrightarrow \mathcal{C}^2(\bar\Omega_{-,n})$$

is compact and the point-wise limit (4.76) unique, we obtain that

$$\lim_{\lambda\uparrow d\sigma_{j+1}} \|L_{[\lambda,\Omega_j]}^{\min} - L\|_{\mathcal{C}^2(\bar\Omega_{-,n})} = 0$$

for all $n \ge n_0$. As

$$\Omega_- = \bigcup_{n\ge n_0} \Omega_{-,n},$$

it becomes apparent that $L \in \mathcal{C}^2(\Omega_-)$ solves (4.43) in Ω_-. By elliptic regularity, we actually have that $L \in \mathcal{C}^{2+\nu}(\Omega_-)$.

Now, for every $\lambda \in \mathbb{R}$ and $M > m > 0$, we consider the singular boundary value problem

$$\begin{cases} -d\Delta u = \lambda u + a(x)f(x,u)u & \text{in } \Omega_{j+1}, \\ u = M & \text{on } (\partial\Omega_{j+1} \setminus \partial\Omega) \setminus \partial\Omega_{0,j+1}, \\ u = m & \text{on } \partial\Omega_{0,j+1} \setminus \partial\Omega, \\ u = 0 & \text{on } \partial\Omega_{j+1} \cap \partial\Omega. \end{cases} \tag{4.82}$$

By adapting the proof of Corollary 4.4 with $\Omega = \Omega_{j+1}$, it becomes apparent that (4.82) possesses a positive solution if and only if $\lambda < d\sigma_{j+1}$. Moreover, it is unique if it exists. Let us denote it by $\Theta_{[\lambda,\Omega_{j+1},M,m]}$. Then, adapting the proof of Theorem 4.7, it is apparent that

$$L^{\min}_{[\lambda,\Omega_{j+1}]} := \lim_{m\uparrow\infty} \Theta_{[\lambda,\Omega_{j+1},M,m]}.$$

Fix $\lambda < d\sigma_{j+1}$. Then, according to (4.38) and (4.80), there exists $M(\lambda) > m_\lambda$ such that, for every $M \geq M(\lambda)$,

$$L^{\min}_{[\lambda,\Omega_j]} \geq \Theta_{[\lambda,\Omega_j,M(\lambda)]} \geq \Theta_{[\lambda,\Omega_{j+1},M(\lambda),m_\lambda]} \qquad \text{in} \quad \Omega_{j+1}. \qquad (4.83)$$

By (4.80), letting $\lambda \uparrow d\sigma_{j+1}$, it is easy to see that (4.83) implies

$$L \geq L^{\min}_{[d\sigma_{j+1},\Omega_{j+1}]} \qquad \text{in} \quad \Omega_{j+1}.$$

Consequently, by the minimality of $L^{\min}_{[d\sigma_{j+1},\Omega_{j+1}]}$,

$$L = L^{\min}_{[d\sigma_{j+1},\Omega_{j+1}]} \qquad \text{in} \quad \Omega_{j+1},$$

which ends the proof.   □

4.8   Limiting behavior as $\lambda \downarrow -\infty$ of the large solutions

The main result of this section can be stated as follows.

Theorem 4.10 *Suppose $a(x)$ and f satisfy* (Ha) *and* (KO)*, respectively. Then,*

$$\lim_{\lambda\downarrow-\infty} L^{\min}_{[\lambda,\Omega_j]} = 0 \qquad \text{uniformly in compact subsets of } \Omega_j \qquad (4.84)$$

for every $j \in \{1, ..., q_0\}$.

Proof: Fix $j \in \{1, ..., q_0\}$. We begin by establishing that

$$\lim_{\lambda\downarrow-\infty} L^{\min}_{[\lambda,\Omega_j]} = 0 \qquad \text{uniformly in compact subsets of } \Omega_- \subset \Omega_j. \qquad (4.85)$$

Let $x_0 \in \Omega_-$ and $R > 0$ be such that $\bar{B}_R(x_0) \subset \Omega_-$, and set

$$D := B_{\frac{R}{2}}(x_0), \quad \tilde{D} := B_{\frac{R}{4}}(x_0), \quad A := \max_{\bar{B}_R(x_0)} a < 0, \quad \xi := \|L^{\min}_{[\lambda,\Omega_j]}\|_{\mathcal{C}(\bar{B}_R(x_0))}.$$

Then, in $B_R(x_0)$, we have that

$$-d\Delta L^{\min}_{[\lambda,\Omega_j]} = \lambda L^{\min}_{[\lambda,\Omega_j]} + af(\cdot, L^{\min}_{[\lambda,\Omega_j]})L^{\min}_{[\lambda,\Omega_j]} \leq \lambda L^{\min}_{[\lambda,\Omega_j]} + Ag(L^{\min}_{[\lambda,\Omega_j]})L^{\min}_{[\lambda,\Omega_j]}$$

and hence, for every $\lambda < d\sigma_{j+1}$, with $\sigma_{q_0+1} := \infty$, $L_{[\lambda,\Omega_j]}^{\min}$ provides us with a positive subsolution of

$$\begin{cases} -d\Delta u = \lambda u + Ag(u)u & \text{in } D, \\ u = \xi & \text{on } \partial D. \end{cases} \tag{4.86}$$

By Theorem 2.4, (4.86) possesses a unique positive solution, $\vartheta_{[\lambda,D,\xi]}$. Moreover, for every $\lambda < d\sigma_{j+1}$,

$$L_{[\lambda,\Omega_j]}^{\min} \leq \vartheta_{[\lambda,D,\xi]} \qquad \text{in } D. \tag{4.87}$$

According to the proof of Theorem 3.2, the function $\vartheta_{[\lambda,D,\xi]}$ is radially symmetric and, for each $\lambda \in \mathbb{R}$, the (increasing) point-wise limit

$$L_\lambda := \lim_{\zeta \uparrow \infty} \vartheta_{[\lambda,D,\zeta]}$$

provides us with the minimal large positive solution of

$$-d\Delta u = \lambda u + Ag(u)u \tag{4.88}$$

in D. Moreover, L_λ is radially symmetric and

$$L_\lambda(x_0) = \min_D L_\lambda. \tag{4.89}$$

On the other hand, by Theorem 2.4, we have that, for every $\lambda < \mu$,

$$\vartheta_{[\lambda,D,\xi]} \ll \vartheta_{[\mu,D,\xi]} \qquad \text{for all } \xi > 0$$

and hence, letting $\xi \uparrow \infty$, Theorem 3.4 guarantees that

$$L_\lambda \leq L_\mu \qquad \text{in } D \quad \text{if } \lambda < \mu. \tag{4.90}$$

Let $\varphi \gg 0$ be any principal eigenfunction associated to $\lambda_1[-\Delta, \tilde{D}]$. Multiplying the differential equation

$$-d\Delta L_\lambda = \lambda L_\lambda + Ag(L_\lambda)L_\lambda \qquad \text{in } D,$$

by φ and integrating by parts in $\tilde{D}$ yields

$$(\lambda_1[-d\Delta, \tilde{D}] - \lambda) \int_{\tilde{D}} L_\lambda \varphi \, dx - A \int_{\tilde{D}} g(L_\lambda)L_\lambda \varphi \, dx = -d \int_{\partial \tilde{D}} L_\lambda \frac{\partial \varphi}{\partial n} \, dS.$$

Thus, since $A < 0$ and $g \geq 0$, we find from (4.90) that

$$0 < (\lambda_1[-d\Delta, \tilde{D}] - \lambda) \int_{\tilde{D}} L_\lambda \varphi \, dx \leq -d \int_{\partial \tilde{D}} L_\lambda \frac{\partial \varphi}{\partial n} \, dS \leq -d \int_{\partial \tilde{D}} L_0 \frac{\partial \varphi}{\partial n} \, dS$$

for all $\lambda < 0$, because $\frac{\partial \varphi}{\partial n} < 0$ on $\partial \tilde{D}$. Therefore, letting $\lambda \downarrow -\infty$ yields

$$\lim_{\lambda \downarrow -\infty} \int_{\tilde{D}} L_\lambda \varphi \, dx = 0. \tag{4.91}$$

As (4.89) implies

$$L_\lambda(x_0) \int_{\tilde{D}} \varphi \, dx \le \int_{\tilde{D}} L_\lambda \varphi \, dx,$$

it becomes apparent from (4.91) that

$$\lim_{\lambda \downarrow -\infty} L_\lambda(x_0) = 0. \qquad (4.92)$$

By (4.87), we have that

$$L_{[\lambda,\Omega_j]}^{\min} \le \vartheta_{[\lambda,D,\xi]} \le L_\lambda \qquad \text{in } D.$$

Thus, we find from (4.92) that

$$\lim_{\lambda \downarrow -\infty} L_{[\lambda,\Omega_j]}^{\min}(x_0) = 0.$$

Similarly, for every $x \in D = B_{R/2}(x_0)$, we have that

$$D_x := B_{R/2}(x) \subset B_R(x_0)$$

and that $L_{[\lambda,\Omega_j]}^{\min}$ is a positive subsolution of

$$\begin{cases} -d\Delta u = \lambda u + Ag(u)u & \text{in } D_x, \\ u = \xi & \text{on } \partial D_x, \end{cases} \qquad (4.93)$$

for all $\lambda < d\sigma_{j+1}$. Consequently, reasoning as above yields

$$L_{[\lambda,\Omega_j]}^{\min} \le \vartheta_{[\lambda,D_x,\xi]} \le L_{\lambda,x} \qquad \text{in } D_x,$$

where $L_{\lambda,x}$ stands for the minimal positive large solution of (4.88) in D_x.

On the other hand, by the radial symmetry of the minimal large solutions of (4.88) in these balls, necessarily

$$L_{\lambda,x}(y) = L_\lambda(x_0 - x + y) \qquad \text{for all } y \in D_x,$$

and hence,

$$L_{[\lambda,\Omega_j]}^{\min}(x) \le L_{\lambda,x}(x) = L_\lambda(x_0) \qquad \text{for all } x \in D.$$

Therefore, according to (4.92),

$$\lim_{\lambda \downarrow -\infty} L_{[\lambda,\Omega_j]}^{\min} = 0 \qquad \text{uniformly in } D.$$

By compactness, (4.85) holds.

As $\Omega_{q_0} = \Omega_-$, (4.85) is (4.84) when $j = q_0$. Hence, the proof is completed in this case. So, suppose $j \le q_0 - 1$. Let $\Omega_{0,k}^i$ be, with $j + 1 \le k \le q_0$ and $1 \le i \le m_k$, an arbitrary component of $\Omega_0 \cap \Omega_j$. For sufficiently small $\delta > 0$, consider

$$\Omega_{\delta,k}^i := \{x \in \Omega \ : \ \text{dist}\,(x, \Omega_{0,k}^i) < \delta\}.$$

The boundary $\partial\Omega^i_{\delta,k}$ consists of the components of $\partial\Omega^i_{0,k} \cap \partial\Omega$ plus the components of $\partial\Omega^i_{\delta,k} \setminus \partial\Omega$, which are compact subsets of Ω_-. As $a \leq 0$ and $f \geq 0$, for each $\lambda < 0$ we have that

$$-d\Delta L^{\min}_{[\lambda,\Omega_j]} \leq 0 \quad \text{in} \ \ \Omega^i_{\delta,k}$$

and hence, $L^{\min}_{[\lambda,\Omega_j]}$ is subharmonic in $\Omega^i_{\delta,k}$. Thus, by the maximum principle,

$$\max_{\bar{\Omega}^i_{\delta,k}} L^{\min}_{[\lambda,\Omega_j]} = \max_{\partial\Omega^i_{\delta,k}\setminus\partial\Omega} L^{\min}_{[\lambda,\Omega_j]} \tag{4.94}$$

because $L^{\min}_{[\lambda,\Omega_j]} = 0$ on $\partial\Omega^i_{0,k} \cap \partial\Omega$. As, according to (4.85),

$$\lim_{\lambda\downarrow-\infty} \max_{\partial\Omega^i_{\delta,k}\setminus\partial\Omega} L^{\min}_{[\lambda,\Omega_j]} = 0,$$

the identity (4.94) ends the proof. $\quad\square$

4.9 Comments on Chapter 4

The most pioneering result characterizing the existence of positive solutions of (4.1) is Theorem 1 of H. Brézis and L. Oswald [31], received by the editors of *Nonlinear Analysis* on November 7, 1984. By using some classical minimization methods, it was shown that (4.1) possesses a positive solution if, and only if,

$$\lambda_1[-d\Delta - V_0(x),\Omega] < 0 < \lambda_1[-d\Delta - V_\infty(x),\Omega], \tag{4.95}$$

where

$$V_0(x) := \lim_{u\downarrow 0}(\lambda + a(x)f(x,u)) = \lambda$$

and

$$V_\infty(x) := \lim_{u\uparrow\infty}(\lambda + a(x)f(x,u)) = \begin{cases} -\infty & \text{if } x \in \Omega_-, \\ \lambda & \text{if } x \in \Omega_0. \end{cases}$$

Consequently, though the first estimate of (4.95) becomes $\lambda > d\sigma_0$, it is far from evident that the second one should become $\lambda < d\sigma_1$, as required by Theorem 4.1. Actually, to get the equivalence between Theorem 4.1 and [31, Th. 1] one must invoke the fundamental property that

$$\lim_{u\uparrow\infty} \lambda_1[-d\Delta - a(x)f(x,u),\Omega] = d\sigma_1, \tag{4.96}$$

going back to J. López-Gómez [141]. Indeed, thanks to (4.96), the estimate (4.95) can be equivalently written as

$$d\sigma_0 < \lambda < d\sigma_1.$$

Another significant pioneering contribution was made by T. Ouyang [203], on this occasion working with Neumann boundary conditions, as a part of his PhD thesis on Yamabe's problem [240] under the supervision of W. M. Ni. By combining a global continuation argument with the existence of a priori bounds for the positive solutions of (4.1) in compact subsets of $(-\infty, d\sigma_1)$, T. Ouyang [203] established Theorem 4.1 for the special case $f(x, u) = u^{p-1}$. Moreover, he also established that

$$\lim_{\lambda \uparrow d\sigma_1} \|\theta_{[\lambda, \Omega]}\|_{L^2(\Omega)} = \infty. \tag{4.97}$$

A. Ambrosetti and J. L. Gámez [14], and M. A. del Pino [70] slightly refined some of the previous findings of T. Ouyang [203]. However, none of these works stressed the tremendous importance of ordering the several components of Ω_0 according to the size of the principal eigenvalues $\lambda_1[-\Delta, \Omega_{0,j}^i]$, $1 \leq j \leq q_0$, $1 \leq i \leq m_j$, in order to describe the dynamics of (1.1), as was done in [109].

The proof of Theorem 4.1 given here goes back to J. M. Fraile et al. [90], where the method of subsolutions and supersolutions was incorporated for the first time to the world of the degenerate diffusive logistic equations. The reader should be aware that prior to the publication of J. López-Gómez [141, 144] and J. M. Fraile et al. [90], the only available supersolutions were the large positive constants, which fail to be supersolutions if $a^{-1}(0) \neq \emptyset$. Probably, this might explain why H. Brézis and L. Oswald [25] and T. Ouyang [203] did not use the method of subsolutions and supersolutions to obtain their pioneering findings. It is worth emphasizing that, according to J. M. Fraile et al. [90], Theorem 4.1 is valid for general semilinear elliptic equations of the form

$$\mathfrak{L}u = \lambda u + a(x) f(x, u) u$$

under general mixed boundary conditions, with $\mathfrak{L}$ of type (1.48).

The point-wise blow up in $\Omega_{0,1}$ of $\theta_{[\lambda, \Omega]}$ as $\lambda \uparrow d\sigma_1$ was observed by the first time in J. López-Gómez [148, Th. 2.4], received for publication on August 26, 1996, though it was not published electronically until October 29, 1999. As a consequence of this delay, it appeared first in J. López-Gómez and J. C. Sabina de Lis [181, Th. 4.2], which was received for publication on May 8, 1997. Undoubtedly, the point-wise blow-up in all the components of $\Omega_{0,1}$ is substantially sharper than the L^2 estimate (4.97).

The results of Sections 4.1, 4.2 and 4.3 are attributable to R. Gómez-Reñasco and J. López-Gómez [109]. Later, they were substantially refined by J. López-Gómez [147]. Many of these results were part of the PhD thesis of R. Gómez-Reñasco [105], completed in February 1999 under the supervision of the author. Most of these results were found during the summer of 1998. The resulting monograph [109] was submitted to H. Brézis for publication in the *Archive of Rational Mechanics and Analysis* in September 1988. But it was rejected on March 4, 1999. Then, on April 5, 1999, it was submitted to P. H. Rabinowitz for publication in *Nonlinear Analysis*, where it appeared in 2002, almost four years after it was written. The results of [109] were communicated personally by R. Gómez-Reñasco at the Conference on Operator Theory and Its Applications held in Winnipeg (Canada) on October 1998, and by J. López-Gómez at the Conference on Differential Equations honoring A. C. Lazer held in Miami (Florida, USA) on January 8 and 9, 1999.

The stabilization of $\theta_{[\lambda,\Omega]}$ to the minimal large solution in

$$\Omega_1 = \Omega \setminus \bar{\Omega}_{0,1}$$

established by Theorem 4.8 goes back to J. García-Melián et al. [100], whose work was accepted on February 17, 1998, by Professor P. H. Rabinowitz. The divergence to infinity of $\theta_{[\lambda,\Omega]}$ on $\partial\Omega_{0,1} \setminus \partial\Omega$ as $\lambda \uparrow d\sigma_1$ goes back to Theorem 4.3 of J. López-Gómez and J. C. Sabina de Lis [181] when $a(x)$ is a function of class $\mathcal{C}^1$ around $\partial\Omega_{0,1} \setminus \partial\Omega$, which was received by the editors of the *Journal of Differential Equations* on May 8, 1997. The proof given in Section 4.6 is the original one of [181]. The proof of this property in the general case when $a(x)$ is continuous, which is the one adopted in the proof of Theorem 4.8, goes back to Y. Du and Q. Huang [80], which was received by the editors of *SIAM Journal of Mathematical Analysis* on February 22, 1999, and published electronically on November 4, 1999. The proof of Y. Du and Q. Huang [80] uses a very classical device originated by the proof of the boundary lemma of E. Hopf [113] and O. A. Oleinik [201] which was later refined by M. W. Protter and H. F. Weinberger [210] and J. López-Gómez [163]. Naturally, the list of references of [80] includes [90], [100] and [181].

Recently, J. López-Gómez and P. H. Rabinowitz [178] have given a short elementary proof of Theorem 4.8 in the one-dimensional setting and have extended a number of results of this chapter to cover the case of nodal solutions.

Theorem 4.8 is an extremely sharp result establishing that the Harnack inequality is an utterly linear property, because it can fail in nonlinear problems when the parameters involved in their setting change. Indeed, let K be the compact subset of Ω defined by

$$K := \{x \in \Omega \ : \ \text{dist}\,(x, \partial\Omega) \geq \varepsilon\}$$

for sufficiently small $\varepsilon > 0$. Thanks to the Harnack inequality, for every $\lambda \in (d\sigma_0, d\sigma_1)$, there exists a positive constant $H_\lambda > 0$ such that

$$\max_K \theta_{[\lambda,\Omega]} \leq H_\lambda \min_K \theta_{[\lambda,\Omega]}. \tag{4.98}$$

Thus, if there are $\delta > 0$ and $H > 0$ such that

$$\sup_{\lambda \in [d\sigma_1 - \delta, d\sigma_1)} H_\lambda \leq H,$$

then, according to (4.98), we should have

$$\lim_{\lambda \uparrow d\sigma_1} \theta_{[\lambda,\Omega]} = \infty \quad \text{uniformly in } K,$$

because

$$\lim_{\lambda \uparrow d\sigma_1} \theta_{[\lambda,\Omega]} = \infty \quad \text{uniformly in compact subsets of} \quad \bar{\Omega}_{0,1} \setminus \partial\Omega,$$

which contradicts the stabilization of $\theta_{[\lambda,\Omega]}$ to the minimal large solution in Ω_1 established by Theorem 4.8. Therefore,

$$\lim_{\lambda \uparrow d\sigma_1} H_\lambda = \infty.$$

Consequently, the Harnack inequality cannot be applied uniformly in compact intervals of $\lambda \in \mathbb{R}$ in a degenerate nonlinear problem like (4.1).

More recently, F. C. Cirstea and V. Radulescu [57] considered the following singular problem

$$\begin{cases} -\Delta u = au - b(x)f(u) & \text{in } \Omega, \\ u = \infty & \text{on } \partial\Omega, \end{cases} \tag{4.99}$$

where Ω is a regular domain of $\mathbb{R}^N$, $N \geq 1$, a is a constant, $b \in \mathcal{C}(\bar{\Omega})$ satisfies $b \geq 0$, $b \neq 0$, and $f \in \mathcal{C}^1$ is a positive function satisfying (KO) and such that $f(u)/u$ is increasing in $\mathbb{R}_+$. Setting

$$\sigma_1 := \lambda_1[-\Delta, \text{int } b^{-1}(0)],$$

the main result of [57] can be stated as follows.

Theorem 4.11 *Problem (4.99) has a solution if and only if $a \in (-\infty, \sigma_1)$. Moreover, in this case, the solution is unique.*

F. C. Cirstea and V. Radulescu [57] acknowledged that

"We point out that our framework in the above result includes the case when b vanishes at some points of $\partial\Omega$, or even if $b = 0$ on $\partial\Omega$. In this sense, our result responds to a question raised to one of us by Professor Haim Brézis in Paris, May 2001."

The existence result had been already found by R. Gómez-Reñasco and J. López-Gómez [109]. The uniqueness was also known, but the uniqueness discussion is postponed to Part II.

Theorem 4.10 refines a result going back to the proof of Part (b) of J. López-Gómez and M. Molina-Meyer [166, Th. 1.1].

This chapter grew from R. Gómez-Reñasco and J. López-Gómez [109], and J. López-Gómez [147, 151, 160].

All the results of this paper are valid for a slightly more general family of differential equations of the form

$$(-d\Delta + V(x))u = \lambda u + a(x)f(x,u)u$$

with $V \in \mathcal{C}^\nu(\bar{\Omega})$, or simply $V \in L^\infty(\Omega)$ if we seek for solutions in $W^{2,p}(\Omega)$, $p > N$.

Chapter 5

Dynamics: Metasolutions

The main goal of this chapter is ascertaining the asymptotic behavior of the solutions of (1.1) as $t \uparrow \infty$ according to the each of the different ranges of values of the parameter $\lambda \in \mathbb{R}$. From the point of view of population dynamics, these behaviors can be briefly sketched as follows:

- If $\lambda \leq d\sigma_0$, then the inhabiting region Ω cannot support the species u.

- If $d\sigma_0 < \lambda < d\sigma_1$, then the species u exhibits logistic growth in Ω.

- If $q_0 \geq 2$ and $d\sigma_j \leq \lambda < d\sigma_{j+1}$ for some $j \in \{1, ..., q_0 - 1\}$, then u has Malthusian growth in

$$\left(\bar{\Omega}_{0,1} \cup \cdots \cup \bar{\Omega}_{0,j} \right) \setminus \partial\Omega,$$

  and logistic growth in the complement

$$\Omega_j := \Omega \setminus \left(\bar{\Omega}_{0,1} \cup \cdots \cup \bar{\Omega}_{0,j} \right).$$

- If $\lambda \geq d\sigma_{q_0}$, then the species u exhibits Malthusian growth in $\Omega \setminus \bar{\Omega}_-$ and logistic growth in Ω_-.

These findings are extremely relevant from the point of view of the applications of the abstract mathematical theory developed in this book to population dynamics, since they provide simultaneous effects of the growth laws of Malthus and Verhulst within the same territory Ω, which seems extremely realistic in applications. Rather naturally, the growth of the species should be severely limited in the regions with a serious shortcoming of resources; however, growth may be huge within patches where resources are abundant, which might help show why agriculture facilitated the emergence of human groups whose size gradually increased over the last ten thousands years; almost nothing on an evolutionary scale, until humans developed the extremely populated areas they inhabit today. Simultaneously, in unfavorable areas, where agriculture was not facilitated by the intricate nature of the territory, as occurs in most rain forest areas, the human population numbers did not vary substantially.

The distribution of this chapter is the following. Section 5.1 states the main theorem of the chapter, Section 5.2 applies it to discuss the dynamics of (1.1) in a special case where $q_0 = 2$, Section 5.3 studies a numerical example within the setting of Section 5.2, and, finally, Section 5.4 gives the detailed proof of the main theorem.

5.1   Concept of metasolution: The main theorem

The concept of *metasolution* will substantially shorten the statement of the main theorem of this chapter. It should be remembered that the Ω_j's, $1 \leq j \leq q_0$, were already defined by (4.3). Roughly defined, a *metasolution* is a generalized solution whose set of singularities might have positive measure. In the context of this book, they are the extensions by ∞ of the large solutions in the Ω_j's.

Definition 5.1 (Metasolution) *For every $1 \leq j \leq q_0$, a function*

$$\mathfrak{M} : \bar{\Omega} \to [0, \infty]$$

is said to be a positive metasolution of

$$-d\Delta u = \lambda u + a(x)f(x, u)u \tag{5.1}$$

supported in Ω_j if there exists a positive solution, L, of

$$\begin{cases} -d\Delta u = \lambda u + a(x)f(x,u)u & in \quad \Omega_j, \\ u = 0 & on \quad \partial\Omega_j \cap \partial\Omega, \\ u = \infty & on \quad \partial\Omega_j \setminus \partial\Omega, \end{cases}$$

such that

$$\mathfrak{M} = \begin{cases} L & in \quad \Omega_j, \\ \infty & in \quad \Omega \setminus \Omega_j, \\ 0 & on \quad \partial\Omega. \end{cases}$$

Thanks to Theorem 4.7, if $a(x)$ satisfies (Ha) and f satisfies (KO), then, for every $1 \leq j \leq q_0$, (5.1) possesses a positive metasolution supported in Ω_j if and only if $\lambda < d\sigma_{j+1}$. Moreover, should it be the case, (5.1) admits minimal and maximal positive metasolutions. Namely,

$$\mathfrak{M}^{\min}_{[\lambda, \Omega_j]} := \begin{cases} L^{\min}_{[\lambda, \Omega_j]} & in \quad \Omega_j, \\ \infty & in \quad \Omega \setminus \Omega_j, \\ 0 & on \quad \partial\Omega, \end{cases} \qquad \mathfrak{M}^{\max}_{[\lambda, \Omega_j]} := \begin{cases} L^{\max}_{[\lambda, \Omega_j]} & in \quad \Omega_j, \\ \infty & in \quad \Omega \setminus \Omega_j, \\ 0 & on \quad \partial\Omega. \end{cases}$$

The next result, which is the main theorem of this chapter and of Part I, provides us with the dynamics of (1.1) according to each of the values of the parameter $\lambda \in \mathbb{R}$.

Theorem 5.2 *Suppose $a(x)$ and f satisfy* (Ha) *and* (KO), *respectively, and denote by*

$$u(x,t) := u_{[\lambda,\Omega]}(x,t;u_0)$$

the unique solution of (1.1). *Then, the following properties hold:*

(a) *If $\lambda \le d\sigma_0$, then*

$$\lim_{t \uparrow \infty} u(\cdot, t) = 0 \quad in \quad \mathcal{C}(\bar{\Omega}).$$

(b) *If $d\sigma_0 < \lambda < d\sigma_1$, then*

$$\lim_{t \uparrow \infty} u(\cdot, t) = \theta_{[\lambda,\Omega]} \quad in \quad \mathcal{C}(\bar{\Omega}),$$

where $\theta_{[\lambda,\Omega]}$ is the unique positive solution of

$$\begin{cases} -d\Delta u = \lambda u + a(x)f(x,u)u & in \quad \Omega, \\ u = 0 & on \quad \partial\Omega. \end{cases} \tag{5.2}$$

(c) *If $d\sigma_j \le \lambda < d\sigma_{j+1}$, $1 \le j \le q_0$, then*

$$\mathfrak{M}_{[\lambda,\Omega_j]}^{\min} \le \liminf_{t \uparrow \infty} u(\cdot, t) \le \limsup_{t \uparrow \infty} u(\cdot, t) \le \mathfrak{M}_{[\lambda,\Omega_j]}^{\max} \quad in \quad \bar{\Omega}. \tag{5.3}$$

If, in addition, $u_0 > 0$ is a subsolution of (5.2), *then*

$$\lim_{t \uparrow \infty} u(\cdot, t) = \mathfrak{M}_{[\lambda,\Omega_j]}^{\min} \quad in \quad \bar{\Omega}. \tag{5.4}$$

According to (5.3) and the definition of the maximal and the minimal metasolutions, it becomes apparent that

$$\lim_{t \uparrow \infty} u(\cdot, t) = \infty \quad \text{uniformly in compact subsets of} \quad \Omega \setminus \bar{\Omega}_j = \bigcup_{i=1}^{j} \bar{\Omega}_{0,i},$$

while in Ω_j the following estimate holds

$$L_{[\lambda,\Omega_j]}^{\min} \le \liminf_{t \uparrow \infty} u(\cdot, t) \le \limsup_{t \uparrow \infty} u(\cdot, t) \le L_{[\lambda,\Omega_j]}^{\max}.$$

If, in addition, u_0 is a subsolution of (5.2) in Ω, then

$$\lim_{t \uparrow \infty} u(\cdot, t) = L_{[\lambda,\Omega_j]}^{\min} \quad in \quad \Omega_j.$$

As a result of Theorem 5.2, the dynamic of (1.1) is governed by the maximal non-negative classical solution of (5.2) if $\lambda < d\sigma_1$, while it is regulated by the minimal and the maximal metasolutions of (5.1) in Ω_j if $d\sigma_j \leq \lambda < d\sigma_{j+1}$ for some $1 \leq j \leq q_0$. It should be remembered that $\sigma_{q_0+1} := \infty$. Consequently, from the point of view of population dynamics, for every $\lambda \in [d\sigma_j, d\sigma_{j+1})$, the species u exhibits Malthusian growth in

$$\left(\bar{\Omega}_{0,1} \cup \cdots \bar{\Omega}_{0,j}\right) \setminus \partial\Omega,$$

whereas it has logistic growth in

$$\Omega_j = \Omega \setminus \left(\bar{\Omega}_{0,1} \cup \cdots \bar{\Omega}_{0,j}\right), \quad 1 \leq j \leq q_0.$$

Naturally, unless $\mathfrak{M}^{\min}_{[\lambda,\Omega_j]} = \mathfrak{M}^{\max}_{[\lambda,\Omega_j]}$ or, equivalently, $L^{\min}_{[\lambda,\Omega_j]} = L^{\max}_{[\lambda,\Omega_j]}$, (5.3) does not provide us with the exact behavior of $u(x,t)$ for large t. This is why the second part of this book will focus attention on the uniqueness of the large positive solutions of (5.1).

Remark 5.3 All the results of Part I and, in particular, Theorem 5.2 are also valid if, instead of $-d\Delta$, we consider the more general differential operator

$$\mathfrak{L} = -d\Delta + V$$

for some $V \in \mathcal{C}^\nu(\bar{\Omega})$. Even the function V might be taken in $L^\infty(\Omega)$ if solutions are regarded in $W^{2,p}(\Omega)$ with $p > N$. Naturally, in such case, the eigenvalues

$$d\sigma_0 = \lambda_1[-d\Delta, \Omega], \quad d\sigma_j = \lambda_1[-d\Delta, \Omega^i_{0,j}], \quad 1 \leq j \leq q_0,$$

should be inter-exchanged by

$$\Sigma_0 := \lambda_1[-d\Delta + V, \Omega], \quad \Sigma_j := \lambda_1[-d\Delta + V, \Omega^i_{0,j}], \quad 1 \leq j \leq q_0.$$

In this context, q_0, m_j, and even the components of Ω, $\Omega^i_{0,j}$, might depend on V and d. Although this was the case dealt with in [153], in this book we have refrained from providing the most general setting available. Nevertheless, in Chapter 10 we will need to work with this slightly more general setting.

5.2   Paradigmatic bifurcation diagram with $q_0 = 2$

In Figure 5.1 we have represented the dynamics of (1.1) under the assumptions of Theorem 5.2 in the special case when $q_0 = 2$. Then,

$$\Omega_1 := \Omega \setminus \bar{\Omega}_{0,1}, \quad \Omega_2 = \Omega \setminus \left(\bar{\Omega}_{0,1} \cup \bar{\Omega}_{0,2}\right) = \Omega_-.$$

Figure 1.1 shows an admissible nodal configuration of $a(x)$ respecting this situation. Precisely, Figure 5.1 shows the dynamics of (1.1) when

$$\mathfrak{M}_{[\lambda,\Omega_j]} := \mathfrak{M}^{\min}_{[\lambda,\Omega_j]} = \mathfrak{M}^{\max}_{[\lambda,\Omega_j]} \tag{5.5}$$

for all $j = 1, 2$ and $\lambda < d\sigma_{j+1}$, where $\sigma_3 := \infty$. In such case, thanks to Theorem 4.7, it is easy to see that the maps

$$\lambda \in (-\infty, d\sigma_2) \mapsto \mathfrak{M}_{[\lambda,\Omega_1]}, \qquad \lambda \in \mathbb{R} \mapsto \mathfrak{M}_{[\lambda,\Omega_2]} \tag{5.6}$$

are continuous and increasing. Moreover, by Theorems 4.8 and 4.9,

$$\lim_{\lambda\uparrow d\sigma_1} \theta_{[\lambda,\Omega]} = \mathfrak{M}_{[d\sigma_1,\Omega_1]}, \qquad \lim_{\lambda\uparrow d\sigma_2} \mathfrak{M}_{[\lambda,\Omega_1]} = \mathfrak{M}_{[d\sigma_2,\Omega_2]}.$$

This explains why in Figure 5.1 the curves of metasolutions $(\lambda, \mathfrak{M}_{[\lambda,\Omega_1]})$, defined for $\lambda < d\sigma_1$, and $(\lambda, \mathfrak{M}_{[\lambda,\Omega_2]})$, defined for $\lambda \in \mathbb{R}$, are increasing and meet at $\lambda = d\sigma_2$. Note that, due to Theorems 4.1 and 4.8, the curve of classical positive solutions $(\lambda, \theta_{[\lambda,\Omega]})$, $d\sigma_0 < \lambda < d\sigma_1$, bifurcates from $u = 0$ at $\lambda = d\sigma_0$ and meets $\mathfrak{M}_{[d\sigma_1,\Omega_1]}$ at $\lambda = d\sigma_1$.

In Figure 5.1 we are representing the value of the parameter λ versus the values of the classical positive solutions and metasolutions, $u(x)$, at some distinguished point $x \in \Omega_- = \Omega_2$, where all of them are bounded, because the large solution of (5.1) in Ω_- is defined for all $\lambda \in \mathbb{R}$. The global bifurcation diagram shows four different types of solutions. The λ-axis represents $u = 0$, which, according to Theorem 5.2(a), is a global attractor if $\lambda \leq d\sigma_0$. As the

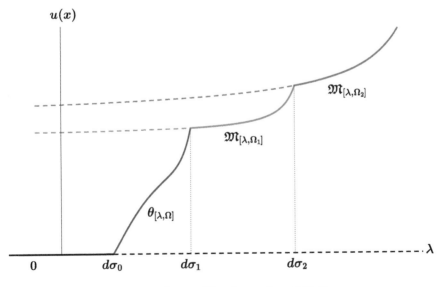

FIGURE 5.1: The dynamics of (1.1).

linearization of (1.1) at $u = 0$ is the linear problem

$$\begin{cases} \frac{\partial u}{\partial t} - d\Delta u = \lambda u & \text{in } \Omega \times (0, \infty), \\ u = 0 & \text{on } \partial\Omega \times (0, \infty), \\ u(\cdot, 0) = u_0 > 0 & \text{in } \Omega, \end{cases}$$

whose unique solution is given by

$$u(x, t) = e^{t(d\Delta + \lambda)} u_0,$$

when $\lambda > d\sigma_0$ the trivial solution $u = 0$ is linearly unstable, because

$$u(x, t) = e^{t(d\Delta + \lambda)} \varphi = e^{t(\lambda - d\sigma_0)} \varphi \to \infty \qquad \text{as } t \uparrow \infty,$$

where $\varphi \gg 0$ stands for any principal eigenfunction of $-\Delta$ associated to σ_0. This is why in Figure 5.1 the u axis has been represented with a dashed line for $\lambda \geq d\sigma_0$. As usual, continuous lines are filled in by stable steady states, or stable metasolutions, while dashed lines are filled in by unstable steady states, or unstable metasolutions. Then, we have represented the positive solution $\theta_{[\lambda,\Omega]}$, $d\sigma_0 < \lambda < d\sigma_1$, which is a global attractor of (1.1) in that range, the curve of metasolutions $\mathfrak{M}_{[\lambda,\Omega_1]}$, $\lambda < d\sigma_2$, supported in Ω_1 and, finally, the curve of metasolutions $\mathfrak{M}_{[\lambda,\Omega_2]}$ supported in Ω_2, $\lambda \in \mathbb{R}$.

According to Theorem 5.2(a), $u = 0$ is a global attractor for the positive solutions of (1.1) if $\lambda \leq d\sigma_0$, by Theorem 5.2(b), $\theta_{[\lambda,\Omega]}$ is a global attractor for the positive solutions if $d\sigma_0 < \lambda < d\sigma_1$, and, due to Theorem 5.2(c), $\mathfrak{M}_{[\lambda,\Omega_1]}$ is a global attractor for $d\sigma_1 \leq \lambda < d\sigma_2$, and $\mathfrak{M}_{[\lambda,\Omega_2]}$ is a global attractor if $\lambda \geq d\sigma_2$. This describes the dynamics of (1.1) for all values of $\lambda \in \mathbb{R}$.

As for $\lambda < d\sigma_1$ the dynamics of (1.1) are governed by the classical non-negative steady states, the metasolutions $\mathfrak{M}_{[\lambda,\Omega_j]}$ must be unstable with respect to these solutions. Consequently, $\lambda = d\sigma_1$ provides us with the critical value of λ where $\mathfrak{M}_{[\lambda,\Omega_1]}$ becomes an attractor for all classical positive solutions. Similarly, $\mathfrak{M}_{[\lambda,\Omega_2]}$ is unstable if $\lambda < d\sigma_2$, while it becomes a global attractor for all classical positive solutions if $\lambda \geq d\sigma_2$.

The monotonicity of the maps (5.6) is an easy consequence from Theorem 4.7, taking into account that, owing to Theorem 4.5(b),

$$\Theta_{[\lambda,\Omega_j,M]} \ll \Theta_{[\mu,\Omega_j,M]}$$

for all $\lambda < \mu < d\sigma_{j+1}$, $M > 0$ and $j = 1, 2$. Indeed, letting $M \uparrow \infty$ implies

$$L_{[\lambda,\Omega_j]} < L_{[\mu,\Omega_j]}$$

provided $\lambda < \mu < d\sigma_{j+1}$. The continuity of the mapping

$$\lambda \mapsto L_{[\lambda,\Omega_j]}, \qquad \lambda < d\sigma_{j+1},$$

under the uniqueness condition (5.5) can be established with the next argument. Suppose λ_n, $n \geq 1$, is a sequence of λ's, with $\lambda_n < d\sigma_{j+1}$, such that

$$\lambda_\omega := \lim_{n \to \infty} \lambda_n < d\sigma_{j+1}.$$

Then, for sufficiently small $\varepsilon > 0$, there exists $n_0 \in \mathbb{N}$ such that

$$L_{[\lambda_n,\Omega_j]} \leq L_{[\lambda_\omega + \varepsilon,\Omega_j]} \qquad \text{for all } n \geq n_0.$$

By adapting the compactness arguments of Chapter 4, it is easy to infer that, along some subsequence relabeled by n, $L_{[\lambda_n,\Omega_j]}$ approximates uniformly in compact subsets of Ω_j to a positive solution of (4.2). By uniqueness, the limit must be $L_{[\lambda_\omega,\Omega_j]}$. As this is true along any subsequence, the continuity holds. Naturally, the continuity might fail if (4.2) has multiple solutions.

Obviously, the general discussion carried out in this section can be adapted almost *mutatis mutandis* to cover the general case when $q_0 \geq 1$.

5.3   Numerical example with $q_0 = 2$

In this section we consider (1.1) in the special case when

$$N = 2, \quad d = 1, \quad \Omega = B_{0.5} = \{x \in \mathbb{R}^2 \; : \; |x| < 0.5\}, \quad a(x) = \rho(r), \quad r = |x|,$$

with

$$\rho(r) := \begin{cases} \sin(5\pi(r + 0.5)), & 0.1 < r < 0.3, \\ 0, & r \in [0, 0.1] \cup [0.3, 0.5], \end{cases}$$

and

$$f(x, u) = u^3, \qquad u \geq 0, \quad x \in \bar{\Omega}.$$

Figure 5.2 shows a plot of $-a(x)$.

By Lemma 3.6, f satisfies the conditions (Hf), (Hg) and (KO). Moreover, $a(x)$ satisfies (Ha) with

$$\Omega_0 = B_{0.1} \cup A_{0.3,0.5}, \qquad \Omega_- = A_{0.1,0.3},$$

where we are denoting

$$A_{a,b} := \{\, x \in \mathbb{R}^N \; : \; a < |x| < b \,\}$$

for all $0 < a < b$. As, due to R. Gómez-Reñasco and J. López-Gómez [109, Sect. 5],

$$\lambda_1[-\Delta, B_{0.5}] \simeq 23.13 < \lambda_1[-\Delta, A_{0.3,0.5}] \simeq 245.14 < \lambda_1[-\Delta, B_{0.1}] \simeq 578.31,$$

it becomes apparent that, actually, $a(x)$ satisfies Hypothesis (Ha) with

$$q_- = 1, \qquad q_0 = 2, \qquad \Omega_{0,1} = A_{0.3,0.5}, \qquad \Omega_{0,2} = B_{0.1},$$

$$\sigma_0 \simeq 23.13 < \sigma_1 \simeq 245.14 < \sigma_2 \simeq 578.31.$$

According to Theorems 4.1 and 4.8, the problem

$$\begin{cases} -\Delta u = \lambda u + a(x)u^4 & \text{in } \Omega, \\ u = 0 & \text{on } \partial\Omega, \end{cases} \tag{5.7}$$

has a (unique) positive solution, $\theta_{[\lambda,\Omega]}$, if and only if $\sigma_0 < \lambda < \sigma_1$. Moreover, by uniqueness, $\theta_{[\lambda,\Omega]}$ is radially symmetric, the mapping $\lambda \mapsto \theta_{[\lambda,\Omega]}(r)$ is increasing for all $r = |x| \in [0, 0.5)$,

$$\lim_{\lambda \downarrow \sigma_0} \theta_{[\lambda,\Omega]} = 0 \quad \text{uniformly in } \Omega = B_{0.5},$$

and

$$\lim_{\lambda \uparrow \sigma_1} \theta_{[\lambda,\Omega]} = \infty \quad \text{uniformly in compact subsets of } \bar{\Omega}_{0,1} \setminus \partial\Omega. \tag{5.8}$$

In particular, the mapping

$$\lambda \mapsto \|\theta_{[\lambda,\Omega]}\|_2 := \|\theta_{[\lambda,\Omega]}\|_{L^2(\Omega)} = \left(\int_\Omega \theta_{[\lambda,\Omega]}^2 \right)^{\frac{1}{2}}, \qquad \sigma_0 < \lambda < \sigma_1,$$

is increasing and it satisfies

$$\lim_{\lambda \downarrow \sigma_0} \|\theta_{[\lambda,\Omega]}\|_2 = 0, \qquad \lim_{\lambda \uparrow \sigma_1} \|\theta_{[\lambda,\Omega]}\|_2 = \infty.$$

Figure 5.3 shows the graph of this curve computed in [109] through a path-following solver (see [109] for technical details). It plots the L^2-norm of the

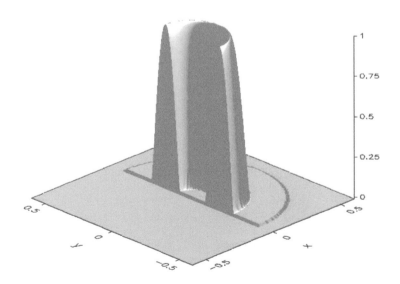

FIGURE 5.2: A plot of $-a(x)$.

non-negative solutions of (5.7) versus the parameter λ in the interval $(0, 250)$. Note that the L^2-norm of $\theta_{[\lambda,\Omega]}$ indeed blows up as $\lambda \uparrow \sigma_1 \simeq 245.14$. Continuous lines are filled in by stable solutions and dashed lines by unstable ones. Each point on the continuous line represents a positive solution of (5.7). According to Theorem 5.2, $u = 0$ is a global attractor for the parabolic problem

$$\begin{cases} \frac{\partial u}{\partial t} - \Delta u = \lambda u + a(x)u^4 & \text{in } \Omega \times (0, \infty), \\ u = 0 & \text{on } \partial\Omega \times (0, \infty), \\ u(\cdot, 0) = u_0 > 0 & \text{in } \Omega, \end{cases} \qquad (5.9)$$

if $\lambda \leq \sigma_0$, whereas $\theta_{[\lambda,\Omega]}$ is a global attractor of (5.9) for all $\lambda \in (\sigma_0, \sigma_1)$.

Figure 5.4 shows the plots of $\theta_{[\lambda,\Omega]}$ for $\lambda = 101.229218$, $\lambda = 191.845246$, $\lambda = 217.71643$ and $\lambda = 232.193199$, respectively. In agreement with Theorems 4.1 and 4.8, these solutions are point-wise increasing with respect to λ and blow-up in

$$\bar{\Omega}_{0,1} \setminus \partial\Omega = \{x \in \mathbb{R}^2 \ : \ 0.3 \leq |x| < 0.5\}$$

as $\lambda \uparrow \sigma_1$, while in the region

$$\Omega_1 := \Omega \setminus \bar{\Omega}_{0,1} = B_{0.3},$$

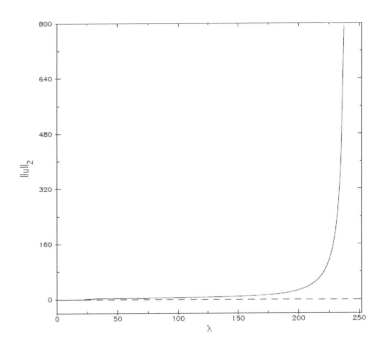

FIGURE 5.3: The curve $\lambda \mapsto \|\theta_{[\lambda,\Omega]}\|_2$.

they stabilize to the unique positive solution, $L_{[\sigma_1,\Omega_1]}$, of the singular problem

$$\begin{cases} -\Delta u = \sigma_1 u + a(x)u^4 & \text{in} \quad \Omega_1, \\ u = \infty & \text{on} \quad \partial\Omega_1, \end{cases} \tag{5.10}$$

as $\lambda \uparrow \sigma_1$. The uniqueness of $L_{[\sigma_1,\Omega_1]}$ and the uniqueness of $L_{[\lambda,\Omega_1]}$ for each $\lambda < \sigma_2$ are guaranteed by Theorem 8.4.

Note that in Figure 5.4 the scales on the vertical axis are different in each of the four plots. This is why the plots of the positive steady states do not seem to grow as λ increases, though they actually do. Furthermore, by (5.8), they do it at a much faster rate in $\Omega_{0,1}$ than in $\Omega \setminus \Omega_{0,1}$.

Figure 5.5 shows a bifurcation diagram computed in [109] of stable positive steady states, $\theta_{[\lambda,\Omega]}$, and stable metasolutions supported in Ω_1, $\mathfrak{M}_{[\lambda,\Omega_1]}$, for the range $\sigma_0 < \lambda < \sigma_2$. For a given $x \in \Omega_- = A_{0.1,0.3}$ with $r = |x| = 0.2$, it plots

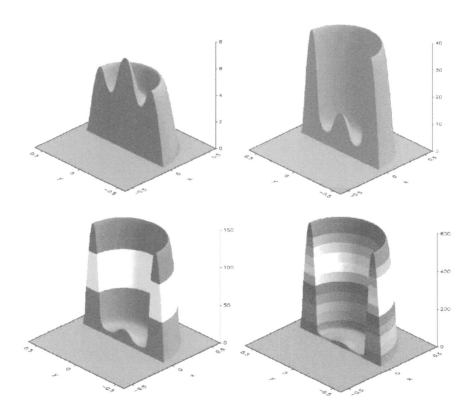

FIGURE 5.4: The classical positive steady states $\theta_{[\lambda,\Omega]}$ for $\lambda = 101.229218$ (top left), $\lambda = 191.845246$ (top right), $\lambda = 217.71643$ (bottom left) and $\lambda = 232.193199$ (bottom right). They are point-wise increasing and blow up to infinity in $A_{0.3,0.5}$ as λ approximates $\sigma_1 \simeq 245.14$.

the value $\theta_{[\lambda,\Omega]}(x)$ versus the parameter λ for $\sigma_0 < \lambda < \sigma_1$, where, according to Theorem 5.2(b), $\theta_{[\lambda,\Omega]}$ is a global attractor of (5.9), and the value $\mathfrak{M}_{[\lambda,\Omega]}(x)$ versus λ for $\sigma_1 < \lambda < \sigma_2$, where, according to Theorem 5.2(c), $\mathfrak{M}_{[\lambda,\Omega]}$ is a global attractor of (5.9). According to Theorem 4.7, $\mathfrak{M}_{[\lambda,\Omega_1]}$ exists if and only if $\lambda < \sigma_2$ and, thanks to Theorem 5.2, it is unstable for $\lambda < \sigma_1$ and stable for $\sigma_1 \leq \lambda < \sigma_2$. The continuous line plots $(\lambda, \theta_{[\lambda,\Omega]}(x))$ for $\sigma_0 < \lambda < \sigma_1$, while the dashed-dotted line plots $(\lambda, \mathfrak{M}_{[\lambda,\Omega_1]}(x))$ for $\sigma_1 \leq \lambda < \sigma_2$. Only in this isolated situation, in order to differentiate classical solutions from metasolutions, a curve of generalized stable steady states will be represented through a dashed line.

Figure 5.6 shows the plots of $L_{[\lambda,\Omega_1]}$ for $\lambda = 300$, $\lambda = 450$, $\lambda = 500$ and $\lambda = 525$, respectively. They equal infinity on $\partial B_{0.3}$ and are point-wise increasing with respect to λ up to $\lambda = \sigma_2$, where, according to Theorem 4.9, they blow up to infinity in $\bar{\Omega}_{0,2} = \bar{B}_{0.1}$ reaching $\mathfrak{M}_{[\sigma_2,\Omega_-]}$. As in Figure 5.4, one should take into account that the scales of the vertical axis in the two rows of Figure 5.6 are different, though in this occasion the increasing character of the metasolution in λ is well differentiated.

As a consequence of Theorem 5.2, when the intrinsic growth rate of the species u, measured by λ, is below the threshold σ_0, the inhabiting area Ω cannot support the species, which is driven to extinction. Contrarily, when $\lambda \in (\sigma_0, \sigma_1)$, the territory Ω is able to maintain u at the critical level $\theta_{[\lambda,\Omega]}$, independently of the size of the initial population u_0. So, as in this range of values of λ the habitat Ω cannot maintain an arbitrarily large population, (5.9) exhibits a genuine logistic behavior.

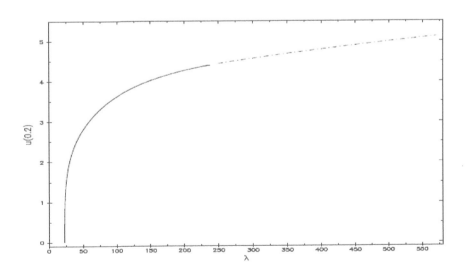

FIGURE 5.5: Classical solutions and stable metasolutions supported in Ω_1.

Rather astonishingly, when $\lambda \in [\sigma_1, \sigma_2)$ the population must be limited by $L_{[\lambda,\Omega_1]}$ in the region $\Omega_1 := \Omega \setminus \bar{\Omega}_{0,1}$, while, as an effect of the random dispersion of the individuals of the species u to the biggest protection zone, $\Omega_{0,1}$, the population can grow arbitrarily in $\Omega_{0,1}$. So, (5.9) exhibits a genuine logistic growth in $\Omega \setminus \bar{\Omega}_{0,1}$ while it exhibits a typical exponential growth in $\Omega_{0,1}$ as $t \uparrow \infty$. Indeed, thanks to the parabolic maximum principle,

$$u_1 := u_{[\lambda,\Omega]}(\cdot, 1; u_0) \gg 0$$

and hence, for every $t \geq 0$,

$$u_{[\lambda,\Omega]}(\cdot, t+1; u_0) = u_{[\lambda,\Omega]}(\cdot, t; u_1) > u_{[\lambda,\Omega_{0,1}]}(\cdot, t; u_1) \quad \text{in } \Omega_{0,1}, \qquad (5.11)$$

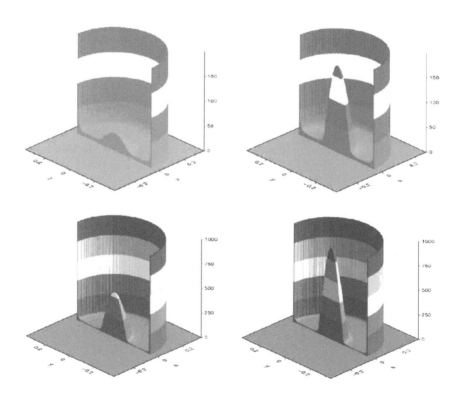

FIGURE 5.6: The large solutions $L_{[\lambda,\Omega_1]}$ for $\lambda = 300$ (top left), $\lambda = 450$ (top right), $\lambda = 500$ (bottom left) and $\lambda = 525 < \sigma_2$ (bottom right). They are point-wise increasing and blow up to infinity in $B_{0.1}$ as λ approximates $\sigma_2 \simeq 578.31$.

where $u_{[\lambda,\Omega_{0,1}]}(x, t; u_1)$ stands for the unique solution of the linear problem

$$\begin{cases} \frac{\partial u}{\partial t} - \Delta u = \lambda u & \text{in} \quad \Omega_{0,1} \times (0, \infty), \\ u = 0 & \text{on} \quad \partial\Omega_{0,1} \times (0, \infty), \\ u(\cdot, 0) = u_1 & \text{in} \quad \Omega_{0,1}, \end{cases}$$

which is given by

$$u_{[\lambda,\Omega_{0,1}]}(x, t; u_1) = e^{t(\lambda+\Delta)}u_1. \tag{5.12}$$

Suppose $\lambda > \sigma_1$ and let $\varphi_{0,1}$ be a principal eigenfunction associated with σ_1. Then, since $u_1 \gg 0$, there exists $\xi > 0$ such that

$$u_1 > \xi\varphi_{0,1}$$

and hence, (5.11) and (5.12) imply that

$$u_{[\lambda,\Omega]}(\cdot, t+1; u_0) > \xi e^{t(\lambda+\Delta)}\varphi_{0,1} = \xi e^{(\lambda-\sigma_1)t}\varphi_{0,1},$$

which shows the exponential growth of the population in $\Omega_{0,1}$ for all $\lambda > \sigma_1$.

Similarly, when $\lambda > \sigma_2$ the population must be bounded in Ω_-, as a result of the limited resources therein, though the individual of the species can disperse towards the most favorable areas $\Omega_{0,1}$ and $\Omega_{0,2}$, where the population density grows arbitrarily at the rates $e^{t(\lambda-\sigma_1)}$ and $e^{t(\lambda-\sigma_2)}$, respectively. Naturally, in the applications, the principal eigenvalues σ_1 and σ_2 measure the quality of the *protection zones* $\Omega_{0,1}$ and $\Omega_{0,2}$.

5.4 Proof of Theorem 5.2

Part (a) is a direct consequence of the last assertion of Theorem 1.7 and Part (b) is a direct consequence of Theorem 1.7(b), because, thanks to Theorem 4.1, (1.1) possesses a positive steady state for each $\lambda \in (\sigma_0, \sigma_1)$. It remains to prove Part (c). The proof will proceed by induction. First, we will establish Part (c) in the special case when

$$d\sigma_1 \leq \lambda < d\sigma_2.$$

Since $\lambda > d\sigma_1 - \varepsilon$ for each $\varepsilon > 0$, by the parabolic maximum principle,

$$u_{[\lambda,\Omega]}(\cdot, t; u_0) \geq u_{[d\sigma_1-\varepsilon,\Omega]}(\cdot, t; u_0)$$

for all $\varepsilon > 0$ and $t \geq 0$. Thus, by Part (b), letting $t \uparrow \infty$ yields

$$\liminf_{t\uparrow\infty} u_{[\lambda,\Omega]}(\cdot, t; u_0) \geq \lim_{t\uparrow\infty} u_{[d\sigma_1-\varepsilon,\Omega]}(\cdot, t; u_0) = \theta_{[d\sigma_1-\varepsilon,\Omega]} \tag{5.13}$$

for sufficiently small $\varepsilon > 0$. Thus, letting $\varepsilon \downarrow 0$, Theorem 4.8 implies that

$$\liminf_{t \uparrow \infty} u_{[\lambda,\Omega]}(\cdot, t; u_0) \geq \lim_{\varepsilon \downarrow 0} \theta_{[d\sigma_1 - \varepsilon, \Omega]} = \mathfrak{M}^{\min}_{[d\sigma_1, \Omega_1]}. \tag{5.14}$$

Actually, as

$$\lim_{\varepsilon \downarrow 0} \theta_{[d\sigma_1 - \varepsilon, \Omega]} = \infty$$

uniformly in compact subsets of $\bar{\Omega}_{0,1} \setminus \partial\Omega$, (5.13) implies that

$$\liminf_{t \uparrow \infty} u_{[\lambda,\Omega]}(\cdot, t; u_0) = \infty \quad \text{unif. in compact subsets of} \quad \bar{\Omega}_{0,1} \setminus \partial\Omega. \tag{5.15}$$

In particular, for every $M > 0$ there exists $T_M > 0$ such that

$$u_{[\lambda,\Omega]}(x, t; u_0) \geq M \qquad \forall \ (x,t) \in (\partial\Omega_{0,1} \setminus \partial\Omega) \times [T_M, \infty).$$

So, $u_{[\lambda,\Omega]}(\cdot, t; u_0)$ is a supersolution of

$$\begin{cases} \frac{\partial u}{\partial t} - d\Delta u = \lambda u + af(\cdot, u)u & \text{in } \Omega_1 \times (T_M, \infty), \\ u = M & \text{on } (\partial\Omega_1 \setminus \partial\Omega) \times (T_M, \infty), \\ u = 0 & \text{on } (\partial\Omega_1 \cap \partial\Omega) \times (T_M, \infty), \\ u(\cdot, T_M) = u_{[\lambda,\Omega]}(\cdot, T_M; u_0) & \text{in } \Omega_1. \end{cases}$$

Consequently, by the parabolic maximum principle,

$$u_{[\lambda,\Omega]}(x, t; u_0) \geq u_{[\lambda,\Omega_1,M]}\left(x, t - T_M; u_{[\lambda,\Omega]}(\cdot, T_M; u_0)\right) \tag{5.16}$$

for all $(x,t) \in \Omega_1 \times (T_M, \infty)$, where $u_{[\lambda,\Omega_1,M]}(x, t; \tilde{u}_0)$ stands for the unique solution of

$$\begin{cases} \frac{\partial u}{\partial t} - d\Delta u = \lambda u + af(\cdot, u)u & \text{in } \Omega_1 \times (T_M, \infty), \\ u = M & \text{on } (\partial\Omega_1 \setminus \partial\Omega) \times (T_M, \infty), \\ u = 0 & \text{on } (\partial\Omega_1 \cap \partial\Omega) \times (T_M, \infty), \\ u(\cdot, 0) = \tilde{u}_0 & \text{in } \Omega_1, \end{cases}$$

which is the parabolic counterpart of (4.33) with $j = 1$. By Theorem 4.5, (4.33) possesses a unique positive solution $\Theta_{[\lambda,\Omega_1,M]}$ for each $\lambda < d\sigma_2$. Thus, thanks to Theorem 1.7(b),

$$\lim_{t \uparrow \infty} u_{[\lambda,\Omega_1,M]}(x, t; \tilde{u}_0) = \Theta_{[\lambda,\Omega_1,M]}$$

for all $\tilde{u}_0 > 0$. So, letting $t \uparrow \infty$ in (5.16) shows that

$$\liminf_{t \uparrow \infty} u_{[\lambda,\Omega]}(x, t; u_0) \geq \Theta_{[\lambda,\Omega_1,M]} \qquad \text{for all } M > 0.$$

Therefore, letting $M \uparrow \infty$, we find from Theorem 4.7 that

$$\liminf_{t \uparrow \infty} u_{[\lambda,\Omega]}(x, t; u_0) \geq L^{\min}_{[\lambda,\Omega_1]} \qquad \text{in} \quad \Omega_1.$$

Bringing together this estimate with (5.15) it becomes apparent that

$$\liminf_{t\uparrow\infty} u_{[\lambda,\Omega]}(x,t;u_0) \geq \mathfrak{M}^{\min}_{[\lambda,\Omega_1]} \quad \text{in} \quad \Omega, \tag{5.17}$$

which is the lower estimate of (5.3) for $j = 1$.

Suppose, in addition, that u_0 is a subsolution of (5.2) in Ω. Then, according to D. Sattinger [220], the function

$$x \mapsto u_{[\lambda,\Omega]}(x,t;u_0), \qquad x \in \bar{\Omega},$$

is a subsolution of (4.1) for all $t > 0$, because $t \mapsto u_{[\lambda,\Omega]}(\cdot,t;u_0)$ is non-decreasing. Fix $t > 0$ and set

$$M_t := \max_{\bar{\Omega}_1} u_{[\lambda,\Omega]}(\cdot,t;u_0).$$

As $u_{[\lambda,\Omega]}(\cdot,t;u_0)$ is a subsolution of (5.2), it is also a subsolution of (4.33) with $j = 1$ for all $M \geq M_t$. Thus, due to Theorem 4.5(a),

$$u_{[\lambda,\Omega]}(\cdot,t;u_0) \leq \Theta_{[\lambda,\Omega_1,M]} \quad \text{in} \quad \Omega_1 \quad \text{for all} \quad M \geq M_t.$$

Hence, owing to (4.51),

$$u_{[\lambda,\Omega]}(\cdot,t;u_0) \leq \lim_{M\uparrow\infty} \theta_{[\lambda,\Omega_1,M]} = L^{\min}_{[\lambda,\Omega_1]} \quad \text{in} \quad \Omega_1$$

and therefore, letting $t \uparrow \infty$ yields

$$\limsup_{t\uparrow\infty} u_{[\lambda,\Omega]}(\cdot,t;u_0) \leq L^{\min}_{[\lambda,\Omega_1]} \quad \text{in} \quad \Omega_1.$$

Consequently, according to (5.17), we can conclude that

$$\lim_{t\uparrow\infty} u_{[\lambda,\Omega]}(\cdot,t;u_0) = \mathfrak{M}^{\min}_{[\lambda,\Omega_1]} \quad \text{in} \quad \Omega, \tag{5.18}$$

which ends the proof of (5.4) for $j = 1$.

To obtain the upper estimates of (5.3), we need to get some a priori bounds in Ω_- for the solutions of (1.1). Fix $\lambda \geq d\sigma_1$, set

$$u_1 := u_{[\lambda,\Omega]}(\cdot,1;u_0) \gg 0$$

and let $\varphi \gg 0$ be a principal eigenfunction associated to σ_0. Then, since $u_1 \gg 0$, there exists $\kappa > 1$ such that

$$u_1 < \kappa\varphi. \tag{5.19}$$

We claim that there exists

$$\Lambda > \max\{\lambda, d\sigma_{q_0}\} \tag{5.20}$$

such that $\kappa\varphi$ is a subsolution of

$$\begin{cases} -d\Delta u = \Lambda u + af(\cdot, u)u & \text{in } \Omega, \\ u = 0 & \text{on } \partial\Omega. \end{cases} \tag{5.21}$$

Indeed, since $\kappa\varphi = 0$ on $\partial\Omega$, $\kappa\varphi$ is a subsolution of (5.21) if, and only if,

$$-d\Delta(\kappa\varphi) \le \Lambda\kappa\varphi + af(\cdot, \kappa\varphi)\kappa\varphi \quad \text{in } \Omega,$$

or, equivalently,

$$-af(\cdot, \kappa\varphi) \le \Lambda - d\sigma_0 \quad \text{in } \Omega,$$

which is true for sufficiently large $\Lambda > \max\{\lambda, d\sigma_{q_0}\}$.

By the semigroup property and the parabolic maximum principle, it follows from (5.19) that

$$u_{[\lambda,\Omega]}(x, t+1; u_0) = u_{[\lambda,\Omega]}(x, t; u_1) \le u_{[\lambda,\Omega]}(\cdot, t; \kappa\varphi)$$

for all $x \in \Omega$ and $t > 0$. Similarly, (5.20) implies

$$u_{[\lambda,\Omega]}(\cdot, t; \kappa\varphi) < u_{[\Lambda,\Omega]}(\cdot, t; \kappa\varphi) \quad \text{in } \Omega$$

for all $t > 0$. Hence,

$$u_{[\lambda,\Omega]}(\cdot, t+1; u_0) \le u_{[\Lambda;\Omega]}(\cdot, t; \kappa\varphi) \quad \text{in } \Omega. \tag{5.22}$$

As $\kappa\varphi$ is a subsolution of (5.21), it follows from D. Sattinger [220] that $u_{[\Lambda;\Omega]}(\cdot, t; \kappa\varphi)$ is a subsolution of (5.21) for all $t > 0$. Fix $t > 0$ and set

$$\tilde{M}_t := \max_{\bar{\Omega}_-} u_{[\Lambda,\Omega]}(\cdot, t; \kappa\varphi).$$

Note that $u_{[\Lambda,\Omega]}(\cdot, t; \kappa\varphi)$ is a subsolution of (5.21) and also a subsolution of (4.33) for $j = q_0$ and any $M \ge \tilde{M}_t$, since $\Omega_{q_0} = \Omega_-$. Thus, due to Theorem 4.5(a),

$$u_{[\Lambda,\Omega]}(\cdot, t; \kappa\varphi) \le \Theta_{[\lambda,\Omega_-,M]} \quad \text{in } \Omega_- \quad \text{for all } M \ge \tilde{M}_t$$

and hence, by (4.38),

$$u_{[\Lambda,\Omega]}(\cdot, t; \kappa\varphi) \le \lim_{M\uparrow\infty} \Theta_{[\Lambda,\Omega_-,M]} = L_{[\Lambda,\Omega_-]}^{\min} \quad \text{in } \Omega_-.$$

Therefore, by (5.22), letting $t \uparrow \infty$ yields

$$\limsup_{t\uparrow\infty} u_{[\lambda,\Omega]}(\cdot, t; u_0) \le L_{[\Lambda,\Omega_-]}^{\min} \quad \text{in } \Omega_-. \tag{5.23}$$

Consequently, $u_{[\lambda,\Omega]}(\cdot, t; u_0)$ is uniformly bounded above, in any compact subset of Ω_-, for all $t > 0$.

Suppose $d\sigma_1 \leq \lambda < d\sigma_2$ and for sufficiently large $n \in \mathbb{N}$, say $n \geq n_0$, consider the open sets defined by (4.35) with $j = 1$, i.e.,

$$\Omega_{1,n} := \{x \in \Omega_1 \; : \; \text{dist}\,(x, \partial\Omega_1 \setminus \partial\Omega) > 1/n\}.$$

Pick $n \geq n_0$. As $\partial\Omega_{1,n} \subset \Omega_- \cup \partial\Omega$, it follows from (5.23) that there exists $M_0 > 0$ such that

$$u_{[\lambda,\Omega]}(\cdot, t; u_0) \leq M \qquad \text{on} \quad \partial\Omega_{1,n} \setminus \partial\Omega$$

for all $M \geq M_0$ and $t > 0$. Thus, by the parabolic maximum principle,

$$u_{[\lambda,\Omega]}(\cdot, t; u_0) \leq u_{[\lambda,\Omega_{1,n},M]}(\cdot, t; u_0) \qquad \text{in} \quad \Omega_{1,n} \tag{5.24}$$

for all $t > 0$, where $u_{[\lambda,\Omega_{1,n},M]}$ stands for the unique solution of

$$\begin{cases} \frac{\partial u}{\partial t} - d\Delta u = \lambda u + af(\cdot, u)u & \text{in } \Omega_{1,n} \times (0, \infty), \\ u = M & \text{on } (\partial\Omega_{1,n} \setminus \partial\Omega) \times (0, \infty), \\ u = 0 & \text{on } (\partial\Omega_{1,n} \cap \partial\Omega) \times (0, \infty), \\ u(\cdot, 0) = u_0 & \text{in } \Omega_{1,n}. \end{cases}$$

By Theorem 1.7,

$$\lim_{t\uparrow\infty} u_{[\lambda,\Omega_{1,n},M]}(\cdot, t; u_0) = \Theta_{[\lambda,\Omega_{1,n},M]} \qquad \text{in} \quad \Omega_{1,n},$$

where $\Theta_{[\lambda,\Omega_{1,n},M]}$ is the unique positive solution of (4.36) with $j = 1$. Hence, letting $t \uparrow \infty$ in (5.24) shows that, for every $n \geq n_0$,

$$\limsup_{t\uparrow\infty} u_{[\lambda,\Omega]}(\cdot, t; u_0) \leq \Theta_{[\lambda,\Omega_{1,n},M]} \qquad \text{in} \quad \Omega_{1,n}. \tag{5.25}$$

Consequently, letting $M \uparrow \infty$ in (5.25), we find that

$$\limsup_{t\uparrow\infty} u_{[\lambda,\Omega]}(\cdot, t; u_0) \leq L_{[\lambda,\Omega_{1,n}]}^{\min} \qquad \text{in} \quad \Omega_{1,n} \tag{5.26}$$

for all $n \geq n_0$. As (5.26) holds for all $n \geq n_0$ and, due to (4.39), we already know that

$$L_{[\lambda,\Omega_1]}^{\max} = \lim_{n\uparrow\infty} L_{[\lambda,\Omega_{1,n}]}^{\min}$$

we can conclude from (5.26) that

$$\limsup_{t\uparrow\infty} u_{[\lambda,\Omega]}(\cdot, t; u_0) \leq L_{[\lambda,\Omega_1]}^{\max} \qquad \text{in} \quad \Omega_1,$$

which ends the proof of (5.3) for $j = 1$. Naturally, this ends the proof of the theorem if $q_0 = 1$, because in such case $\sigma_2 = \infty$.

Subsequently, we suppose $q_0 \geq 2$ and, fixing $1 \leq j \leq q_0 - 1$, assume that

Part (c) holds for all $\lambda < d\sigma_{j+1}$. It remains to prove that in such case Part (c) also holds for all $\lambda \in [d\sigma_{j+1}, d\sigma_{j+2})$. Suppose

$$d\sigma_{j+1} \leq \lambda < d\sigma_{j+2}.$$

Since $\lambda > d\sigma_{j+1} - \varepsilon$ for all $\varepsilon > 0$, by the parabolic maximum principle,

$$u_{[\lambda,\Omega]}(\cdot, t; u_0) \geq u_{[d\sigma_{j+1}-\varepsilon,\Omega]}(\cdot, t; u_0).$$

Thus, as soon as $d\sigma_j < d\sigma_{j+1} - \varepsilon < d\sigma_{j+1}$, by the induction hypothesis, we can infer that

$$\liminf_{t \uparrow \infty} u_{[\lambda,\Omega]}(\cdot, t; u_0) \geq \liminf_{t \uparrow \infty} u_{[d\sigma_{j+1}-\varepsilon,\Omega]}(\cdot, t; u_0) = \mathfrak{M}^{\min}_{[d\sigma_{j+1}-\varepsilon,\Omega_j]}. \quad (5.27)$$

Thus, letting $\varepsilon \downarrow 0$, Theorem 4.9 implies that

$$\liminf_{t \uparrow \infty} u_{[\lambda,\Omega]}(\cdot, t; u_0) \geq \lim_{\varepsilon \downarrow 0} \mathfrak{M}^{\min}_{[d\sigma_{j+1}-\varepsilon,\Omega_j]} = \mathfrak{M}^{\min}_{[d\sigma_{j+1},\Omega_{j+1}]}. \quad (5.28)$$

Actually, as

$$\lim_{\varepsilon \downarrow 0} \mathfrak{M}^{\min}_{[d\sigma_{j+1}-\varepsilon,\Omega_j]} = \infty$$

uniformly in compact subsets of $\bar{\Omega}_{0,j+1} \setminus \partial\Omega$, (5.27) implies that

$$\liminf_{t \uparrow \infty} u_{[\lambda,\Omega]}(\cdot, t; u_0) = \infty \quad \text{unif. in compact subsets of } \bigcup_{i=1}^{j+1} \bar{\Omega}_{0,i} \setminus \partial\Omega. \quad (5.29)$$

In particular, for every $M > 0$ there exists $T_M > 0$ such that

$$u_{[\lambda,\Omega]}(x, t; u_0) \geq M \qquad \forall \ (x, t) \in (\partial\Omega_{j+1} \setminus \partial\Omega) \times [T_M, \infty).$$

So, $u_{[\lambda,\Omega]}(\cdot, t; u_0)$ is a supersolution of

$$\begin{cases} \frac{\partial u}{\partial t} - d\Delta u = \lambda u + af(\cdot, u)u & \text{in } \Omega_{j+1} \times (T_M, \infty), \\ u = M & \text{on } (\partial\Omega_{j+1} \setminus \partial\Omega) \times (T_M, \infty), \\ u = 0 & \text{on } (\partial\Omega_{j+1} \cap \partial\Omega) \times (T_M, \infty), \\ u(\cdot, T_M) = u_{[\lambda,\Omega]}(\cdot, T_M; u_0) & \text{in } \Omega_{j+1}. \end{cases}$$

Consequently, by the parabolic maximum principle,

$$u_{[\lambda,\Omega]}(x, t; u_0) \geq u_{[\lambda,\Omega_{j+1},M]}\left(x, t - T_M; u_{[\lambda,\Omega]}(\cdot, T_M; u_0)\right) \quad (5.30)$$

for all $(x, t) \in \Omega_{j+1} \times (T_M, \infty)$, where $u_{[\lambda,\Omega_{j+1},M]}(x, t; \tilde{u}_0)$ stands for the unique solution of

$$\begin{cases} \frac{\partial u}{\partial t} - d\Delta u = \lambda u + af(\cdot, u)u & \text{in } \Omega_{j+1} \times (T_M, \infty), \\ u = M & \text{on } (\partial\Omega_{j+1} \setminus \partial\Omega) \times (T_M, \infty), \\ u = 0 & \text{on } (\partial\Omega_{j+1} \cap \partial\Omega) \times (T_M, \infty), \\ u(\cdot, 0) = \tilde{u}_0 & \text{in } \Omega_{j+1}, \end{cases}$$

which is the parabolic counterpart of (4.33) at $j + 1$. By Theorem 4.5, this parabolic problem possesses a unique positive steady-state solution, $\Theta_{[\lambda,\Omega_{j+1},M]}$, for each $\lambda < d\sigma_{j+2}$. Thus, thanks to Theorem 1.7(b),

$$\lim_{t\uparrow\infty} u_{[\lambda,\Omega_{j+1},M]}(x,t;\tilde{u}_0) = \Theta_{[\lambda,\Omega_{j+1},M]}$$

for all $\tilde{u}_0 > 0$. So, letting $t \uparrow \infty$ in (5.30) shows that

$$\liminf_{t\uparrow\infty} u_{[\lambda,\Omega]}(x,t;u_0) \geq \Theta_{[\lambda,\Omega_{j+1},M]} \qquad \text{for all } M > 0.$$

Therefore, letting $M \uparrow \infty$, we find from Theorem 4.7 that

$$\liminf_{t\uparrow\infty} u_{[\lambda,\Omega]}(x,t;u_0) \geq L^{\min}_{[\lambda,\Omega_{j+1}]} \qquad \text{in } \Omega_{j+1}.$$

Bringing together this estimate with (5.29) it becomes apparent that

$$\liminf_{t\uparrow\infty} u_{[\lambda,\Omega]}(x,t;u_0) \geq \mathfrak{M}^{\min}_{[\lambda,\Omega_{j+1}]} \qquad \text{in } \Omega, \tag{5.31}$$

which is the lower estimate of (5.3) at $j + 1$.

Suppose, in addition, that u_0 is a subsolution of (5.2) in Ω. Then, arguing as above, the function $x \mapsto u_{[\lambda,\Omega]}(x,t;u_0)$, $x \in \bar{\Omega}$, is a subsolution of (5.2) for all $t > 0$, because $t \mapsto u_{[\lambda,\Omega]}(\cdot,t;u_0)$ is non-decreasing. Fix $t > 0$ and set

$$\hat{M}_t := \max_{\bar{\Omega}_{j+1}} u_{[\lambda,\Omega]}(\cdot,t;u_0).$$

As $u_{[\lambda,\Omega]}(\cdot,t;u_0)$ is a subsolution of (5.2), it is also a subsolution of (4.33) with $j + 1$, instead of j, for all $M \geq \hat{M}_t$. Thus, due to Theorem 4.5(a),

$$u_{[\lambda,\Omega]}(\cdot,t;u_0) \leq \Theta_{[\lambda,\Omega_{j+1},M]} \qquad \text{in } \Omega_{j+1} \quad \text{for all } M \geq \hat{M}_t.$$

Hence,

$$u_{[\lambda,\Omega]}(\cdot,t;u_0) \leq \lim_{M\uparrow\infty} \theta_{[\lambda,\Omega_{j+1},M]} = L^{\min}_{[\lambda,\Omega_{j+1}]} \qquad \text{in } \Omega_{j+1}$$

and therefore, letting $t \uparrow \infty$ yields

$$\limsup_{t\uparrow\infty} u_{[\lambda,\Omega]}(\cdot,t;u_0) \leq L^{\min}_{[\lambda,\Omega_{j+1}]} \qquad \text{in } \Omega_{j+1}.$$

Consequently, according to (5.31), we can conclude that

$$\lim_{t\uparrow\infty} u_{[\lambda,\Omega]}(\cdot,t;u_0) = \mathfrak{M}^{\min}_{[\lambda,\Omega_{j+1}]} \qquad \text{in } \Omega, \tag{5.32}$$

which ends the proof of (5.4) for $j + 1$.

Suppose $d\sigma_{j+1} \leq \lambda < d\sigma_{j+2}$ and for sufficiently large $n \in \mathbb{N}$, say $n \geq n_0$, consider the open sets defined by (4.35)

$$\Omega_{j+1,n} := \{x \in \Omega_{j+1} \; : \; \text{dist}\,(x, \partial\Omega_{j+1} \setminus \partial\Omega) > 1/n\}.$$

Pick $n \geq n_0$. As $\partial\Omega_{j+1,n} \subset \Omega_- \cup \partial\Omega$, it follows from (5.23) that there exists $M_0 > 0$ such that

$$u_{[\lambda,\Omega]}(\cdot, t; u_0) \leq M \qquad \text{on} \quad \partial\Omega_{j+1,n} \setminus \partial\Omega$$

for all $M \geq M_0$ and $t > 0$. Thus, by the parabolic maximum principle,

$$u_{[\lambda,\Omega]}(\cdot, t; u_0) \leq u_{[\lambda,\Omega_{j+1,n},M]}(\cdot, t; u_0) \qquad \text{in} \quad \Omega_{j+1,n} \qquad (5.33)$$

for all $t > 0$, where $u_{[\lambda,\Omega_{j+1,n},M]}$ stands for the unique solution of

$$\begin{cases} \frac{\partial u}{\partial t} - d\Delta u = \lambda u + af(\cdot, u)u & \text{in } \Omega_{j+1,n} \times (0, \infty), \\ u = M & \text{on } (\partial\Omega_{j+1,n} \setminus \partial\Omega) \times (0, \infty), \\ u = 0 & \text{on } (\partial\Omega_{j+1,n} \cap \partial\Omega) \times (0, \infty), \\ u(\cdot, 0) = u_0 & \text{in } \Omega_{j+1,n}. \end{cases}$$

By Theorem 1.7,

$$\lim_{t\uparrow\infty} u_{[\lambda,\Omega_{j+1,n},M]}(\cdot, t; u_0) = \Theta_{[\lambda,\Omega_{j+1,n},M]} \qquad \text{in} \quad \Omega_{j+1,n},$$

where $\Theta_{[\lambda,\Omega_{j+1,n},M]}$ is the unique positive solution of (4.36) at $j + 1$. Hence, letting $t \uparrow \infty$ in (5.33) shows that, for every $n \geq n_0$,

$$\limsup_{t\uparrow\infty} u_{[\lambda,\Omega]}(\cdot, t; u_0) \leq \Theta_{[\lambda,\Omega_{j+1,n},M]} \qquad \text{in} \quad \Omega_{j+1,n}. \qquad (5.34)$$

Consequently, letting $M \uparrow \infty$ in (5.34), we find that

$$\limsup_{t\uparrow\infty} u_{[\lambda,\Omega]}(\cdot, t; u_0) \leq L^{\min}_{[\lambda,\Omega_{j+1,n}]} \qquad \text{in} \quad \Omega_{j+1,n} \qquad (5.35)$$

for all $n \geq n_0$. As (5.35) holds for all $n \geq n_0$ and, due to (4.39), we already know that

$$L^{\max}_{[\lambda,\Omega_{j+1}]} = \lim_{n\uparrow\infty} L^{\min}_{[\lambda,\Omega_{j+1,n}]},$$

we can conclude from (5.35) that

$$\limsup_{t\uparrow\infty} u_{[\lambda,\Omega]}(\cdot, t; u_0) \leq L^{\max}_{[\lambda,\Omega_{j+1}]} \qquad \text{in} \quad \Omega_{j+1},$$

which ends the proof of the theorem.

5.5 Approximating metasolutions by classical solutions

In this section we will compare the dynamics of the parabolic problem

$$\begin{cases} \frac{\partial u}{\partial t} - d\Delta u = \lambda u + (a(x) + \varepsilon b(x))f(x,u)u & \text{in } \Omega \times (0,\infty), \\ u = 0 & \text{on } \partial\Omega \times (0,\infty), \\ u(\cdot,0) = u_0 > 0 & \text{in } \Omega, \end{cases} \quad (5.36)$$

with the dynamics of (1.1) as $\varepsilon \downarrow 0$. Here, $b \in \mathcal{C}^\nu(\bar\Omega)$ satisfies

$$b(x) < 0 \quad \text{for all} \quad x \in \bar\Omega_0 = a^{-1}(0)$$

and $\varepsilon > 0$ is a constant. Naturally, a and f are assumed to satisfy (Ha) and (KO), respectively. Thus,

$$a(x) + \varepsilon b(x) < 0 \quad \text{for all} \quad x \in \bar\Omega$$

and (5.36) is a classical generalized diffusive logistic equation, of the same type as those analyzed in Chapter 2. Consequently, owing to Theorem 2.2, the next result holds.

Theorem 5.4 *Suppose $a(x)$ and f satisfy (Ha) and (Hf)–(Hg), respectively. Then, for every $\varepsilon > 0$, (5.36) possesses a unique positive steady state, $\theta_{[\lambda,\Omega,\varepsilon]}$, if and only if, $\lambda > d\sigma_0$. Moreover, $\theta_{[\lambda,\Omega,\varepsilon]} \gg 0$ and*

$$\lim_{t\uparrow\infty} u_{[\lambda,\Omega,\varepsilon]}(\cdot,t;u_0) = \begin{cases} 0, & \text{if } \lambda \leq d\sigma_0, \\ \theta_{[\lambda,\Omega,\varepsilon]}, & \text{if } \lambda > d\sigma_0, \end{cases} \quad \text{in } \mathcal{C}(\bar\Omega), \quad (5.37)$$

where $u_{[\lambda,\Omega,\varepsilon]}$ stands for the unique solution of (5.36).

The main result of this section can be stated as follows.

Theorem 5.5 *Suppose $a(x)$ and f satisfy (Ha) and (KO), respectively, and set $\sigma_{q_0+1} := \infty$, as usual. Then,*

(a) *For every $\lambda \in (d\sigma_0, d\sigma_1)$,*

$$\lim_{\varepsilon\downarrow 0} \theta_{[\lambda,\Omega,\varepsilon]} = \theta_{[\lambda,\Omega]} \quad \text{in } \mathcal{C}(\bar\Omega), \quad (5.38)$$

where $\theta_{[\lambda,\Omega]}$ is the unique positive steady-state of (1.1).

(b) *For every $1 \leq j \leq q_0$ and $\lambda \in [d\sigma_j, d\sigma_{j+1})$,*

$$\lim_{\varepsilon\downarrow 0} \theta_{[\lambda,\Omega,\varepsilon]} = \mathfrak{M}^{\min}_{[\lambda,\Omega_j]} \quad \text{in } \bar\Omega. \quad (5.39)$$

Proof: Fix $\lambda \in (d\sigma_0, d\sigma_1)$. Then, for every $\varepsilon > 0$, $\theta_{[\lambda,\Omega,\varepsilon]}$ is the unique positive solution of

$$\begin{cases} -d\Delta u = \lambda u + (a(x) + \varepsilon b(x))f(x,u)u & \text{in } \Omega, \\ u = 0 & \text{on } \partial\Omega, \end{cases} \tag{5.40}$$

whereas, according to Theorem 4.1, $\theta_{[\lambda,\Omega]}$ is the unique solution of (5.40) with $\varepsilon = 0$. Subsequently, we will set

$$\theta_\varepsilon := \theta_{[\lambda,\Omega,\varepsilon]}, \qquad \varepsilon > 0, \qquad \theta_0 := \theta_{[\lambda,\Omega]}.$$

As in Chapter 2, the positive solutions of (5.40) are regarded as the positive zeroes of the nonlinear operator $\mathfrak{F} : \mathbb{R} \times \mathcal{C}_0(\bar{\Omega}) \to \mathcal{C}_0(\bar{\Omega})$ defined by

$$\mathfrak{F}(\varepsilon, u) := u - (-d\Delta)^{-1}\left[\lambda u + (a + \varepsilon b)f(\cdot, u)u\right], \qquad \varepsilon \in \mathbb{R}, \ \ u \in \mathcal{C}_0(\bar{\Omega}),$$

where $(-d\Delta)^{-1}$ stands for the resolvent operator of $-d\Delta$ in Ω under homogeneous Dirichlet boundary conditions. The operator $\mathfrak{F}$ is of class $\mathcal{C}^1$ and, by elliptic regularity, $\mathfrak{F}(\varepsilon, \cdot)$ is a nonlinear compact perturbation of the identity map for all $\varepsilon \in \mathbb{R}$. Since $\mathfrak{F}(0, \theta_0) = 0$, by Lemma 2.1, $\theta_0 \gg 0$ and

$$\lambda = \lambda_1[-d\Delta - af(\cdot, \theta_0), \Omega]. \tag{5.41}$$

Moreover, differentiating $\mathfrak{F}$ with respect to u yields

$$D_u\mathfrak{F}(0, \theta_0)u = u - (-d\Delta)^{-1}\left[\lambda u + a\frac{\partial f}{\partial u}(\cdot, \theta_0)\theta_0 u + af(\cdot, \theta_0)u\right]$$

for all $u \in \mathcal{C}_0(\bar{\Omega})$. Hence, $D_u\mathfrak{F}(0, \theta_0)$ is a Fredholm operator of index zero. Moreover, as it is injective, it is also a linear topological isomorphism. Indeed, if

$$u - (-d\Delta)^{-1}\left[\lambda u + a\frac{\partial f}{\partial u}(\cdot, \theta_0)\theta_0 u + af(\cdot, \theta_0)u\right] = 0$$

for some $u \in \mathcal{C}_0(\bar{D})$, then, by elliptic regularity, $u \in \mathcal{C}_0^{2+\nu}(\bar{\Omega})$ and

$$\mathfrak{L}u = 0 \qquad \text{in } \Omega, \tag{5.42}$$

where

$$\mathfrak{L} := -d\Delta - \lambda - a\frac{\partial f}{\partial u}(\cdot, \theta_0)\theta_0 - af(\cdot, \theta_0). \tag{5.43}$$

On the other hand, since $a(x) < 0$ for all $x \in \Omega_-$, by the monotonicity properties of the principal eigenvalue, it follows from (5.41) and (Hf) that

$$\lambda_1[\mathfrak{L}, \Omega] > \lambda_1[-d\Delta - \lambda - af(\cdot, \theta_0), \Omega] = 0.$$

Hence, (5.42) implies $u = 0$. So, $D_u\mathfrak{F}(0, \theta_0)$ is a linear topological isomorphism. Therefore, combining the uniqueness of the positive solutions with the implicit function theorem, it becomes apparent that (5.38) holds, which ends the proof of Part (a).

Pick $1 \leq j \leq q_0$ and $\lambda \in [d\sigma_j, d\sigma_{j+1})$. Let $\varphi \gg 0$ be any principal eigenfunction associated to σ_0. Then, $u_0 := \eta\varphi$ is a subsolution of (5.40) for sufficiently small $\varepsilon \geq 0$ and $\eta > 0$. Moreover, $\eta > 0$ can be shortened, if necessary, so that

$$u_0 < \mathfrak{M}^{\min}_{[\lambda, \Omega_j]}.$$

According to D. Sattinger [220], $u_{[\lambda, \varepsilon]}(\cdot, t; u_0) \gg 0$ also is a subsolution of (5.40) for sufficiently small $\varepsilon > 0$ and all $t > 0$. Thus, by Theorem 5.2(c),

$$\lim_{t \uparrow \infty} u_{[\lambda, \Omega, 0]}(\cdot, t; u_0) = \mathfrak{M}^{\min}_{[\lambda, \Omega_j]}. \tag{5.44}$$

In particular, for any compact subset $K \subset \Omega_j$ and $\delta > 0$ there exists a time $t_\delta > 0$ such that

$$\|u_{[\lambda, \Omega, 0]}(\cdot, t; u_0) - \mathfrak{M}^{\min}_{[\lambda, \Omega_j]}\|_{\mathcal{C}(K)} \leq \delta/2 \qquad \text{for all } t \geq t_\delta. \tag{5.45}$$

On the other hand, by continuous dependence with respect to the parameter ε (e.g., Chapter 8 of A. Lunardi [186]), there exists $\varepsilon_0 := \varepsilon(t_\delta) > 0$ such that, for every $\varepsilon \in (0, \varepsilon_0]$,

$$\|u_{[\lambda, \Omega, \varepsilon]}(\cdot, t; u_0) - u_{[\lambda, \Omega, 0]}(\cdot, t; u_0)\|_{\mathcal{C}(K)} \leq \delta/2 \qquad \text{for all } t \leq t_\delta. \tag{5.46}$$

Hence, by (5.45) and (5.46), we find that

$$\|u_{[\lambda, \Omega, \varepsilon]}(\cdot, t_\delta; u_0) - \mathfrak{M}^{\min}_{[\lambda, \Omega_j]}\|_{\mathcal{C}(K)} \leq \delta \qquad \text{for all } 0 < \varepsilon \leq \varepsilon_0. \tag{5.47}$$

Shortening ε_0, if necessary, we can assume that u_0 also is a subsolution of (5.40) and hence,

$$u_{[\lambda, \Omega, \varepsilon]}(\cdot, t_\delta; u_0) \leq u_{[\lambda, \Omega, \varepsilon]}(\cdot, t; u_0) < \theta_{[\lambda, \Omega, \varepsilon]} < \mathfrak{M}^{\min}_{[\lambda, \Omega_j]} \tag{5.48}$$

for all $t \geq t_\delta$. According to (5.47) and (5.48), we find that

$$\|u_{[\lambda, \Omega, \varepsilon]}(\cdot, t; u_0) - \mathfrak{M}^{\min}_{[\lambda, \Omega_j]}\|_{\mathcal{C}(K)} \leq \delta \qquad \text{for all } t \geq t_\delta$$

and $0 < \varepsilon \leq \varepsilon_0$. Therefore, letting $t \uparrow \infty$, it follows from Theorem 5.4 that

$$\|\theta_{[\lambda, \Omega, \varepsilon]} - \mathfrak{M}^{\min}_{[\lambda, \Omega_j]}\|_{\mathcal{C}(K)} \leq \delta \qquad \text{for all } 0 < \varepsilon \leq \varepsilon_0.$$

Consequently,

$$\lim_{\varepsilon \downarrow 0} \theta_{[\lambda, \Omega, \varepsilon]} = \mathfrak{M}^{\min}_{[\lambda, \Omega_j]} \quad \text{uniformly on compact subsets of } \Omega_j. \tag{5.49}$$

To complete the proof of (5.39) it remains to prove that

$$\lim_{\varepsilon \downarrow 0} \theta_{[\lambda, \Omega, \varepsilon]} = \infty \qquad \text{in } (\Omega \setminus \bar{\Omega}_j) \cup (\partial\Omega_j \cap \Omega). \tag{5.50}$$

Indeed, by (5.44), for any positive constant $C > 0$, there exists a time $t_C > 0$ such that, for every $t \geq t_C$,

$$u_{[\lambda,\Omega,0]}(\cdot,t;u_0) \geq u_{[\lambda,\Omega,0]}(\cdot,t_C;u_0) > C \qquad \text{in } (\Omega \setminus \bar{\Omega}_j) \cup (\partial\Omega_j \cap \Omega). \quad (5.51)$$

On the other hand, by continuous dependence with respect to ε, we find from (5.51) that there exists $\varepsilon_0 > 0$ such that

$$u_{[\lambda,\Omega,\varepsilon]}(\cdot,t_C;u_0) > C \qquad \text{in } (\Omega \setminus \bar{\Omega}_j) \cup (\partial\Omega_j \cap \Omega)$$

for all $0 \leq \varepsilon \leq \varepsilon_0$. Consequently, for every $t \geq t_C$ and $0 \leq \varepsilon \leq \varepsilon_0$,

$$u_{[\lambda,\Omega,\varepsilon]}(\cdot,t;u_0) \geq u_{[\lambda,\Omega,\varepsilon]}(\cdot,t_C;u_0) > C \qquad \text{in } (\Omega \setminus \bar{\Omega}_j) \cup (\partial\Omega_j \cap \Omega),$$

since $t \mapsto u_{[\lambda,\Omega,\varepsilon]}(\cdot,t;u_0)$ is increasing. Thus, letting $t \uparrow \infty$, it follows from Theorem 5.4 that

$$\theta_{[\lambda,\Omega,\varepsilon]} \geq C \qquad \text{in } (\Omega \setminus \bar{\Omega}_j) \cup (\partial\Omega_j \cap \Omega)$$

for all $0 \leq \varepsilon \leq \varepsilon_0$, which shows (5.39) and ends the proof of the theorem. $\qquad \square$

Remark 5.6 By the Schauder interior estimates, it is easily seen that (5.49) implies the convergence of $\theta_{[\lambda,\Omega,\varepsilon]}$ to $\mathfrak{M}_{[\lambda,\Omega_j]}^{\min}$ as $\varepsilon \downarrow 0$ in $\mathcal{C}^{2+\nu}(K)$ for every compact subset K of Ω_j.

Theorem 5.5(b) provides us with a reasonable numerical scheme to approximate the minimal metasolutions supported in Ω_j in the interval of λ's where they are attractors from below, i.e., $d\sigma_j \leq \lambda < d\sigma_{j+1}$. By obvious reasons, this scheme cannot be used to compute them outside these ranges of λ.

5.6   Pattern formation in classical logistic problems

Let Ω_j^i, $1 \leq j \leq q$, $1 \leq i \leq m_j$, be an arbitrary family of smooth subdomains of Ω satisfying

$$\bar{\Omega}_j^i \cap \bar{\Omega}_{\hat{j}}^{\hat{i}} = \emptyset \qquad \text{if } (i,j) \neq (\hat{i},\hat{j})$$

and

$$\sigma_j := \lambda_1[-\Delta,\Omega_j^i], \qquad 1 \leq i \leq m_j, \ 1 \leq j \leq q,$$
$$\sigma_j < \sigma_{j+1}, \qquad 1 \leq j \leq q-1,$$

and consider the open subset of Ω defined by

$$\mathcal{O} := \bigcup_{j=1}^{q} \bigcup_{i=1}^{m_j} \Omega_j^i.$$

Suppose $a(x) > 0$ for all $x \in \bar{\Omega}$. Then, (1.1) is a classical diffusive logistic parabolic problem, whose dynamics are governed by its maximal non-negative steady state solution. Suppose

$$\max_{\bar{\mathcal{O}}} a \simeq \varepsilon > 0$$

while, for any compact subset $K \subset \Omega \setminus \bar{\mathcal{O}}$, $\min_K a$ is separated away from zero, then, by continuous dependence, Theorem 5.5 provides us with the shape of $u(x, t; u_0)$, for sufficiently large t, according to the different ranges of values of the parameter λ considered in Theorem 5.5. So, Theorems 5.2 and 5.5 provide us with all the admissible patterns exhibited by the classical diffusive logistic equations through the dynamics of the associated degenerated situations.

5.7 Biological discussion

From the perspective of population dynamics, according to Theorem 5.2, a species u becomes extinct if $\lambda \leq d\sigma_0$. For a given species with an intrinsic growth rate λ this occurs if $d \geq d_0 := \lambda/\sigma_0$, which measures the critical size of the dispersion rate d, d_0, so that the habitat Ω can support it. If $d \geq d_0$, then the individuals of the species approximate the habitat edges at a sufficiently high rate as to drive the species to extinction as a consequence of the hostility of the surroundings of the inhabiting area Ω.

Adopting another perspective, for any given $\lambda > 0$ and $d > 0$, the extinction can be avoided by placing the species u in a sufficiently large habitat Ω. Indeed, according to the Faber–Krahn inequality, we already know that there is a universal constant $C > 0$ such that

$$\sigma_0 = \lambda_1[-\Delta, \Omega] \geq C|\Omega|^{-2/N}, \qquad C := \lambda_1[-\Delta, B_1]|B_1|^{2/N},$$

where N is the spatial dimension. Therefore, as soon as the measure of the habitat satisfies

$$|\Omega| \leq (Cd/\lambda)^{N/2},$$

Ω cannot support a population u with an intrinsic growth rate λ dispersing at the rate d. Conversely, if Ω is sufficiently large to contain some ball of radius $R > 0$, $B_R(x_0)$, then, by the monotonicity of $\lambda_1[-\Delta, D]$ with respect to D, it is apparent that

$$\lambda_1[-\Delta, \Omega] \leq \lambda_1[-\Delta, B_R(x_0)] = \lambda_1[-\Delta; B_R(0)] = \lambda_1[-\Delta; B_1]R^{-2}$$

and hence,

$$\lim_{R\uparrow\infty} \sigma_0 = \lim_{R\uparrow\infty} \lambda_1[-\Delta, \Omega] = 0.$$

Thus, $\lambda > d\sigma_0$ for sufficiently large R. Consequently, in sufficiently large habitats the species u will be always permanent. Naturally, once we have fixed λ, the bigger the dispersion rate, measured by d, the larger should be the inhabiting area Ω to avoid extinction.

It should be also noted that even the simplest linear problem

$$\begin{cases} \frac{\partial u}{\partial t} - d\Delta u = \lambda u & \text{in } \Omega \times (0, \infty), \\ u = 0 & \text{on } \partial\Omega \times (0, \infty), \\ u(\cdot, 0) = u_0 > 0 & \text{in } \Omega, \end{cases} \tag{5.52}$$

predicts extinction if $\lambda < d\sigma_0$, as we have already discussed in Chapter 1. So, we should not expect Ω to be able to support any species u with the same intrinsic growth rate, λ, and dispersion rate, d, if, in addition, we are incorporating interspecific competition of the individuals for the available resources, measured by $af(\cdot, u)u \leq 0$.

Theorem 5.2 predicts that the behavior of the species u should be of classical logistic type if $d\sigma_0 < \lambda < d\sigma_1$. The condition $\lambda > d\sigma_0$ allows Ω to support the species u independently of the level of intraspecific competition, measured by $af(\cdot, u)u$. In the context of population dynamics, the condition

$$\lambda < d\sigma_1 = d\lambda_1[-\Delta, \Omega_{0,1}^i], \quad 1 \leq i \leq m_1,$$

means that the *largest protection zones*, measured by the size of the associated principal eigenvalues, cannot support the species u in the absence of intraspecific competition effects, because the solutions of the linear problem

$$\begin{cases} \frac{\partial u}{\partial t} - d\Delta u = \lambda u & \text{in } \Omega_{0,1}^i \times (0, \infty), \\ u = 0 & \text{on } \partial\Omega_{0,1}^i \times (0, \infty), \\ u(\cdot, 0) = u_0 > 0 & \text{in } \Omega_{0,1}^i, \end{cases} \tag{5.53}$$

approximate 0 as $t \uparrow \infty$, for all $1 \leq i \leq m_1$. This might be attributable to the lack of room in these protection zones to accumulate the necessary resources to maintain the species u therein. Note that we are measuring the size of the protection zones through the size of the principal eigenvalue of $-\Delta$. Theorem 5.2 establishes that the previous situation changes dramatically when

$$d\sigma_1 < \lambda < d\sigma_2 = d\lambda_1[-\Delta, \Omega_{0,2}^i], \quad 1 \leq i \leq m_2,$$

as, in such case, the largest protection zones $\Omega_{0,1}^i$, $1 \leq i \leq m_1$, can support the species u, in the sense that the solutions of (5.53) exhibit a genuine Malthusian growth therein if $\lambda > d\sigma_1$, though the smaller protection zones, $\Omega_{0,j}^i$, $2 \leq j \leq q_0$, $1 \leq i \leq m_j$, cannot support the species u because $\lambda < d\sigma_2$. In such circumstances, Theorem 5.2 predicts exponential growth in the largest protection zones,

$$\Omega_{0,1} = \bigcup_{i=1}^{m_1} \Omega_{0,1}^i$$

and logistic growth in their complement

$$\Omega_1 := \Omega \setminus \bar{\Omega}_{0,1}.$$

More generally, when $q_0 \geq 3$ and $d\sigma_j < \lambda < d\sigma_{j+1}$ for some $2 \leq j \leq q_0 - 1$, any protection zone of the form $\Omega_{0,k}^i$, $1 \leq k \leq j$, $1 \leq i \leq m_k$, can support u, in the sense that the solutions of the linear problems

$$\begin{cases} \frac{\partial u}{\partial t} - d\Delta u = \lambda u & \text{in} \quad \Omega_{0,k}^i \times (0, \infty), \\ u = 0 & \text{on} \quad \partial\Omega_{0,k}^i \times (0, \infty), \\ u(\cdot, 0) = u_0 > 0 & \text{in} \quad \Omega_{0,k}^i, \end{cases} \tag{5.54}$$

grow exponentially for all $1 \leq k \leq j$ and $1 \leq i \leq m_k$. However, the smaller protection zones $\Omega_{0,k}^i$, $j + 1 \leq k \leq q_0$, $1 \leq i \leq m_k$, cannot do it, in the sense that the solutions of (5.54) become extinct for all $j + 1 \leq k \leq q_0$, $1 \leq i \leq m_k$. In these circumstances, Theorem 5.2 predicts exponential growth of the species u in the largest protection zones

$$\bigcup_{k=1}^{j} \bigcup_{i=1}^{m_k} \Omega_{0,k}^i$$

and logistic growth in the complement

$$\Omega_j := \Omega_- \cup \bigcup_{k=j+1}^{q_0} \bigcup_{i=1}^{m_k} \Omega_{0,k}^i.$$

When $\lambda > d\sigma_{q_0}$, there is exponential growth in all the protection zones and logistic growth in Ω_-, the region where the behavior of the species has been assumed to be of logistic type for the non-spatial model.

5.8   Comments on Chapter 5

The first two parts of Theorem 5.2 go back to J. M. Fraile et al. [90] and are valid for general second order elliptic operators like (1.48) under mixed boundary conditions (see Theorem 3.7(a)(b) of [90]). Actually, according to the proof of Theorem 3.7(c) of [90], it is apparent that

$$\lim_{t \uparrow \infty} u_{[\lambda, \Omega]}(x, t; u_0) = \infty \tag{5.55}$$

for all $x \in \Omega_{0,1}$ and $\lambda \geq d\sigma_1$. The problem of the asymptotic behavior of the solutions of (1.1) was addressed in J. M. Fraile et al. [90].

The first occasion that Theorem 5.2 was stated in its greatest generality was in the four summarizing items on pages 573 and 574 of R. Gómez-Reñasco and J. López-Gómez [109], submitted in September 1998. The pictures included in Section 5.3 were taken from [109]. The detailed proofs of these results were given in J. López-Gómez [147], which was submitted for publication

Part II

Uniqueness of the large solution

Chapter 6

A canonical one-dimensional problem

The main goal of the second part of this book is to obtain general uniqueness results for the large solutions constructed in the first part. The whole program adopted to accomplish this task will be divided into three steps delimited by each of the chapters included in this part. This chapter establishes the existence and the uniqueness of the positive solution, $\ell(x)$, of the singular boundary value problem

$$\begin{cases} u''(x) = a(x)h(u(x)), & x > 0, \\ u(0) = \infty, \ u(\infty) = 0, \end{cases} \tag{6.1}$$

under the following conditions on $a(x)$ and $h(u)$:

(A1) $a \in \mathcal{C}[0, \infty)$ satisfies $a(x) \geq a(y) > 0$ whenever $x \geq y > 0$.

(A2) $h \in \mathcal{C}^1[0, \infty)$ satisfies $h(0) = h'(0) = 0$, $h'(u) > 0$ if $u > 0$, as well as the Keller–Osserman condition

$$I(u) := \int_u^\infty \frac{dx}{\sqrt{\int_u^x h}} < \infty, \quad u > 0, \qquad \lim_{u \to \infty} I(u) = 0. \tag{6.2}$$

Moreover, it will ascertain the blow-up rate of $\ell(x)$ at $x = 0$ when, besides (A1) and (A2), the next conditions holds.

(A3) The quotient $g(u) := h(u)/u$, $u > 0$, satisfies $g'(u) \geq 0$ for all $u > 0$ and the limit

$$H := \lim_{u \uparrow \infty} \frac{h(u)}{u^p} > 0 \tag{6.3}$$

is well defined for some $p > 1$.

Note that (6.3) implies (6.2). Indeed, by (6.3), there exists $z > 0$ such that

$$h(u) \geq \frac{H}{2} u^p, \qquad u \geq z.$$

Hence, since $p > 1$, we have that

$$\int_z^\infty \left[\int_z^x h(s)\, ds \right]^{-\frac{1}{2}} dx \leq \sqrt{\frac{2(p+1)}{H}} \int_z^\infty \frac{dx}{\sqrt{x^{p+1} - z^{p+1}}} < \infty.$$

Naturally, by $u(0) = \infty$ and $u(\infty) = 0$ it is meant that

$$\lim_{x \downarrow 0} u(x) = \infty, \qquad \lim_{x \uparrow \infty} u(x) = 0.$$

It should be noted that, according to the notations introduced in Part I,

$$\mathfrak{a}(x) := -a(x)/d \qquad \text{and} \qquad h(u) := g(u)u.$$

Moreover, as discussed in the final section of Chapter 3, condition (6.2) for $u > u^*$ (see (3.46)) entails (KO) when $h(u)$ is defined as in the proof of Theorem 3.2. Consequently, as (6.2) is stronger than (KO), all the theory developed in Chapter 3 can be applied here.

This chapter is organized as follows. Section 6.2 shows the existence and the uniqueness of the positive solution of (6.1), and of some closely related problems whose significance will become apparent later. For any given $R > 0$, Section 6.3 studies some of the most basic properties of the function

$$\mathfrak{b}(x) = \int_x^R \left(\int_0^y \mathfrak{a}^{\frac{1}{p+1}} \right)^{-\frac{p+1}{p-1}} dy, \qquad x \in (0, R].$$

Section 6.4 shows that $(\varepsilon\mathfrak{b}, \kappa\mathfrak{b})$ provides us with a sub-supersolution pair of (6.1) in $(0, \delta]$ for sufficiently small $\delta > 0$ and $\varepsilon > 0$ and sufficiently large κ, in order to establish that

$$\lim_{x \downarrow 0} \frac{\ell(x)}{\mathfrak{b}(x)} = I_0^{\frac{-p}{p-1}} \left(\frac{p+1}{p-1} \right)^{\frac{p+1}{p-1}} H^{\frac{-1}{p-1}},$$

where

$$I_0 := \lim_{x \downarrow 0} \frac{\mathfrak{b}(x)\mathfrak{b}''(x)}{[\mathfrak{b}'(x)]^2} \in (0, \infty).$$

Some sufficient conditions for $0 < I_0 < \infty$ are collected in Section 6.2. Note that if $\mathfrak{b}(x)$ is logarithmically convex, then $I_0 \geq 1$ (see the end of Section 3.4). Precisely, if there exists $\varepsilon > 0$ such that $\mathfrak{a} \in C^3(0, \varepsilon]$, $\mathfrak{a}'(x) > 0$ and $(\ln \mathfrak{a})''(x) < 0$ for all $x \in (0, \varepsilon)$, and the function $\frac{\mathfrak{a}\mathfrak{a}''}{(\mathfrak{a}')^2}$ is non-oscillating in $(0, \varepsilon)$, in the sense discussed in Definition 6.6, then $I_0 \in [1, \infty)$.

6.1 Existence and uniqueness

The following result shows the existence and the uniqueness of the positive solution of (6.1), $\ell(x)$, and establishes some of its basic properties.

Theorem 6.1 *Suppose* $\mathfrak{a}(x)$ *and* $h(u)$ *satisfy* (A1) *and* (A2). *Then, the problem* (6.1) *possesses a unique positive solution,* $\ell(x)$, $x > 0$. *Moreover,*

$$\ell(x) > 0, \qquad \ell'(x) < 0, \qquad \ell''(x) > 0, \qquad x > 0, \tag{6.4}$$

and

$$\lim_{x \downarrow 0} \ell'(x) = -\infty, \qquad \lim_{x \uparrow \infty} \ell'(x) = 0. \tag{6.5}$$

Consequently, the graph of $\ell(x)$ looks like Figure 6.1.

Proof: Subsequently, for every $M > 0$ and $L > 0$, we will consider the problem

$$\begin{cases} u''(x) = \mathfrak{a}(x)h(u(x)), & 0 < x < L, \\ u(0) = M, \ u(L) = 0. \end{cases} \tag{6.6}$$

By our assumptions on $\mathfrak{a}(x)$ and $h(u)$, it is apparent that $(\underline{u}, \overline{u}) := (0, M)$ is an ordered sub-supersolution pair of (6.6). Thus, by Theorem 1.2, (6.6) possesses a solution $u \in \mathcal{C}^2[0, L]$ with $u(x) \in [0, M]$ for all $x \in [0, L]$. Since $u(0) = M > 0$, $u \neq 0$. Thus,

$$u(x) > 0 \quad \text{for all } x \in [0, L). \tag{6.7}$$

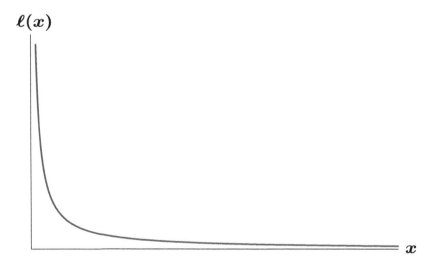

FIGURE 6.1: The graph of $\ell(x)$.

Indeed, if there is $x_0 \in (0, L)$ such that $u(x_0) = 0$, then $u'(x_0) = 0$ and (u, u') solves the Cauchy problem

$$u' = v, \quad v' = \mathfrak{a}(x)h(u), \quad u(x_0) = v(x_0) = 0,$$

whose unique solution is $(u, v) = (0, 0)$, by the Cauchy–Lipschitz theorem. This contradiction shows (6.7) and establishes that $u'(L) < 0$. Moreover, since

$$u''(x) = \mathfrak{a}(x)h(u(x)) > 0$$

for all $x \in (0, L)$, $u'(x)$ is increasing. So, for every $x \in [0, L]$

$$u'(x) \leq u'(L) < 0.$$

Therefore, u is decreasing.

Now, we will show the uniqueness of the solution of (6.6). Our proof will proceed by contradiction. If $u_1 \neq u_2$ are two positive solutions of (6.6) such that $u_1(x_0) > u_2(x_0)$ for some $x_0 \in (0, L)$, then there are $x_1 \in [0, x_0)$ and $x_2 \in (x_0, L]$ such that

$$u_1(x_1) = u_2(x_1), \quad u_1(x_2) = u_2(x_2),$$

and

$$u_1(x) > u_2(x) \quad \forall\, x \in (x_1, x_2).$$

Pick $x_m \in (x_1, x_2)$ such that

$$u_1(x_m) - u_2(x_m) = \max_{x \in [x_1, x_2]} (u_1(x) - u_2(x)) > 0.$$

Then,

$$0 \geq (u_1 - u_2)''(x_m) = \mathfrak{a}(x_m) \left[h(u_1(x_m)) - h(u_2(x_m)) \right] > 0,$$

by (A1) and (A2). This contradiction shows the uniqueness. Subsequently, we will denote by $u_{[M,L]}$ the unique positive solution of (6.6).

Next, we will show that $M_1 \geq M_2 > 0$ and $L_1 \geq L_2 > 0$ imply

$$M_1 \geq u_{[M_1, L_1]} \geq u_{[M_2, L_2]} \quad \text{in} \quad [0, L_2]. \tag{6.8}$$

Indeed, the function

$$\bar{u} := u_{[M_1, L_1]}\big|_{[0, L_2]}$$

is a supersolution of

$$\begin{cases} u''(x) = \mathfrak{a}(x)h(u(x)), & 0 < x < L_2, \\ u(0) = M_2, \ u(L_2) = 0, \end{cases} \tag{6.9}$$

in the interval $[0, L_2]$, because it solves the differential equation and

$$\bar{u}(0) = M_1 \geq M_2, \quad \bar{u}(L_2) \geq 0.$$

Clearly, $\underline{u} := 0 \leq \bar{u}$ is a subsolution. Therefore, (6.8) follows from Theorem 1.2, since $u_{[M_2, L_2]}$ is the unique positive solution of (6.9). The first estimate of (6.8) follows from the fact that $u_{[M,L]}$ is decreasing.

According to (6.8), for any $M > 0$, the point-wise limit

$$u_M(x) := \lim_{L \uparrow \infty} u_{[M,L]}(x) \qquad (6.10)$$

is well defined for all $x \in [0, \infty)$. Moreover, $0 \leq u_M \leq M$ in $[0, \infty)$. Actually, as the set of prolonged functions

$$\mathfrak{F} := \{\tilde{u}_{[M,L]}, \ L > 0\}, \qquad \text{where} \quad \tilde{u}_{[M,L]} := \begin{cases} u_{[M,L]}, & \text{in } [0, L], \\ 0, & \text{in } (L, \infty), \end{cases}$$

is bounded in $\mathcal{C}[0, \infty)$, one can infer from the Ascoli–Arzela theorem that $u_M \geq 0$ provides us with a solution of

$$u''(x) = \mathfrak{a}(x)h(u(x)), \qquad x > 0,$$

such that $u(0) = M$. Arguing as above, we find that $u_M(x) > 0$ for all $x > 0$. Moreover, by (6.10), u_M is non-increasing. Hence, $u'_M(x) \leq 0$ for all $x > 0$. Also, $u''_M(x) > 0$ for all $x > 0$. Thus, u'_M is increasing in $[0, \infty)$. Consequently, $u'_M(x) < 0$ for all $x > 0$ and the next limit is well defined:

$$\omega := \lim_{x \uparrow \infty} u_M(x) \geq 0.$$

Suppose $\omega > 0$, Then, for every $x > 1$, we find from (A1) and (A2) that

$$u''_M(x) = \mathfrak{a}(x)h(u_M(x)) \geq \mathfrak{a}(1)h(\omega) > 0$$

and hence, for every $x > 1$, we find that

$$u'_M(x) = u'_M(1) + \int_1^x u''_M(s)\, ds \geq u'_M(1) + \mathfrak{a}(1)h(\omega)(x - 1),$$

which is impossible because this estimate implies $u'_M(x) > 0$ for sufficiently large x and we already know that u_M is decreasing. Therefore, $\omega = 0$ and u_M is a positive solution of

$$\begin{cases} u''(x) = \mathfrak{a}(x)h(u(x)), & x > 0, \\ u(0) = M, \ u(\infty) = 0. \end{cases} \qquad (6.11)$$

Now, we will show that u_M is the unique positive solution of (6.11). Suppose that $u_1 \neq u_2$ are two solutions of (6.11) with $u_1(x_0) > u_2(x_0)$ for some $x_0 > 0$. Then, arguing as in the proof of the uniqueness for (6.6), there are $x_1 \in [0, x_0)$ and $x_2 \in (x_0, \infty]$ with $u_1(x_1) = u_2(x_1)$, $u_1(x_2) = u_2(x_2)$, and $u_1(x) > u_2(x)$ for all $x \in (x_1, x_2)$. Pick $x_m \in (x_1, x_2)$ such that

$$u_1(x_m) - u_2(x_m) = \max_{x \in [x_1, x_2]} (u_1(x) - u_2(x)) > 0.$$

Then, by (A1) and (A2),

$$0 \geq (u_1 - u_2)''(x_m) = \mathfrak{a}(x_m)\left[h(u_1(x_m)) - h(u_2(x_m))\right] > 0,$$

which is impossible. Therefore, u_M is the unique positive solution of (6.11).

Owing to (6.8), for every $0 < M < \tilde{M}$ and $L > 0$ we have that $u_{[M,L]} \leq u_{[\tilde{M},L]}$. Hence, letting $L \uparrow \infty$ implies $u_M \leq u_{\tilde{M}}$. Actually,

$$u_M(x) < u_{\tilde{M}}(x) \quad \text{for all} \ x \geq 0. \tag{6.12}$$

Indeed, if $u_M(x_0) = u_{\tilde{M}}(x_0)$ for some $x_0 > 0$, then $u'_M(x_0) = u'_{\tilde{M}}(x_0)$ because $u_M \leq u_{\tilde{M}}$, and hence, by the uniqueness of solution for the Cauchy problem,

$$u' = v, \qquad v' = \mathfrak{a}(x)h(u), \qquad u(x_0) = u_M(x_0), \quad u'(x_0) = u'_M(x_0),$$

necessarily $u_M = u_{\tilde{M}}$, which implies $M = \tilde{M}$, a contradiction. Consequently, (6.12) holds.

Subsequently, for every $L > 0$ we will consider the singular problem

$$\begin{cases} u''(x) = \mathfrak{a}(x)h(u(x)), & 0 < x < L, \\ u(0) = \infty, \ u(L) = \infty. \end{cases} \tag{6.13}$$

As h satisfies (A2), it follows from Theorem 3.4 that (6.13) possesses minimal and maximal positive solutions. Let denote by $\ell_{\min}$ the minimal one.

Fix $M > 0$ and $L > 0$. Then, there is $\varepsilon > 0$ for which

$$u_M \leq M \leq \ell_{\min} \quad \text{in} \quad [0, \varepsilon] \cup [L - \varepsilon, L].$$

Thus, $\ell_{\min}|_{[\varepsilon, L-\varepsilon]}$ is a supersolution of

$$\begin{cases} u''(x) = \mathfrak{a}(x)h(u(x)), & \varepsilon < x < L - \varepsilon, \\ u(\varepsilon) = u_M(\varepsilon), \ u(L - \varepsilon) = u_M(L - \varepsilon). \end{cases} \tag{6.14}$$

The proof of the uniqueness of the positive solution for (6.6) and (6.11) can be easily adapted to show that u_M is the unique positive solution of (6.14). As $\underline{u} := 0$ is a subsolution of (6.14) with $\underline{u} \leq \ell_{\min}|_{[\varepsilon, L-\varepsilon]}$, we find from Theorem 1.2 that (6.14) possesses a solution between 0 and $\ell_{\min}$, necessarily u_M. Thus, $u_M \leq \ell_{\min}$ in $[\varepsilon, L - \varepsilon]$. Therefore,

$$u_M \leq \ell_{\min} \quad \text{in} \quad [0, L]. \tag{6.15}$$

So, by the monotonicity of u_M with respect to M, the point-wise limit

$$\ell := \lim_{M \uparrow \infty} u_M \tag{6.16}$$

is well defined in $[0, \infty)$. Moreover, by (6.15), the Ascoli–Arzela theorem guarantees that ℓ solves (6.1). By (6.16), $\ell(x) > 0$ for every $x > 0$ since $u_M(x) > 0$

and the mapping $M \mapsto u_M(x)$ is increasing. Moreover, ℓ is non-increasing, because u_M is decreasing for all $M > 0$. Thus, $\ell' \leq 0$. As for all $x > 0$

$$\ell''(x) = \mathfrak{a}(x)h(\ell(x)) > 0,$$

the derivative ℓ' is increasing. Therefore, $\ell'(x) < 0$ for all $x > 0$ and (6.4) holds. Actually, ℓ is the minimal positive solution of (6.1), because any other solution, $u(x)$, satisfies $u \geq u_M$ for all $M > 0$ and so, $u \geq \ell$. Consequently, we will subsequently denote it by

$$\ell_{\min} := \ell.$$

To show the existence of a maximal solution for the problem (6.1) we consider the problems

$$\begin{cases} u''(x) = \mathfrak{a}(x)h(u(x)), & x > \varepsilon, \\ u(\varepsilon) = \infty, \ u(\infty) = 0, \end{cases} \tag{6.17}$$

where $\varepsilon > 0$. By making the change of variable $y := x - \varepsilon$, (6.17) fits into the abstract setting of (6.1). Thus, for every $\varepsilon > 0$, (6.17) possesses a minimal positive solution, denoted by $\ell^\varepsilon_{\min}$. Similarly, for every $M > 0$ and $\varepsilon > 0$,

$$\begin{cases} u''(x) = \mathfrak{a}(x)h(u(x)), & x > \varepsilon, \\ u(\varepsilon) = M, \ u(\infty) = 0, \end{cases} \tag{6.18}$$

has a unique positive solution, denoted by u^ε_M. Moreover,

$$\lim_{M \uparrow \infty} u^\varepsilon_M = \ell^\varepsilon_{\min}.$$

Pick $0 < \varepsilon < \tilde{\varepsilon}$. In the interval $[\tilde{\varepsilon}, \infty)$, u^ε_M is a supersolution of

$$\begin{cases} u''(x) = \mathfrak{a}(x)h(u(x)), & x > \tilde{\varepsilon}, \\ u(\tilde{\varepsilon}) = u^\varepsilon_M(\tilde{\varepsilon}), \ u(\infty) = 0, \end{cases} \tag{6.19}$$

because

$$u^\varepsilon_M(\tilde{\varepsilon}) < u^\varepsilon_M(\varepsilon) = M = u^{\tilde{\varepsilon}}_M(\tilde{\varepsilon}).$$

Moreover, 0 is a subsolution. Thus, by uniqueness,

$$u^\varepsilon_M \leq u^{\tilde{\varepsilon}}_M \quad \text{in } [\tilde{\varepsilon}, \infty).$$

Thus, letting $M \uparrow \infty$ yields

$$\ell^\varepsilon_{\min} \leq \ell^{\tilde{\varepsilon}}_{\min} \quad \text{in } [\tilde{\varepsilon}, \infty).$$

From these estimates, it is easily seen that the point-wise limit

$$\ell_{\max} := \lim_{\varepsilon \downarrow 0} \ell^\varepsilon_{\min}$$

is well defined and it provides us with a positive solution of (6.1). As for any solution u of (6.1) and $\varepsilon > 0$, we have that $u \leq \ell^\varepsilon_{\min}$ in $[\varepsilon, \infty)$, letting $\varepsilon \downarrow 0$

yield $u \leq \ell_{\max}$. Therefore, $\ell_{\max}$ is the maximal solution of (6.1). To establish the uniqueness of the solution of (6.1) it remains to prove that

$$\ell_{\min} = \ell_{\max}. \tag{6.20}$$

Note that all previous solutions are non-negative. So, by the uniqueness of the associated Cauchy problems, they must be positive.

Let ℓ be an arbitrary positive solution of (6.1) and, for any fixed $\varepsilon > 0$, consider

$$\bar{\ell}(x) := \ell(x - \varepsilon), \qquad x > \varepsilon.$$

According to (6.1), for every $x > \varepsilon$ we have that

$$\bar{\ell}''(x) = \ell''(x - \varepsilon) = \mathfrak{a}(x - \varepsilon)h(\ell(x - \varepsilon)) \leq \mathfrak{a}(x)h(\bar{\ell}(x)).$$

Hence, $\bar{\ell}$ is a supersolution of (6.18) for all $M > 0$. Thus, for every $M > 0$ and $\varepsilon > 0$, we have that

$$\bar{\ell} \geq u_M^\varepsilon \quad \text{in} \quad [\varepsilon, \infty) \quad \text{for all} \quad M > 0, \ \varepsilon > 0. \tag{6.21}$$

To prove (6.21) we will proceed by contradiction. Suppose (6.21) fails. As $\bar{\ell}(\varepsilon) = \infty$, (6.21) holds for sufficiently small $x - \varepsilon > 0$. Thus, there exist $x_1 > \varepsilon$ and $x_2 \in (x_1, \infty]$ such that

$$\bar{\ell}(x_1) = u_M^\varepsilon(x_1), \qquad \bar{\ell}(x_2) = u_M^\varepsilon(x_2),$$

and

$$u_M^\varepsilon(x) > \bar{\ell}(x) = \ell(x - \varepsilon), \qquad \forall \, x \in (x_1, x_2).$$

Let $x_m \in (x_1, x_2)$ be such that

$$u_M^\varepsilon(x_m) - \bar{\ell}(x_m) = \max_{x \in [x_1, x_2]} \left(u_M^\varepsilon(x) - \bar{\ell}(x) \right).$$

Then,

$$0 \geq (u_M^\varepsilon - \bar{\ell})''(x_m) = \mathfrak{a}(x_m) \left[h(u_M^\varepsilon(x_m)) - h(\bar{\ell}(x_m)) \right] > 0,$$

which is impossible. Thus, (6.21) holds. Now, letting $M \uparrow \infty$ in (6.21) yields

$$\ell_{\min}^\varepsilon = \lim_{M \uparrow \infty} u_M^\varepsilon \leq \bar{\ell} = \ell(\cdot - \varepsilon) \quad \text{in} \quad [\varepsilon, \infty),$$

where $\ell_{\min}^\varepsilon$ stands for the minimal positive solution of (6.17). Consequently, letting $\varepsilon \downarrow 0$, we find that, for every $x > 0$,

$$\ell(x) \geq \lim_{\varepsilon \downarrow 0} \ell_{\min}^\varepsilon(x) = \ell_{\max}(x).$$

In particular, $\ell_{\min} \geq \ell_{\max}$. Therefore, (6.20) holds, which concludes the proof of the uniqueness.

To complete the proof of the theorem, it remains to show (6.5), where ℓ stands for the unique positive solution of (6.1). Since $\ell''(x) > 0$ for all $x > 0$, ℓ' is increasing. Thus, since $\ell' < 0$, the limit

$$\ell_0' := \lim_{x \downarrow 0} \ell'(x) \in [-\infty, 0)$$

is well defined. Suppose $\ell_0' \in (-\infty, 0)$. Then, for every $x > y > 0$,

$$\ell(x) - \ell(y) = \int_y^x \ell' \geq \ell_0'(x - y).$$

Hence,

$$\ell(y) \leq \ell(x) - \ell_0'(x - y),$$

which is impossible, because

$$\lim_{y \downarrow 0} \ell(y) = \infty.$$

Therefore, $\ell_0' = -\infty$, which is the first limit of (6.5).

Lastly, for each integer $n \geq 2$, there exists $x_n \in [n-1, n]$ such that

$$\ell(n) = \ell(n-1) + \int_{n-1}^n \ell' = \ell(n-1) + \ell'(x_n).$$

Thus, since $\lim_{x \uparrow \infty} \ell(x) = 0$, we find that

$$\lim_{n \to \infty} \ell'(x_n) = 0.$$

Therefore, since $\ell' \leq 0$ is increasing, it becomes apparent that

$$\lim_{x \uparrow \infty} \ell'(x) = 0.$$

This establishes (6.5) and ends the proof of the theorem. □

Similarly, the next result holds.

Theorem 6.2 *Suppose $\mathfrak{a}(x)$ and $h(u)$ satisfy (A1) and (A2), and there are $L > 0$ and $M \in [0, \infty)$ for which*

$$\begin{cases} u''(x) = \mathfrak{a}(x)h(u(x)), & x \in (0, L), \\ u(0) = \infty, \quad u(L) = M, \end{cases} \tag{6.22}$$

has a positive solution $U(x)$, $0 < x \leq L$, with $U'(L) < 0$. Then, $U(x)$ is the unique positive solution of (6.22).

Remark 6.3 (a) When $M = 0$, any positive solution, u, of (6.22) must satisfy $u'(L) < 0$. Indeed, necessarily $u'(L) \leq 0$. So, if, in addition, $u'(L) = 0$, then $u(L) = u'(L) = 0$, which implies $u = 0$, because, due to the Cauchy-Lipschitz theorem, $(u, v) = (0, 0)$ is the unique solution of

$$u' = v, \quad v' = \mathfrak{a}(x)h(u), \quad u(L) = v(L) = 0.$$

(b) As a direct consequence of Theorem 6.2, for every $L > 0$, the function $\ell(x)$ is the unique positive solution of

$$\begin{cases} u''(x) = \mathfrak{a}(x)h(u(x)), & x \in (0, L), \\ u(0) = \infty, \quad u(L) = \ell(L), \end{cases}$$

because $\ell'(L) < 0$, by (6.4).

(c) For sufficiently large $M > 0$, (6.22) might not admit a solution U with $U'(L) < 0$, but, instead, $U'(L) > 0$. Indeed, as $M \uparrow \infty$, these solutions approximate a solution of

$$\begin{cases} u''(x) = \mathfrak{a}(x)h(u(x)), & x \in (0, L), \\ u(0) = \infty, \quad u(L) = \infty. \end{cases}$$

In such cases, whether or not uniqueness occurs might depend on some hidden growth properties of $h(u)$. Therefore, $U'(L) < 0$ might be really necessary for the validity of Theorem 6.2.

Proof of Theorem 6.2: Fix $L > 0$ and $M \geq 0$, and consider the problem (6.22). The existence of minimal and maximal solutions follows the general scheme of Theorem 6.1. Indeed, for every $N > 0$, the problem

$$\begin{cases} u''(x) = \mathfrak{a}(x)h(u(x)), & x \in (0, L), \\ u(0) = N, \quad u(L) = M, \end{cases}$$

possesses a unique positive solution, u_N, and

$$\ell_{\min} := \lim_{N \uparrow \infty} u_N$$

provides us with the minimal positive solution of (6.22). Similarly, for sufficiently small $\varepsilon > 0$, we denote by $\ell_{\min}^\varepsilon$ the minimal positive solution of

$$\begin{cases} u''(x) = \mathfrak{a}(x)h(u(x)), & x \in (\varepsilon, L), \\ u(\varepsilon) = \infty, \quad u(L) = M. \end{cases} \tag{6.23}$$

Then, the point-wise limit

$$\ell_{\max} := \lim_{\varepsilon \downarrow 0} \ell_{\min}^\varepsilon$$

provides us with the maximal positive solution of (6.22).

It remains to prove the uniqueness. Let $L > 0$ and $M \geq 0$ for which (6.22) admits a positive solution $U(x)$ with $U'(L) < 0$. Then, for sufficiently small $\varepsilon > 0$,

$$U(L - \varepsilon) \geq U(L).$$

Hence, the auxiliary function

$$\tilde{U}(x) := U(x - \varepsilon), \qquad \varepsilon \leq x \leq L,$$

satisfies

$$\tilde{U}(\varepsilon) = U(0) = \infty, \qquad \tilde{U}(L) = U(L - \varepsilon) \geq U(L) = M,$$

and, for each $x \in (\varepsilon, L)$,

$$\tilde{U}''(x) = U''(x - \varepsilon) = \mathfrak{a}(x - \varepsilon)h(U(x - \varepsilon))$$
$$= \mathfrak{a}(x - \varepsilon)h(\tilde{U}(x)) \leq \mathfrak{a}(x)h(\tilde{U}(x)).$$

Thus, $\tilde{U}(x)$ is a supersolution of (6.23). Hence, arguing as in the proof of Theorem 6.1 yields

$$\tilde{U}(x) = U(x - \varepsilon) \geq \ell_{\min}^\varepsilon(x) \quad \text{for all } x \in (\varepsilon, L].$$

Consequently, letting $\varepsilon \downarrow 0$, we find that

$$U(x) \geq \lim_{\varepsilon \downarrow 0} \ell_{\min}^\varepsilon(x) = \ell_{\max}(x)$$

for all $x \in (0, L]$. Therefore,

$$U = \ell_{\max}.$$

In other words, $U = \ell_{\max}$ is the unique positive solution of (6.22) with $U'(L) < 0$. To prove that $\ell_{\min} = \ell_{\max}$ we will proceed by contradiction. Suppose

$$\ell_{\min} < \ell_{\max} = U.$$

Then, $\ell_{\min}'(L) \geq 0$ and, for every

$$v_0 \in (\ell_{\max}'(L), \ell_{\min}'(L)),$$

the unique solution of the Cauchy problem

$$u' = v, \quad v' = \mathfrak{a}(x)h(u), \quad u(L) = M, \quad u'(L) = v_0,$$

must be globally defined in $(0, L]$ and satisfy

$$\ell_{\min}(x) < u(x) < \ell_{\max}(x) \qquad \text{for all } x \in (0, L). \tag{6.24}$$

This is impossible, since u would provide us with a solution of (6.1) such that $u'(L) = v_0 < 0$ but $u \neq U$, which contradicts the previous uniqueness result. The estimate (6.24) holds true because for every $\alpha \in (0, L)$ and $N > 0$, the boundary value problem

$$\begin{cases} u''(x) = \mathfrak{a}(x)h(u(x)), & x \in (\alpha, L), \\ u(\alpha) = N, \quad u(L) = M, \end{cases}$$

possesses a unique positive solution. This ends the proof. $\qquad \square$

6.2 Auxiliary function $\mathfrak{b}$

In this section we suppose that $\mathfrak{a}$ satisfies (A1), fix $R > 0$ and $p > 1$, and consider the function $\mathfrak{b}$ defined through

$$\mathfrak{b}(x) := \int_x^R \frac{1}{A}, \qquad A(x) := \left(\int_0^x \mathfrak{a}^{\frac{1}{p+1}} \right)^{\frac{p+1}{p-1}}, \qquad x \in (0, R], \qquad (6.25)$$

i.e.,

$$\mathfrak{b}(x) = \int_x^R \left(\int_0^y \mathfrak{a}^{\frac{1}{p+1}} \right)^{-\frac{p+1}{p-1}} dy, \qquad x \in (0, R].$$

The following result demonstrates the main properties of this function.

Lemma 6.4 *The function $\mathfrak{b}$ defined by (6.25) satisfies $\mathfrak{b} \in C^2(0, R]$ and*

$$\mathfrak{b}(x) > 0, \quad \mathfrak{b}'(x) < 0, \quad \mathfrak{b}''(x) > 0, \qquad \forall\ x \in (0, R). \qquad (6.26)$$

Moreover,

$$\lim_{x \downarrow 0} \mathfrak{b}(x) = \infty, \quad \lim_{x \downarrow 0} \mathfrak{b}'(x) = -\infty, \qquad (6.27)$$

and

$$\lim_{x \downarrow 0} \frac{-\mathfrak{b}'(x)}{\mathfrak{b}(x)} = \lim_{x \downarrow 0} \frac{\mathfrak{b}''(x)}{-\mathfrak{b}'(x)} = \lim_{x \downarrow 0} \frac{\mathfrak{b}''(x)}{\mathfrak{b}(x)} = \infty. \qquad (6.28)$$

Proof: By (A1), $A \in C^1[0, R]$, $A(0) = 0$ and $A(x) > 0$ for all $x \in (0, R]$. Thus, $1/A \in C^1(0, R]$ and $\mathfrak{b} \in C^2(0, R]$. Clearly, $\mathfrak{b}(x) > 0$ for each $x \in (0, R)$ and $\mathfrak{b}(R) = 0$. Moreover,

$$A'(x) = \frac{p+1}{p-1} A(x) \left(\int_0^x \mathfrak{a}^{\frac{1}{p+1}} \right)^{-1} \mathfrak{a}^{\frac{1}{p+1}}(x) > 0 \qquad (6.29)$$

for all $x \in (0, R]$. Hence,

$$\mathfrak{b}'(x) = -\frac{1}{A(x)} < 0, \qquad \mathfrak{b}''(x) = \frac{A'(x)}{A^2(x)} > 0, \qquad (6.30)$$

which concludes the proof of (6.26).

Next, we will prove (6.27). As $\mathfrak{a}$ is non-decreasing and positive,

$$A(x) = \left(\int_0^x \mathfrak{a}^{\frac{1}{p+1}} \right)^{\frac{p+1}{p-1}} \leq \mathfrak{a}^{\frac{1}{p-1}}(R) \, x^{\frac{p+1}{p-1}}$$

for all $x \in (0, R]$. Thus,

$$\mathfrak{b}(x) \geq \mathfrak{a}^{\frac{-1}{p-1}}(R) \int_x^R y^{-\frac{p+1}{p-1}} \, dy = \frac{p-1}{2} \mathfrak{a}^{\frac{-1}{p-1}}(R) \left(x^{\frac{-2}{p-1}} - R^{\frac{-2}{p-1}} \right).$$

Consequently, the first limit of (6.27) holds true. The validity of the second limit can be shown from the first identity of (6.30) taking into account that $A(0) = 0$. This ends the proof of (6.27).

Lastly, we will show (6.28). By (A1), (6.29) implies that

$$A'(x) \geq \frac{p+1}{p-1} A(x) x^{-1}, \qquad x \in (0, R]. \tag{6.31}$$

Thus, by (6.30) and (6.31), it becomes apparent that

$$\lim_{x \downarrow 0} \frac{\mathfrak{b}''(x)}{-\mathfrak{b}'(x)} = \lim_{x \downarrow 0} \frac{A'(x)}{A(x)} = \infty.$$

This establishes the second limit of (6.28).

On the other hand, since

$$\frac{-\mathfrak{b}'(x)}{\mathfrak{b}(x)} = \left(A(x) \int_x^R \frac{1}{A} \right)^{-1}, \qquad x \in (0, R],$$

the first limit of (6.28) is equivalent to

$$\lim_{x \downarrow 0} \left(A(x) \int_x^R \frac{1}{A} \right) = 0. \tag{6.32}$$

To prove (6.32), let $\varepsilon \in (0, R)$ and $x \in (0, \varepsilon)$. Then, since A is increasing,

$$A(x) \int_x^R \frac{1}{A} = \int_x^\varepsilon \frac{A(x)}{A(s)} \, ds + A(x) \int_\varepsilon^R \frac{1}{A} \leq \varepsilon - x + A(x) \int_\varepsilon^R \frac{1}{A}.$$

Thus, for every $\varepsilon > 0$,

$$\limsup_{x \downarrow 0} \left(A(x) \int_x^R \frac{1}{A} \right) \leq \varepsilon,$$

because $A(0) = 0$. Consequently, (6.32) holds. Finally, since

$$\lim_{x \downarrow 0} \frac{\mathfrak{b}''(x)}{\mathfrak{b}(x)} = \lim_{x \downarrow 0} \frac{\mathfrak{b}''(x)}{-\mathfrak{b}'(x)} \cdot \lim_{x \downarrow 0} \frac{-\mathfrak{b}'(x)}{\mathfrak{b}(x)} = \infty,$$

the proof is complete. ☐

According to (6.29) and (6.30), we also have

$$\mathfrak{b}'(R) < 0, \qquad \mathfrak{b}''(R) > 0.$$

The next lemma reveals an extremely important property of $\mathfrak{b}$ that will be very useful later.

Lemma 6.5 *Suppose*

$$\lim_{x \downarrow 0} \frac{b(x) b''(x)}{[b'(x)]^2} = I_0 \in (0, \infty). \tag{6.33}$$

Then, the auxiliary function

$$c(x) := \begin{cases} \dfrac{a(x) b^p(x)}{b''(x)}, & 0 < x \leq R, \\[3mm] \left(\dfrac{p-1}{p+1}\right)^{p+1} I_0^p, & x = 0, \end{cases} \tag{6.34}$$

lies in $C[0, R]$ and it is bounded away from zero in $[0, R - \varepsilon]$ for all $0 < \varepsilon < R$. Note that $c(R) = 0$, because $b(R) = 0$, $a(R) > 0$ and $b''(R) > 0$.

Proof: Due to Lemma 6.4, $c \in C(0, R]$ and $c(x) > 0$ for all $x \in (0, R)$. Thus, it suffices to prove that

$$\lim_{x \downarrow 0} c(x) = \left(\frac{p-1}{p+1}\right)^{p+1} I_0^p. \tag{6.35}$$

By (6.29) and (6.30), we find that

$$b''(x) = \frac{p+1}{p-1} \frac{1}{A(x)} \left(\int_0^x a^{\frac{1}{p+1}} \right)^{-1} a^{\frac{1}{p+1}}(x) > 0, \qquad x \in (0, R).$$

Thus, for every $x \in (0, R)$,

$$\begin{aligned}
c(x) &= \frac{p-1}{p+1} a(x) b^p(x) a^{\frac{-1}{p+1}}(x) \left(\int_0^x a^{\frac{1}{p+1}} \right)^{\frac{p+1}{p-1}+1} \\[2mm]
&= \frac{p-1}{p+1} b^p(x) a^{\frac{p}{p+1}}(x) \left(\int_0^x a^{\frac{1}{p+1}} \right)^{\frac{2p}{p-1}} \\[2mm]
&= \frac{p-1}{p+1} \left[b(x) a^{\frac{1}{p+1}}(x) \left(\int_0^x a^{\frac{1}{p+1}} \right)^{\frac{2}{p-1}} \right]^p.
\end{aligned}$$

On the other hand, by (6.29), we have that

$$A'(x) = \frac{p+1}{p-1} a^{\frac{1}{p+1}}(x) \left(\int_0^x a^{\frac{1}{p+1}} \right)^{\frac{2}{p-1}}, \qquad x \in [0, R].$$

Therefore, taking into account (6.30), we find that

$$c(x) = \left(\frac{p-1}{p+1}\right)^{p+1} [A'(x) b(x)]^p = \left(\frac{p-1}{p+1}\right)^{p+1} \left\{ \frac{b(x) b''(x)}{[b'(x)]^2} \right\}^p \tag{6.36}$$

for all $x \in (0, R)$. By (6.33), letting $x \downarrow 0$ in (6.36) shows (6.35), which ends the proof. $\quad \square$

Subsequently, in order to discuss the meaning of condition (6.33), we will introduce the quotient function

$$q_b(x) := \frac{b(x)}{b'(x)} = \frac{1}{(\ln b)'(x)}, \qquad x \in (0, R].$$

By Lemma 6.4, $q_b \in \mathcal{C}^1(0, R]$ and

$$\lim_{x \downarrow 0} q_b(x) = 0.$$

Thus, if we extend q_b to $[0, R]$ by $q_b(0) := 0$, then $q_b \in \mathcal{C}[0, R]$. Moreover, for every $x \in (0, R]$, we have that

$$q_b'(x) = 1 - \frac{b''(x)b(x)}{[b'(x)]^2}.$$

Consequently, $q_b \in \mathcal{C}^1[0, R]$ if and only if the following limit exists

$$I_0 := \lim_{x \downarrow 0} \frac{b(x)b''(x)}{[b'(x)]^2}.$$

In such case, by (6.26), $I_0 \geq 0$. Thus, (6.33) can be equivalently expressed as

$$q_b \in \mathcal{C}^1[0, R] \quad \text{with} \quad q_b'(0) < 1. \tag{6.37}$$

Note that if $q_b \in \mathcal{C}^1[0, R]$ but $q_b'(0) = 1$, then $I_0 = 0$. In such case, the function c defined by (6.34) satisfies $c(0) = 0$ and the conclusions of Lemma 6.5 fail. The next result will provide us with some sufficient conditions for (6.33), or, equivalently, (6.37). To state it, we need the next concept.

Definition 6.6 *A function $\xi \in \mathcal{C}^1(0, \varepsilon]$, $\varepsilon > 0$, is said to be non-oscillating in $(0, \varepsilon]$ if, either $\xi'(x) > 0$ for each $x \in (0, \varepsilon]$, or $\xi'(x) < 0$ for each $x \in (0, \varepsilon]$, or $\xi = \xi_0$ in $(0, \varepsilon]$ for some constant ξ_0. The function ξ is said to be non-oscillating at $\xi = 0$ if it is non-oscillating in $(0, \delta]$ for some $0 < \delta \leq \varepsilon$.*

Proposition 6.7 *Suppose $p > 1$, a satisfies (A1), $a(0) = 0$, and there exists $0 < \varepsilon < R$ such that $a \in \mathcal{C}^2(0, \varepsilon]$,*

$$a'(x) > 0 \quad and \quad (\ln a)''(x) < 0 \quad for\ all \quad x \in (0, \varepsilon]. \tag{6.38}$$

Then, there exists $M > 1$ such that

$$1 < A'(x) \int_x^R \frac{1}{A} \leq M \quad for\ all\ x \in (0, \varepsilon]. \tag{6.39}$$

Suppose, in addition, that the function

$$Q_{\mathfrak{a}}(x) := \frac{\mathfrak{a}'(x)}{\mathfrak{a}(x)} \cdot \frac{\int_0^x \mathfrak{a}^{\frac{1}{p+1}}}{\mathfrak{a}^{\frac{1}{p+1}}(x)}, \qquad x \in (0, \varepsilon], \tag{6.40}$$

is non-oscillating in $(0, \varepsilon]$*, as discussed in Definition 6.6. Then,*

$$I_0 := \lim_{x \downarrow 0} \left(A'(x) \int_x^R \frac{1}{A} \right) = \lim_{x \downarrow 0} \frac{\mathfrak{b}(x) \mathfrak{b}''(x)}{[\mathfrak{b}'(x)]^2} \in [1, M]. \tag{6.41}$$

In particular, (6.33)*, or equivalently* (6.37)*, holds.*

The local logarithmic concavity (6.38) is rather natural, since $\mathfrak{a}(0) = 0$ implies

$$\lim_{x \downarrow 0} \ln \mathfrak{a}(x) = -\infty.$$

Obviously, it allows most of the decaying rates for $\mathfrak{a}$ at $x = 0$. Indeed, it holds if there are two constants, $\beta > 0$ and $\gamma > 0$, such that

$$\mathfrak{a}(x) \sim \beta x^\gamma \quad \text{for} \quad x \sim 0,$$

or

$$\mathfrak{a}(x) \sim -x \ln x \quad \text{for} \quad x \sim 0,$$

or

$$\mathfrak{a}(x) \sim e^{-1/x} \quad \text{for} \quad x \sim 0.$$

More generally, since

$$(\ln \mathfrak{a})''(x) = \frac{\mathfrak{a}''(x)\mathfrak{a}(x) - [\mathfrak{a}'(x)]^2}{\mathfrak{a}^2(x)} \qquad \text{for all} \quad x \in (0, \varepsilon),$$

a simple sufficient condition for (6.38) is

$$\mathfrak{a}'(x) > 0 \quad \text{and} \quad \mathfrak{a}''(x) \leq 0 \quad \text{for all} \quad x \in (0, \varepsilon].$$

After we complete the proof of Proposition 6.7, we will give a very simple criterion based on the quotient $\mathfrak{a}/\mathfrak{a}'$, to guarantee that $Q_{\mathfrak{a}}(x)$ is non-oscillating for all $p \geq 1$.

Proof of Proposition 6.7: Subsequently, we consider the function $\varphi(x)$ defined through

$$\varphi(x) := A(x) \int_x^R \frac{1}{A} = -\frac{\mathfrak{b}(x)}{\mathfrak{b}'(x)}, \qquad x \in (0, R].$$

By (6.28), $\varphi(0) = 0$. Moreover, $\varphi(x) > 0$ for all $x \in (0, R)$. Thus, differentiating and rearranging terms, we find that

$$\varphi'(x) = A'(x) \int_x^R \frac{1}{A} - 1 \tag{6.42}$$

and

$$\varphi''(x) = \frac{A'(x)}{A(x)} \left\{ \frac{A(x)A''(x)}{[A'(x)]^2} [\varphi'(x) + 1] - 1 \right\}. \tag{6.43}$$

On the other hand, after some straightforward manipulations, from (6.25) it becomes apparent that

$$\frac{A(x)A''(x)}{[A'(x)]^2} = \frac{1}{p+1} \left(2 + \frac{p-1}{p+1} Q_\mathfrak{a}(x) \right), \qquad 0 < x < R, \tag{6.44}$$

where $Q_\mathfrak{a}(x)$ is the function introduced in (6.40). Note that setting

$$\mathfrak{q}_\mathfrak{a}(x) := \frac{\mathfrak{a}(x)}{\mathfrak{a}'(x)} \ (> 0), \qquad x \in (0, \varepsilon],$$

the function $Q_\mathfrak{a}(x)$ satisfies

$$(\mathfrak{q}_\mathfrak{a} Q_\mathfrak{a})' = 1 - \frac{1}{p+1} Q_\mathfrak{a} \tag{6.45}$$

and hence,

$$\mathfrak{q}_\mathfrak{a}(x) Q_\mathfrak{a}'(x) = -Q_\mathfrak{a}(x) \left[\mathfrak{q}_\mathfrak{a}'(x) + \frac{1}{p+1} \right] + 1, \qquad x \in (0, \varepsilon]. \tag{6.46}$$

Moreover, by (6.38),

$$\mathfrak{q}_\mathfrak{a}'(x) = \frac{[\mathfrak{a}'(x)]^2 - \mathfrak{a}(x)\mathfrak{a}''(x)}{[\mathfrak{a}'(x)]^2} = -\mathfrak{q}_\mathfrak{a}^2(x)(\ln \mathfrak{a})''(x) > 0, \quad x \in (0, \varepsilon].$$

Thus, for every $\tilde{x} \in Q_\mathfrak{a}^{-1}(p+1) \cap (0, \varepsilon]$, we find from (6.46) that

$$\mathfrak{q}_\mathfrak{a}(\tilde{x}) Q_\mathfrak{a}'(\tilde{x}) = -(p+1) \left[\mathfrak{q}_\mathfrak{a}'(\tilde{x}) + \frac{1}{p+1} \right] + 1 = -(p+1)\mathfrak{q}_\mathfrak{a}'(\tilde{x}) < 0,$$

because $\mathfrak{q}_\mathfrak{a}'(\tilde{x}) > 0$. Consequently, $Q_\mathfrak{a}'(\tilde{x}) < 0$. Hence, $Q_\mathfrak{a}(x) > p+1$ for all $x \in (0, \tilde{x})$, if there exists $\tilde{x} \in (0, \varepsilon]$ with $Q_\mathfrak{a}(\tilde{x}) = p+1$. Suppose such a $\tilde{x}$ exists. Then,

$$(\mathfrak{q}_\mathfrak{a} Q_\mathfrak{a})' = 1 - \frac{Q_\mathfrak{a}}{p+1} < 0 \quad \text{in } (0, \tilde{x}).$$

So,

$$\frac{d}{dx} \frac{1}{\mathfrak{q}_\mathfrak{a} Q_\mathfrak{a}} > 0 \quad \text{in } (0, \tilde{x}).$$

Thus, since

$$\frac{1}{q_a(x)Q_a(x)} = \frac{a^{\frac{1}{p+1}}(x)}{\int_0^x a^{\frac{1}{p+1}}} = \frac{d}{dx}\left(\ln \int_0^x a^{\frac{1}{p+1}}\right), \qquad x \in (0, \tilde{x}),$$

we find that

$$\frac{d^2}{dx^2}\left(\ln \int_0^x a^{\frac{1}{p+1}}\right) > 0 \qquad \text{for all } x \in (0, \tilde{x}),$$

which is impossible, because

$$\lim_{x \downarrow 0}\left(\ln \int_0^x a^{\frac{1}{p+1}}\right) = -\infty,$$

and consequently the second derivative should be somewhere negative in $(0, x)$ for every $x > 0$. Therefore,

$$Q_a(x) < p + 1 \qquad \text{for all } x \in (0, \varepsilon]. \tag{6.47}$$

Subsequently we will restrict ourselves to consider $x \in (0, \varepsilon]$. First, we will show that there exists $\eta \in (0, \varepsilon]$ such that

$$\varphi'(x) \geq 0, \qquad \forall\, x \in (0, \eta]. \tag{6.48}$$

To prove (6.48) we will argue by contradiction. As $\varphi(0) = 0$ and $\varphi(x) > 0$ for all $x \in (0, R]$, φ' must be somewhere positive in every interval $(0, \delta]$ with $0 < \delta \leq \varepsilon$. Thus, if (6.48) fails for all $\eta \in (0, \varepsilon)$, then $\varphi'(x)$ must change sign in $(0, \delta]$ for all $\delta > 0$. In such case, there exist two sequences, $x_n \downarrow 0$ and $y_n \downarrow 0$, as $n \uparrow \infty$, such that

$$\varphi''(x_n) = \varphi''(y_n) = 0, \qquad \varphi'(x_n) > 0, \quad \varphi'(y_n) < 0, \qquad n \geq 1.$$

Thus, since $A(x) > 0$ and $A'(x) > 0$ for all $x \in (0, \varepsilon]$, (6.43) yields

$$\frac{A(x_n)A''(x_n)}{[A'(x_n)]^2} = \frac{1}{\varphi'(x_n) + 1} < 1, \qquad \frac{A(y_n)A''(y_n)}{[A'(y_n)]^2} = \frac{1}{\varphi'(y_n) + 1} > 1.$$

On the other hand, based on (6.44), we have that

$$\frac{A(x)A''(x)}{[A'(x)]^2} - 1 = \frac{p-1}{p+1}\left(\frac{Q_a(x)}{p+1} - 1\right). \tag{6.49}$$

Hence,

$$Q_a(x_n) < p + 1, \qquad Q_a(y_n) > p + 1,$$

which contradicts (6.47). Consequently, $\varepsilon > 0$ may be shortened if necessary to get (6.48) with $\eta = \varepsilon$.

Suppose there is $\tilde{x} \in (0, \varepsilon]$ such that $\varphi'(\tilde{x}) = 0$. Then, owing to (6.48), we also have that $\varphi''(\tilde{x}) = 0$, because $\tilde{x}$ is a local minimum of φ'. Thus, by (6.43),

$$\frac{A(\tilde{x})A''(\tilde{x})}{[A'(\tilde{x})]^2} = 1.$$

So, by (6.49), $Q_a(\tilde{x}) = p + 1$, which contradicts (6.47). Therefore, $\varepsilon > 0$ can be shortened if necessary so that

$$\varphi'(x) > 0 \quad \text{for all } x \in (0, \varepsilon]. \tag{6.50}$$

According to (6.42), (6.50) establishes the lower estimate of (6.39).

Also, thanks to (6.44), (6.47) and (6.49), it is apparent that

$$\frac{2}{p+1} < \frac{A(x)A''(x)}{[A'(x)]^2} < 1 \qquad \text{for all } x \in (0, \varepsilon]. \tag{6.51}$$

Obviously, ε can be shortened if necessary so that some of the following excluding options occur:

(a) $\varphi'' \geq 0$ on $(0, \varepsilon]$.

(b) $\varphi'' \leq 0$ on $(0, \varepsilon]$.

(c) φ'' changes sign infinitely many times in $(0, \delta]$ for all $0 < \delta < \varepsilon$.

Suppose (a) occurs. Then, $\varphi'(x)$ is non-decreasing and hence, owing to (6.42),

$$A'(x) \int_x^R \frac{1}{A} = \varphi'(x) + 1 \leq \varphi'(\varepsilon) + 1$$

for all $x \in (0, \varepsilon]$, which concludes the proof of (6.39). Note that in this case the next limit exists

$$L := \lim_{x \downarrow 0} \varphi'(x) \geq 0. \tag{6.52}$$

Therefore,

$$\lim_{x \downarrow 0} \left(A'(x) \int_x^R \frac{1}{A} \right) = L + 1 \geq 1,$$

which ends the proof of (6.41).

Suppose (b) occurs. Then, according to (6.42), (6.43) and (6.51), it becomes apparent that, for every $x \in (0, \varepsilon]$,

$$A'(x) \int_x^R \frac{1}{A} = \varphi'(x) + 1 \leq \frac{[A'(x)]^2}{A''(x)A(x)} < \frac{p+1}{2}. \tag{6.53}$$

Consequently, (6.39) also holds in this case. Similarly, in this case the limit (6.52) is well defined, since $\varphi'(x)$ is non-increasing. Therefore, (6.41) holds.

Suppose (c) occurs and pick $x_j \in (0, \varepsilon]$, $j \in \{1, 2, 3\}$, such that $x_1 < x_2 < x_3$ and $\varphi'' \geq 0$ on $[x_1, x_2]$, while $\varphi'' \leq 0$ on $[x_2, x_3]$. By (6.42), (6.43) and (6.51), it is apparent that (6.53) holds on $[x_2, x_3]$. Since $\varphi''(x_2) = 0$, we also find that

$$A'(x_2) \int_{x_2}^{R} \frac{1}{A} = \varphi'(x_2) + 1 = \frac{[A'(x_2)]^2}{A''(x_2)A(x_2)} < \frac{p+1}{2}.$$

Moreover, since $\varphi'' \geq 0$ on $[x_1, x_2]$, the function φ' is non-decreasing on $[x_1, x_2]$. Thus, for every $x \in [x_1, x_2]$,

$$A'(x) \int_{x}^{R} \frac{1}{A} = \varphi'(x) + 1 \leq \varphi'(x_2) + 1 < \frac{p+1}{2}.$$

Therefore,

$$A'(x) \int_{x}^{R} \frac{1}{A} < \frac{p+1}{2} \qquad \text{for all } x \in [x_1, x_3].$$

As this is valid for every x_1, x_2, x_3 satisfying the previous requirements, (6.39) holds with $M = (p+1)/2$. This ends the proof of (6.39) in all possible cases.

To complete the proof of the proposition it remains to show that φ cannot satisfy alternative (c) if Q_a is non-oscillating in $(0, \varepsilon]$, as in the remaining cases the limit (6.41) has been already established. Indeed, by differentiating (6.43) and rearranging terms it is easily seen that

$$\varphi''' = \frac{A'}{A} \left[2\frac{AA''}{(A')^2} - 1 \right] \varphi'' + \frac{A'}{A} \left[\frac{AA''}{(A')^2} \right]' (\varphi' + 1).$$

Thus, by (6.50), for every $\hat{x} \in (0, \varepsilon]$ with $\varphi''(\hat{x}) = 0$, differentiating (6.44) with respect to x yields

$$\operatorname{sign} \varphi'''(\hat{x}) = \operatorname{sign} \left[\frac{AA''}{(A')^2} \right]' (\hat{x}) = \operatorname{sign} Q_a'(\hat{x}).$$

Consequently, for sufficiently small $\varepsilon > 0$, φ'' cannot change sign in $(0, \varepsilon]$ if either $Q_a'(x) > 0$ for all $x \in (0, \varepsilon]$ or $Q_a'(x) < 0$ for all $x \in (0, \varepsilon]$. Note that we are assuming that $Q_a(x)$ is non-oscillating in $(0, \varepsilon]$.

Suppose $Q_a \equiv Q_a(0)$ is constant on $(0, \varepsilon]$. Then, it follows from (6.46) and (6.47) that

$$0 < Q_a(0) < p+1, \qquad \mathfrak{q}_a'(x) = \frac{1}{Q_a(0)} - \frac{1}{p+1}, \qquad x \in (0, \varepsilon].$$

Thus, setting

$$\gamma := \frac{Q_a(0)(p+1)}{p+1 - Q_a(0)} > 0,$$

there exists a constant $C \geq 0$ such that

$$\mathfrak{q}_a(x) = \gamma^{-1} x + C \qquad \text{for all } x \in (0, \varepsilon].$$

Consequently, since $\mathfrak{q}_\mathfrak{a}(x) := \frac{\mathfrak{a}(x)}{\mathfrak{a}'(x)}$, there exists $\beta > 0$ such that

$$\mathfrak{a}(x) = \beta(x + \gamma C)^\gamma, \qquad x \in (0, \varepsilon].$$

Necessarily $C = 0$, because $\mathfrak{a}(0) = 0$. Thus, $\mathfrak{a}(x) = \beta x^\gamma$, $x \in (0, \varepsilon]$. So,

$$A(x) = \beta^{\frac{1}{p-1}} \left(\int_0^x s^{\frac{\gamma}{p+1}} \, ds \right)^{\frac{p+1}{p-1}} = \beta^{\frac{1}{p-1}} \left(\frac{p+1}{\gamma+p+1} \right)^{\frac{p+1}{p-1}} x^{\frac{\gamma+p+1}{p-1}}$$

for all $x \in (0, \varepsilon]$. Therefore,

$$\lim_{x \downarrow 0} \left(A'(x) \int_x^R \frac{1}{A} \right) = \frac{\gamma + p + 1}{\gamma + 2}, \tag{6.54}$$

which shows (6.41) and ends the proof. $\quad\square$

Remark 6.8 As $\gamma \in [0, \infty)$, the limit (6.54) ranges in $[1, (p+1)/2)$.

The following result provides us with a sufficient condition so that $Q_\mathfrak{a}(x)$ is non-oscillating for all $p \geq 1$.

Proposition 6.9 *Suppose $p > 1$, $\mathfrak{a}$ satisfies (A1), $\mathfrak{a}(0) = 0$, and there exists $0 < \varepsilon < R$ such that $\mathfrak{a} \in C^3(0, \varepsilon]$ and (6.38) holds. Suppose, in addition, that*

$$\mathfrak{q}'_\mathfrak{a} = \left(\frac{\mathfrak{a}}{\mathfrak{a}'} \right)' = \frac{(\mathfrak{a}')^2 - \mathfrak{a}\mathfrak{a}''}{(\mathfrak{a}')^2} = 1 - \frac{\mathfrak{a}\mathfrak{a}''}{(\mathfrak{a}')^2}$$

is non-oscillating in $(0, \varepsilon]$. Then, for each $p \geq 1$, the function $Q_\mathfrak{a}(x)$ defined by (6.40) is non-oscillating. If, in addition,

$$a_1 = \lim_{x \downarrow 0} \mathfrak{q}'_\mathfrak{a}(x) \in [0, \infty),$$

then,

$$Q_\mathfrak{a}(0) := \lim_{x \downarrow 0} Q_\mathfrak{a}(x) = \frac{p+1}{(p+1)a_1 + 1}.$$

Note that $\mathfrak{q}'_\mathfrak{a}$ is non-oscillating if and only if $\frac{\mathfrak{a}\mathfrak{a}''}{(\mathfrak{a}')^2}$ is non-oscillating.

Proof: By the proof of Proposition 6.7, we already know that

$$\mathfrak{q}'_\mathfrak{a}(x) = -\mathfrak{q}^2_\mathfrak{a}(x)(\ln \mathfrak{a})''(x) > 0, \qquad x \in (0, \varepsilon], \tag{6.55}$$

by (6.38). Moreover, differentiating (6.45) shows that, for every $p \geq 1$,

$$\mathfrak{q}''_\mathfrak{a}(x)Q_\mathfrak{a}(x) + 2\mathfrak{q}'_\mathfrak{a}(x)Q'_\mathfrak{a}(x) + \mathfrak{q}_\mathfrak{a}(x)Q''_\mathfrak{a}(x) = -\frac{Q'_\mathfrak{a}(x)}{p+1}, \qquad x \in (0, \varepsilon].$$

Suppose there is a $p \geq 1$ for which $Q_{\mathfrak{a}}(x)$ is not non-oscillating. Then, there are two sequences, x_n, y_n, $n \geq 1$, such that

$$\lim_{n \to \infty} x_n = 0 = \lim_{n \to \infty} y_n$$

and, for every $n \geq 1$,

$$Q'_{\mathfrak{a}}(x_n)Q'_{\mathfrak{a}}(y_n) \leq 0, \qquad Q''_{\mathfrak{a}}(x_n) = Q''_{\mathfrak{a}}(y_n) = 0. \qquad (6.56)$$

Thus, for every $n \geq 1$, we find that

$$\begin{aligned}
\left(2q'_{\mathfrak{a}}(x_n) + \frac{1}{p+1}\right) Q'_{\mathfrak{a}}(x_n) &= -q''_{\mathfrak{a}}(x_n)Q_{\mathfrak{a}}(x_n), \\
\left(2q'_{\mathfrak{a}}(y_n) + \frac{1}{p+1}\right) Q'_{\mathfrak{a}}(y_n) &= -q''_{\mathfrak{a}}(y_n)Q_{\mathfrak{a}}(y_n).
\end{aligned} \qquad (6.57)$$

Since $q'_{\mathfrak{a}}$ is non-oscillating in $(0, \varepsilon]$, some of the following options occur. Either $q''_{\mathfrak{a}}(x) > 0$ for all $x \in (0, \varepsilon]$, or $q''_{\mathfrak{a}}(x) < 0$ for all $x \in (0, \varepsilon]$, or $q'_{\mathfrak{a}} \equiv a_1$ for some constant $a_1 > 0$. Suppose $q''_{\mathfrak{a}} > 0$, or $q''_{\mathfrak{a}} < 0$, on $(0, \varepsilon]$. Then, by (6.55), we find from (6.57) that

$$\operatorname{sign} Q'_{\mathfrak{a}}(x_n) = -\operatorname{sign} q''_{\mathfrak{a}}(x_n), \qquad \operatorname{sign} Q'_{\mathfrak{a}}(y_n) = -\operatorname{sign} q''_{\mathfrak{a}}(y_n),$$

for all $n \geq 1$. Hence,

$$Q'_{\mathfrak{a}}(x_n)Q'_{\mathfrak{a}}(y_n) > 0, \qquad n \geq 1,$$

which contradicts (6.56). Thus, $q'_{\mathfrak{a}} \equiv a_1$ on $(0, \varepsilon]$ for some constant $a_1 > 0$. Therefore,

$$q_{\mathfrak{a}}(x) = a_1 x \qquad \text{for all } x \in (0, \varepsilon].$$

Consequently, integrating in $\mathfrak{a}$ yields

$$\mathfrak{a}(x) = Cx^{1/a_1} \qquad \text{for all } x \in (0, \varepsilon]$$

for some constant $C > 0$. So, $Q_{\mathfrak{a}}$ must be constant in $(0, \varepsilon]$, which shows the non-oscillatory behavior of $Q_{\mathfrak{a}}(x)$ in $(0, \varepsilon]$ for all $p \geq 1$.

Note that, thanks to (6.47), $Q_{\mathfrak{a}}(0) \in [0, p+1]$. Moreover,

$$a_1 := \lim_{x \downarrow 0} q'_{\mathfrak{a}}(x) = \infty$$

cannot be a priori excluded, because $q'_{\mathfrak{a}}$ does not need to be bounded if it is non-oscillating.

Suppose $a_1 \in [0, \infty)$. Then, by the L'Hôpital's rule,

$$Q_{\mathfrak{a}}(0) = \lim_{x \downarrow 0} \frac{\mathfrak{a}''(x) \int_0^x \mathfrak{a}^{\frac{1}{p+1}} + \mathfrak{a}'(x) \mathfrak{a}^{\frac{1}{p+1}}(x)}{\left(\frac{1}{p+1} + 1\right) \mathfrak{a}^{\frac{1}{p+1}}(x) \mathfrak{a}'(x)}$$

$$= \frac{p+1}{p+2} \lim_{x \downarrow 0} \left(\frac{\mathfrak{a}''(x) \int_0^x \mathfrak{a}^{\frac{1}{p+1}}}{\mathfrak{a}'(x) \mathfrak{a}^{\frac{1}{p+1}}(x)} + 1 \right)$$

$$= \frac{p+1}{p+2} \lim_{x \downarrow 0} \left(\frac{\mathfrak{a}''(x) \mathfrak{a}(x)}{[\mathfrak{a}'(x)]^2} Q_{\mathfrak{a}}(x) + 1 \right)$$

$$= \frac{p+1}{p+2} \lim_{x \downarrow 0} \left[(1 - \mathfrak{q}_{\mathfrak{a}}'(x)) Q_{\mathfrak{a}}(x) + 1 \right].$$

Therefore,

$$Q_{\mathfrak{a}}(0) = \frac{p+1}{p+2} [(1 - a_1) Q_{\mathfrak{a}}(0) + 1],$$

which ends the proof. □

6.3 Getting sharp estimates for $\ell(x)$ at $x = 0$ through $\mathfrak{b}$

Proposition 6.10 *Suppose* (A1)–(A3) *and* (6.33). *Then, there are* $\delta \in (0, R)$ *and* $0 < \varepsilon_0 < \kappa_0$ *such that for every* $\varepsilon \in (0, \varepsilon_0]$ *and* $\kappa \in [\kappa_0, \infty)$, *the functions* $\underline{u} := \varepsilon \mathfrak{b}$ *and* $\overline{u} := \kappa \mathfrak{b}$ *are a subsolution and a supersolution, respectively, of the problem*

$$\begin{cases} u''(x) = \mathfrak{a}(x) h(u(x)), & x \in (0, \delta), \\ u(0) = \infty, \ u(\delta) = \ell(\delta), \end{cases} \tag{6.58}$$

where ℓ *is the unique positive solution of* (6.1) *given by Theorem 6.1.*

Proof: By Lemma 6.5, there exist two constants, $0 < m \leq M$, such that

$$0 < m \leq \mathfrak{c}(x) \leq M, \qquad \forall \ x \in [0, R/2]. \tag{6.59}$$

Moreover, by (6.3), there exists $\Lambda > 0$ such that

$$\frac{H}{2} u^p < h(u) < \frac{3H}{2} u^p \qquad \text{if } u \geq \Lambda. \tag{6.60}$$

Subsequently, we set

$$C := \max\{M, \Lambda\} + 1 > 0 \tag{6.61}$$

and pick $0 < \tilde{\varepsilon}_0 < \tilde{\kappa}_0$ and $\delta := \delta(\tilde{\varepsilon}_0) \in (0, R/2)$ satisfying

$$0 < \tilde{\varepsilon}_0 < \left(\frac{2}{3HC}\right)^{\frac{1}{p-1}} < \left(\frac{2}{Hm}\right)^{\frac{1}{p-1}} < \tilde{\kappa}_0 \qquad (6.62)$$

and

$$\mathfrak{b}(x) > \frac{C}{\tilde{\varepsilon}_0} > \frac{\Lambda}{\tilde{\kappa}_0} \qquad \text{for all } x \in (0, \delta]. \qquad (6.63)$$

Obviously, since $3C > m$, (6.62) can be accomplished by choosing a sufficiently small $\tilde{\varepsilon}_0 > 0$ and a sufficiently large $\tilde{\kappa}_0$. As, due to Lemma 6.4,

$$\mathfrak{b}(x) > 0, \quad \mathfrak{b}'(x) < 0, \quad \lim_{x\downarrow 0} \mathfrak{b}(x) = \infty,$$

to get (6.63) it suffices to choose δ with $\mathfrak{b}(\delta) > C/\tilde{\varepsilon}_0$.

Now, fix $0 < \varepsilon_0 < \kappa_0$ satisfying

$$0 < \varepsilon_0 < \min\left\{\tilde{\varepsilon}_0, \frac{\ell(\delta)}{\mathfrak{b}(\delta)}\right\} \le \tilde{\varepsilon}_0 < \tilde{\kappa}_0 \le \max\left\{\tilde{\kappa}_0, \frac{\ell(\delta)}{\mathfrak{b}(\delta)}\right\} < \kappa_0. \qquad (6.64)$$

Obviously, (6.64) can be accomplished by choosing a sufficiently small $\varepsilon_0 > 0$ and a sufficiently large κ_0. These are the choices of ε_0, κ_0 and δ for which all the assertions of the proposition hold.

Fix $x \in (0, \delta]$. Then, since $\mathfrak{b}(x) > 0$ and $\tilde{\kappa}_0 > \tilde{\varepsilon}_0$, (6.61) and (6.63) imply

$$\tilde{\kappa}_0 \mathfrak{b}(x) > \tilde{\varepsilon}_0 \mathfrak{b}(x) > C > \Lambda.$$

Thus, by (6.60), the following estimates hold

$$h(\tilde{\varepsilon}_0 \mathfrak{b}(x)) < \frac{3H}{2}\tilde{\varepsilon}_0^p \mathfrak{b}^p(x), \qquad h(\tilde{\kappa}_0 \mathfrak{b}(x)) > \frac{H}{2}\tilde{\kappa}_0^p \mathfrak{b}^p(x). \qquad (6.65)$$

Hence, according to (6.59), it follows from (6.62), (6.64), (6.65) and (A3) that, for every $x \in (0, \delta]$,

$$\frac{\kappa_0 \mathfrak{b}^p(x)}{h(\kappa_0 \mathfrak{b}(x))} < \frac{\tilde{\kappa}_0 \mathfrak{b}^p(x)}{h(\tilde{\kappa}_0 \mathfrak{b}(x))} < \frac{2}{H}\frac{1}{\tilde{\kappa}_0^{p-1}} < m \le \mathfrak{c}(x)$$

$$\le M < C < \frac{2}{3H}\frac{1}{\tilde{\varepsilon}_0^{p-1}} < \frac{\tilde{\varepsilon}_0 \mathfrak{b}^p(x)}{h(\tilde{\varepsilon}_0 \mathfrak{b}(x))} < \frac{\varepsilon_0 \mathfrak{b}^p(x)}{h(\varepsilon_0 \mathfrak{b}(x))}.$$

Therefore, by (A3), for every $0 < \varepsilon \le \varepsilon_0$, $\kappa \ge \kappa_0$, and $x \in (0, \delta]$, the next estimates hold

$$\frac{\kappa \mathfrak{b}^p(x)}{h(\kappa \mathfrak{b}(x))} \le \frac{\kappa_0 \mathfrak{b}^p(x)}{h(\kappa_0 \mathfrak{b}(x))} < m \le \mathfrak{c}(x) \le M < \frac{\varepsilon_0 \mathfrak{b}^p(x)}{h(\varepsilon_0 \mathfrak{b}(x))} \le \frac{\varepsilon \mathfrak{b}^p(x)}{h(\varepsilon \mathfrak{b}(x))}.$$

In particular, by the definition of $\mathfrak{c}$, it becomes apparent that

$$\frac{\kappa \mathfrak{b}^p(x)}{h(\kappa \mathfrak{b}(x))} < \frac{a(x)\mathfrak{b}^p(x)}{\mathfrak{b}''(x)} < \frac{\varepsilon \mathfrak{b}^p(x)}{h(\varepsilon \mathfrak{b}(x))}$$

for all $x \in (0, \delta]$, $\kappa \geq \kappa_0$ and $\varepsilon \in (0, \varepsilon_0]$, or, equivalently,

$$\frac{\kappa}{h(\kappa b(x))} < \frac{a(x)}{b''(x)} < \frac{\varepsilon}{h(\varepsilon b(x))}.$$

Therefore, using (6.26), we find that

$$\kappa b''(x) < a(x)h(\kappa b(x)), \qquad \varepsilon b''(x) > a(x)h(\varepsilon b(x)), \qquad (6.66)$$

for all $x \in (0, \delta]$. Finally, according to (6.64),

$$0 < \varepsilon_0 < \frac{\ell(\delta)}{b(\delta)} < \kappa_0$$

and, so, for every $\varepsilon \in (0, \varepsilon_0]$ and $\kappa \geq \kappa_0$, it is apparent that

$$\varepsilon b(\delta) \leq \varepsilon_0 b(\delta) < \ell(\delta) < \kappa_0 b(\delta) \leq \kappa b(\delta),$$

which ends the proof. $\square$

Based on Proposition 6.10, the next result holds.

Theorem 6.11 *Suppose* (A1)–(A3) *and* (6.33), *and let* $\delta > 0$, $\varepsilon_0 > 0$, $\kappa_0 > 0$ *be the constants given by Proposition 6.10. Then, for every* $0 < \varepsilon \leq \varepsilon_0$ *and* $\kappa \geq \kappa_0 > 0$,

$$\varepsilon b(x) \leq \ell(x) \leq \kappa b(x), \qquad x \in (0, \delta], \qquad (6.67)$$

where $\ell(x)$ *is the unique positive solution of* (6.1). *Moreover, for every* $\varepsilon \in (0, \varepsilon_0]$ *and* $\kappa \geq \kappa_0$, *the following estimate holds*

$$\kappa \frac{h(\varepsilon b(x))}{h(\kappa b(x))} b''(x) \leq \ell''(x) \leq \varepsilon \frac{h(\kappa b(x))}{h(\varepsilon b(x))} b''(x) \qquad (6.68)$$

for all $x \in (0, \delta]$. *Consequently,* $\ell(x)$ *inherits most of the local properties of* $b(x)$ *at* $x = 0$ *established by Lemma 6.4. In particular,*

$$\lim_{x\downarrow 0} \ell(x) = \infty, \qquad \lim_{x\downarrow 0} \ell'(x) = -\infty, \qquad (6.69)$$

and

$$\lim_{x\downarrow 0} \frac{-\ell'(x)}{\ell(x)} = \lim_{x\downarrow 0} \frac{\ell''(x)}{-\ell'(x)} = \lim_{x\downarrow 0} \frac{\ell''(x)}{\ell(x)} = \infty. \qquad (6.70)$$

Proof: According to Remark 6.3(b), $\ell(x)$ is the unique positive solution of

$$\begin{cases} u''(x) = a(x)h(u(x)), & x \in (0, \delta), \\ u(0) = \infty, \ u(\delta) = \ell(\delta). \end{cases} \qquad (6.71)$$

Subsequently, given $\varepsilon \in (0, \varepsilon_0]$ and $\kappa \geq \kappa_0$, for each natural number $n > \delta^{-1}$ we consider the boundary value problem

$$\begin{cases} u''(x) = a(x)h(u(x)), & x \in (n^{-1}, \delta), \\ u(n^{-1}) = \frac{\varepsilon+\kappa}{2} b(n^{-1}), \ u(\delta) = \ell(\delta). \end{cases} \qquad (6.72)$$

By Proposition 6.10, $(\varepsilon\mathfrak{b}, \kappa\mathfrak{b})$ provides us with an ordered sub-supersolution pair of (6.72). Thus, (6.72) possesses a solution, u_n, such that

$$\varepsilon\mathfrak{b}(x) \leq u_n \leq \kappa\mathfrak{b}(x) \qquad \forall \; x \in [n^{-1}, \delta].$$

Actually, u_n is unique, but this uniqueness is far from necessary in our argument here. By a standard compactness argument, we can extract a subsequence of u_n, say u_{n_m}, $m \geq 1$, approximating to a solution of (6.71) as $m \to \infty$, necessarily ℓ, by the uniqueness of ℓ. Therefore, letting $m \to \infty$ in the estimates

$$\varepsilon\mathfrak{b}(x) \leq u_{n_m} \leq \kappa\mathfrak{b}(x) \qquad \forall \; x \in [n_m^{-1}, \delta],$$

(6.67) holds.

On the other hand, by (A2), it follows from (6.67) that

$$\mathfrak{a}(x)h(\varepsilon\mathfrak{b}(x)) \leq \mathfrak{a}(x)h(\ell(x)) = \ell''(x) \leq \mathfrak{a}(x)h(\kappa\mathfrak{b}(x)), \quad x \in (0, \delta]. \tag{6.73}$$

Moreover, owing to (6.66),

$$\frac{\mathfrak{a}(x)h(\varepsilon\mathfrak{b}(x))}{\varepsilon} < \mathfrak{b}''(x) < \frac{\mathfrak{a}(x)h(\kappa\mathfrak{b}(x))}{\kappa}, \quad x \in (0, \delta]. \tag{6.74}$$

Thus, combining (6.73) and (6.74) yields

$$\kappa\frac{h(\varepsilon\mathfrak{b}(x))}{h(\kappa\mathfrak{b}(x))}\mathfrak{b}''(x) \leq \mathfrak{a}(x)h(\varepsilon\mathfrak{b}(x)) \leq \mathfrak{a}(x)h(\ell(x))$$

$$= \ell''(x) \leq \mathfrak{a}(x)h(\kappa\mathfrak{b}(x)) \leq \varepsilon\frac{h(\kappa\mathfrak{b}(x))}{h(\varepsilon\mathfrak{b}(x))}\mathfrak{b}''(x),$$

for all $x \in (0, \delta]$, which provides us with (6.68).

To complete the proof it remains to establish (6.69) and (6.70). The first limit of (6.69) holds by the definition of $\ell(x)$. For any given $x \in (0, \delta)$, integrating in $[x, \delta]$ the estimate (6.68) yields

$$\kappa\int_x^\delta \frac{h(\varepsilon\mathfrak{b}(s))}{h(\kappa\mathfrak{b}(s))}\mathfrak{b}''(s)\,ds \leq \ell'(\delta) - \ell'(x) \leq \varepsilon\int_x^\delta \frac{h(\kappa\mathfrak{b}(s))}{h(\varepsilon\mathfrak{b}(s))}\mathfrak{b}''(s)\,ds. \tag{6.75}$$

Subsequently, we set

$$Q(s) := \frac{h(\varepsilon\mathfrak{b}(s))}{h(\kappa\mathfrak{b}(s))}, \qquad s \in (0, \delta].$$

The function Q is continuous and $Q(s) > 0$ for all $s \in (0, \delta]$. Moreover, since

$$\lim_{s\downarrow 0}\mathfrak{b}(s) = \infty,$$

it follows from (6.3) that

$$\lim_{s\downarrow 0}Q(s) = \lim_{s\downarrow 0}\frac{h(\varepsilon\mathfrak{b}(s))/(\varepsilon\mathfrak{b}(s))^p}{h(\kappa\mathfrak{b}(s))/(\kappa\mathfrak{b}(s))^p}\left(\frac{\varepsilon}{\kappa}\right)^p = \left(\frac{\varepsilon}{\kappa}\right)^p > 0.$$

Thus, the quantity

$$Q_L := \inf_{s \in (0,\delta]} \frac{h(\varepsilon \mathfrak{b}(s))}{h(\kappa \mathfrak{b}(s))} \in (0,1] \tag{6.76}$$

is well defined. As $\mathfrak{b}''(x) > 0$ for all $x \in (0,\delta]$, substituting (6.76) into (6.75), it becomes apparent that

$$\kappa Q_L(\mathfrak{b}'(\delta) - \mathfrak{b}'(x)) \leq \ell'(\delta) - \ell'(x) \leq \varepsilon Q_L^{-1}(\mathfrak{b}'(\delta) - \mathfrak{b}'(x)), \qquad x \in (0,\delta].$$

Therefore,

$$\ell'(\delta) - \varepsilon Q_L^{-1}(\mathfrak{b}'(\delta) - \mathfrak{b}'(x)) \leq \ell'(x) \leq \ell'(\delta) - \kappa Q_L(\mathfrak{b}'(\delta) - \mathfrak{b}'(x)) \tag{6.77}$$

for all $x \in (0,\delta]$. By Lemma 6.4,

$$\lim_{x \downarrow 0} \mathfrak{b}'(x) = -\infty.$$

Consequently, (6.77) implies

$$\lim_{x \downarrow 0} \ell'(x) = -\infty,$$

which ends the proof of (6.69). Now, for sufficiently small $x > 0$, it follows from (6.77) and (6.67) that

$$\frac{-\ell'(\delta) + \kappa Q_L(\mathfrak{b}'(\delta) - \mathfrak{b}'(x))}{\kappa \mathfrak{b}(x)} \leq \frac{-\ell'(x)}{\ell(x)} \leq \frac{-\ell'(\delta) + \varepsilon Q_L^{-1}(\mathfrak{b}'(\delta) - \mathfrak{b}'(x))}{\varepsilon \mathfrak{b}(x)}.$$

Therefore,

$$\lim_{x \downarrow 0} \frac{-\mathfrak{b}'(x)}{\mathfrak{b}(x)} = \infty \quad \text{implies} \quad \lim_{x \downarrow 0} \frac{-\ell'(x)}{\ell(x)} = \infty.$$

Similarly, (6.67) and (6.68) imply

$$\frac{h(\varepsilon \mathfrak{b}(x))}{h(\kappa \mathfrak{b}(x))} \frac{\mathfrak{b}''(x)}{\mathfrak{b}(x)} \leq \frac{\ell''(x)}{\ell(x)} \leq \frac{h(\kappa \mathfrak{b}(x))}{h(\varepsilon \mathfrak{b}(x))} \frac{\mathfrak{b}''(x)}{\mathfrak{b}(x)}$$

for all $x \in (0,\delta]$. Thus,

$$Q_L \frac{\mathfrak{b}''(x)}{\mathfrak{b}(x)} \leq \frac{\ell''(x)}{\ell(x)} \leq Q_L^{-1} \frac{\mathfrak{b}''(x)}{\mathfrak{b}(x)}.$$

Consequently,

$$\lim_{x \downarrow 0} \frac{\mathfrak{b}''(x)}{\mathfrak{b}(x)} = \infty \quad \text{implies} \quad \lim_{x \downarrow 0} \frac{\ell''(x)}{\ell(x)} = \infty.$$

Analogously, from (6.68) and (6.77), it becomes apparent that

$$\lim_{x \downarrow 0} \frac{\mathfrak{b}''(x)}{-\mathfrak{b}'(x)} = \infty \quad \text{implies} \quad \lim_{x \downarrow 0} \frac{\ell''(x)}{-\ell'(x)} = \infty,$$

which ends the proof. □

6.4   The exact blow-up rate of $\ell(x)$ at $x = 0$

The following result ascertains the exact blow-up rate of $\ell(x)$ at $x = 0$.

Theorem 6.12 *Suppose* (A1)-(A3) *and* (6.33). *Then,*

$$\lim_{x \downarrow 0} \frac{\ell(x)}{b(x)} = I_0^{\frac{-p}{p-1}} \left(\frac{p+1}{p-1} \right)^{\frac{p+1}{p-1}} H^{\frac{-1}{p-1}}, \tag{6.78}$$

where $\ell(x)$ is the unique positive solution of (6.1). *In particular, the limit* (6.78) *is independent of $R > 0$.*

Proof: According to Lemma 6.5,

$$\lim_{x \downarrow 0} \frac{a(x)b^p(x)}{b''(x)} = \left(\frac{p-1}{p+1} \right)^{p+1} I_0^p. \tag{6.79}$$

Subsequently, we will consider the quotient

$$q(x) := \frac{\ell(x)}{b(x)}, \qquad x \in (0, \delta]. \tag{6.80}$$

Owing to (6.67), we already know that

$$\varepsilon \le q(x) \le \kappa, \qquad \forall \ x \in (0, \delta].$$

Hence,

$$0 < \varepsilon \le q_L := \liminf_{x \downarrow 0} q(x) \le q_M := \limsup_{x \downarrow 0} q(x) \le \kappa.$$

To show the existence of $\lim_{x \downarrow 0} q(x)$ we will argue by contradiction. Suppose

$$q_L < q_M.$$

Then, there are two sequences, t_n, s_n, $n \ge 1$, such that

$$\lim_{n \to \infty} t_n = \lim_{n \to \infty} s_n = 0, \qquad \lim_{n \to \infty} q(t_n) = q_M, \qquad \lim_{n \to \infty} q(s_n) = q_L,$$

and, for every $n \ge 1$,

$$q'(t_n) = q'(s_n) = 0, \qquad q''(t_n) \le 0, \qquad q''(s_n) \ge 0. \tag{6.81}$$

Since $\ell'' = ah(\ell)$, we find from (6.80) that

$$q''b + 2q'b' + qb'' = \ell'' = ah(qb) \qquad \text{on } (0, \delta].$$

So,

$$q'' \frac{b}{b''} + 2q' \frac{b'}{b''} + q = \frac{ah(qb)}{b''}.$$

Thus, according to (6.81), we find that, for every $n \geq 1$,

$$\mathfrak{q}(t_n) \geq \mathfrak{q}''(t_n) \frac{\mathfrak{b}(t_n)}{\mathfrak{b}''(t_n)} + \mathfrak{q}(t_n) = \frac{\mathfrak{a}(t_n) h(\mathfrak{q}(t_n)\mathfrak{b}(t_n))}{\mathfrak{b}''(t_n)},$$

$$\mathfrak{q}(s_n) \leq \mathfrak{q}''(s_n) \frac{\mathfrak{b}(s_n)}{\mathfrak{b}''(s_n)} + \mathfrak{q}(s_n) = \frac{\mathfrak{a}(s_n) h(\mathfrak{q}(s_n)\mathfrak{b}(s_n))}{\mathfrak{b}''(s_n)},$$

because, owing to Lemma 6.4, $\mathfrak{b}(x) > 0$ and $\mathfrak{b}''(x) > 0$ for all $x \in (0, \delta]$. Therefore, letting $n \to \infty$ in these inequalities, (6.3) and (6.79) imply that

$$\mathfrak{q}_M \geq \left(\frac{p-1}{p+1}\right)^{p+1} I_0^p \mathfrak{q}_M^p H \qquad \text{and} \qquad \mathfrak{q}_L \leq \left(\frac{p-1}{p+1}\right)^{p+1} I_0^p \mathfrak{q}_L^p H.$$

Therefore,

$$\mathfrak{q}_L = \mathfrak{q}_M = I_0^{\frac{-p}{p-1}} \left(\frac{p+1}{p-1}\right)^{\frac{p+1}{p-1}} H^{\frac{-1}{p-1}}$$

which contradicts $\mathfrak{q}_L < \mathfrak{q}_M$. Consequently, the next limit exists:

$$\mathfrak{q}_0 := \lim_{x \downarrow 0} \frac{\ell(x)}{\mathfrak{b}(x)} \in [\varepsilon, \kappa]. \tag{6.82}$$

Finally, combining Lemma 6.4 and Theorem 6.11 with the L'Hôpital rule, it follows from (6.82), (6.3), (6.79) and (6.80) that

$$\mathfrak{q}_0 = \lim_{x \downarrow 0} \frac{\ell''(x)}{\mathfrak{b}''(x)} = \lim_{x \downarrow 0} \left[\frac{\mathfrak{a}(x)\mathfrak{b}^p(x)}{\mathfrak{b}''(x)} \cdot \frac{h(\ell(x))}{\mathfrak{b}^p(x)}\right]$$

$$= \lim_{x \downarrow 0} \left[\frac{\mathfrak{a}(x)\mathfrak{b}^p(x)}{\mathfrak{b}''(x)} \cdot \frac{h(\ell(x))}{\ell^p(x)} \cdot \mathfrak{q}^p(x)\right] = \left(\frac{p-1}{p+1}\right)^{p+1} I_0^p H \mathfrak{q}_0^p,$$

because $\ell(x) \uparrow \infty$ as $x \downarrow 0$. Consequently, since $\mathfrak{q}_0 > 0$, it is apparent that

$$\mathfrak{q}_0 = I_0^{\frac{-p}{p-1}} \left(\frac{p+1}{p-1}\right)^{\frac{p+1}{p-1}} H^{\frac{-1}{p-1}},$$

which shows (6.78) and ends the proof. ☐

6.5 Comments on Chapter 6

This chapter is a refinement of S. Cano-Casanova and J. López-Gómez [39], going back to J. López-Gómez [161], where, for a given $p > 1$, the special case

$$h(u) = u^p, \qquad u \geq 0, \tag{6.83}$$

was analyzed. In the simplest case when (6.83) holds and there are two constants $H > 0$ and $\gamma \geq 0$ such that

$$\mathfrak{a}(x) = Hx^\gamma, \qquad x \geq 0,$$

a direct calculation shows that actually

$$\ell(x) = \left[\frac{\gamma+2}{p-1} \left(\frac{\gamma+2}{p-1} + 1 \right) \right]^{\frac{1}{p-1}} H^{\frac{-1}{p-1}} x^{-\frac{\gamma+2}{p-1}}, \qquad x > 0. \tag{6.84}$$

The blow-up rate of this function $\ell(x)$ at $x = 0$ provides us with the most classical blow-up rates of C. Loewner and L. Nirenberg [138], V. A. Kondratiev and V. A. Nikishin [125], C. Bandle and M. Marcus [16, 17, 18], M. Marcus and L. Véron [188], L. Véron [230], and A. C. Lazer and P. J. McKenna [129, 130], for the special case when $\gamma = 0$, and Y. Du and Q. Huang [80], as well as J. García-Melián, R. Letelier and J. C. Sabina de Lis [101], for the general case when $\gamma \geq 0$.

Indeed, it turns out that in the general multidimensional case when $N \geq 1$, at least for these particular choices of $\mathfrak{a}(x)$ and $h(u)$, the blow-up rate of $\ell(x)$ given by (6.84), $-\frac{\gamma+2}{p-1}$, actually provides us with the blow-up rates on $\partial\Omega$ of all the positive solutions of

$$\begin{cases} \Delta u(x) = \mathfrak{a}(d(x))u^p, & x \in \Omega, \\ u = \infty, & \text{on } \partial\Omega, \end{cases} \tag{6.85}$$

provided

$$\lim_{d(x)\downarrow 0} \frac{\mathfrak{a}(d(x))}{Hd^\gamma(x)} = 1,$$

where we are denoting

$$d(x) := \operatorname{dist}(x, \partial\Omega), \qquad x \in \Omega.$$

But this particular issue will be discussed in the next two chapters. So, we refrain from giving more details herein.

As a corollary of Propositions 6.7 and 6.9, if there exists $\varepsilon > 0$ such that $\mathfrak{a} \in \mathcal{C}^3(0, \varepsilon]$, $\mathfrak{a}'(x) > 0$ and $(\ln \mathfrak{a})''(x) < 0$ for all $x \in (0, \varepsilon)$, and, in addition, the function $\frac{\mathfrak{a}\mathfrak{a}''}{(\mathfrak{a}')^2}$ is non-oscillating in $(0, \varepsilon)$, then (6.33), or equivalently (6.37), holds for all $p \geq 1$. Although we do not hope that (6.33) will be satisfied in general without imposing some non-oscillation property on $\mathfrak{a}(x)$, we did not try to look for a counterexample yet.

S. Cano-Casanova [34] has recently used the techniques of this chapter to ascertain the decay rate to zero at $x = \infty$ of the positive solutions of

$$\begin{cases} u''(x) = \mathfrak{a}(x)h(u(x)), & x > 0, \\ u(0) = M, \quad u(\infty) = 0, \end{cases} \tag{6.86}$$

where M is a positive constant, $\mathfrak{a}$ satisfies (A1), h satisfies (A2) with $h(u)/u$, $u > 0$, increasing and

$$A := \lim_{u \uparrow 0} \frac{h(u)}{u^q} > 0 \qquad (6.87)$$

for some $q > 1$. According to the proof of Theorem 6.1 we already know that (6.86) possesses a unique positive solution, u_M. Setting

$$\mathfrak{b}_0(x) := \int_x^\infty \left(\int_R^y \mathfrak{a}^{\frac{1}{q+1}} \right)^{-\frac{q+1}{q-1}} dy,$$

the main theorem of S. Cano-Casanova [34] establishes that

$$\lim_{x \to \infty} \frac{u_M(x)}{\mathfrak{b}_0(x)} = \lim_{x \to \infty} \frac{u_M'(x)}{\mathfrak{b}_0'(x)} = \lim_{x \to \infty} \frac{u_M''(x)}{\mathfrak{b}_0''(x)} = J_0^{\frac{-q}{q-1}} \left(\frac{q+1}{q-1} \right)^{\frac{q+1}{q-1}} A^{\frac{-1}{q-1}}$$

provided

$$J_0 := \lim_{x \uparrow \infty} \frac{\mathfrak{b}_0(x) \mathfrak{b}_0''(x)}{[\mathfrak{b}_0'(x)]^2} \in (0, \infty).$$

This result is a substantial improvement of some available results for a class of generalized Thomas-Fermi equations (see, e.g., V. Maric and M. Tomic [190] and S. D. Taliaferro [225], [226]) and it may present a huge number of applications in analyzing *exterior problems*.

Chapter 7

Uniqueness of the large solution under radial symmetry

Throughout this chapter, we consider $x_0 \in \mathbb{R}^N$, $N \geq 1$, $R > 0$, $R_2 > R_1 > 0$,

$$B_R(x_0) := \{x \in \mathbb{R}^N \ : \ |x - x_0| < R\},$$
$$A_{R_1,R_2}(x_0) := \{x \in \mathbb{R}^N \ : \ R_1 < |x - x_0| < R_2\},$$

a continuous function $\mathfrak{a} \in \mathcal{C}[0, \infty)$ such that

$$\mathfrak{a}(t) > 0 \qquad \text{for all} \quad t > 0, \tag{7.1}$$

a function $g \in \mathcal{C}[0, \infty) \cap \mathcal{C}^1(0, \infty)$ such that

$$g(0) = 0, \qquad g'(u) > 0 \quad \text{for all} \quad u > 0, \qquad \lim_{u \uparrow \infty} g(u) = \infty, \tag{7.2}$$

and

$$\Omega \in \{B_R(x_0), A_{R_1,R_2}(x_0)\}. \tag{7.3}$$

Also, we set
$$d(x) := \text{dist}(x, \partial\Omega), \qquad x \in \Omega.$$

Under these assumptions, by Theorem 2.4, the boundary value problem

$$\begin{cases} -\Delta u = \lambda u - \mathfrak{a}(d(x))g(u)u & \text{in } \Omega, \\ u = M & \text{on } \partial\Omega, \end{cases} \tag{7.4}$$

possesses a unique positive solution, $\theta_{[\lambda,\Omega,M]}$, for every $M > 0$. Moreover, the map $M \to \theta_{[\lambda,\Omega,M]}$ is increasing, and, since (7.4) is invariant by rotations, $\theta_{[\lambda,\Omega,M]}$ must be radially symmetric for all $M > 0$, by uniqueness. However, unless the function

$$f(u) := g(u)u, \qquad u \geq 0,$$

satisfies the generalized Keller–Osserman condition (KO), the limit

$$\theta_{[\lambda,\Omega,\infty]} := \lim_{M\uparrow\infty} \theta_{[\lambda,\Omega,M]} \tag{7.5}$$

might be infinity somewhere. According to Theorem 3.4, when f satisfies (KO), (7.5) equals the minimal large positive solution, $L^{\min}_{[\lambda,\Omega]}$, of the singular problem

$$\begin{cases} -\Delta u = \lambda u - \mathfrak{a}(d(x))g(u)u & \text{in } \Omega, \\ u = \infty & \text{on } \partial\Omega. \end{cases} \tag{7.6}$$

Naturally, in such case, $L^{\min}_{[\lambda,\Omega]}$ must be radially symmetric, because it is a limit of radially symmetric functions. By the construction of the maximal large positive solution, $L^{\max}_{[\lambda,\Omega]}$, carried out in the proof of Theorem 3.4, it becomes apparent that $L^{\max}_{[\lambda,\Omega]}$ is radially symmetric. It should be remembered that

$$\theta_{[\lambda,\Omega,\infty]} = L^{\min}_{[\lambda,\Omega]}, \qquad L^{\max}_{[\lambda,\Omega]} = \lim_{\varepsilon\downarrow 0} L^{\min}_{[\lambda,\Omega_\varepsilon]}, \tag{7.7}$$

where

$$\Omega_\varepsilon := \{x \in \Omega \ : \ \text{dist}(x, \partial\Omega) > \varepsilon\}$$

for sufficiently small $\varepsilon > 0$. In general, it is a very appealing open problem to ascertain whether

$$L^{\min}_{[\lambda,\Omega]} = L^{\max}_{[\lambda,\Omega]}, \tag{7.8}$$

which entails the uniqueness of the large solution of (7.6). This chapter gives some rather satisfactory answers to this problem in the radially symmetric case.

7.1   The main uniqueness result

Theorem 7.1 *Suppose* (7.3), $\lambda \geq 0$, $\mathfrak{a} \in \mathcal{C}[0,\infty)$ *satisfies*

$$0 < \mathfrak{a}(t) \leq \mathfrak{a}(s) \qquad \text{if} \ \ 0 < t \leq s, \tag{7.9}$$

g *satisfies* (7.2) *and* (KO), *and there exists* $r := r(g) > 0$ *such that*

$$\gamma^r g(u) \leq g(\gamma u) \quad \text{for all} \ \gamma > 1 \quad \text{and} \quad u > 0. \tag{7.10}$$

Then, (7.8) *holds, i.e.,* (7.6) *admits a unique positive solution,*

$$L_{[\lambda,\Omega]} := L^{\min}_{[\lambda,\Omega]} = L^{\max}_{[\lambda,\Omega]}.$$

Moreover, $L_{[\lambda,\Omega]}$ *is radially symmetric.*

Note that (7.10) also holds if $\gamma = 1$, or $u = 0$, because $g(0) = 0$. Moreover, by (7.2), $g(u) > 0$ for all $u > 0$. Hence, (7.10) holds for every $r \in (0, r(g)]$. Moreover, the change of variable

$$\alpha := \frac{2}{r(g)} > 0, \qquad \gamma = \varrho^\alpha, \tag{7.11}$$

transforms (7.10) into

$$\varrho^2 g(\varrho^{-\alpha} u) \leq g(u) \qquad \text{for all} \quad (\varrho, u) \in (1, \infty) \times [0, \infty). \tag{7.12}$$

The next result provides us with an important sufficient condition for (7.10), or, equivalently, (7.12).

Lemma 7.2 *Suppose* $g \in C[0, \infty) \cap C^2(0, \infty)$, $g(0) = 0$, *and*

$$g''(u) \geq 0 \quad \text{for all} \quad u > 0. \tag{7.13}$$

Then, (7.10) holds with $r = 1$. *If, in addition,* $g(u) > 0$ *for all* $u > 0$, *then, (7.10) also holds for every* $r \in (0, 1]$.

Proof: For every $\gamma \geq 1$, consider the function

$$G_\gamma(u) := g(\gamma u) - \gamma g(u), \qquad u \geq 0.$$

Obviously, $G_\gamma \in C[0, \infty) \cap C^2(0, \infty)$, and, for every $\gamma \geq 1$,

$$G_\gamma(0) = 0. \tag{7.14}$$

Moreover, by (7.13), g' is non-decreasing. Thus,

$$G_\gamma'(u) = \gamma \left(g'(\gamma u) - g'(u)\right) \geq 0, \quad u > 0, \quad \gamma \geq 1. \tag{7.15}$$

Hence, from (7.14) and (7.15), it becomes apparent that

$$G_\gamma(u) \geq 0 \qquad \text{for all} \quad \gamma \geq 1, \ u \geq 0.$$

Therefore,

$$g(\gamma u) \geq \gamma g(u) \quad \text{for all} \quad (\gamma, u) \in [1, \infty) \times [0, \infty).$$

Consequently, (7.10) holds with $r = 1$. The last assertion is obvious, because $g(u) > 0$ if $u > 0$, and $\gamma > \gamma^r$ for all $\gamma > 1$ and $r \in (0, 1)$. $\square$

Although at first glance condition (7.10) seems restrictive, according to Lemma 7.2, it holds for a wide class of g's satisfying (7.2) and

$$H := \lim_{u \uparrow \infty} \frac{g(u)u}{u^p} > 0 \tag{7.16}$$

for some $p > 1$. For example, for any integer $n \geq 1$ and real numbers

$$0 < r_1 < r_2 < \ldots < r_n, \qquad \{a_1, \ldots, a_n\} \subset (0, \infty),$$

the function

$$g(u) := \sum_{k=1}^{n} a_k u^{r_k}$$

satisfies (7.2), (7.16) and (7.10). Indeed, by construction, (7.2) holds. More-over,

$$\lim_{u \uparrow \infty} \frac{g(u)u}{u^{r_n+1}} = a_n > 0.$$

So, g satisfies (7.16). Finally, for every $\gamma > 1$ and $u > 0$,

$$g(\gamma u) = \sum_{k=1}^{n} a_k \gamma^{r_k} u^{r_k} = \gamma^{r_1} \sum_{k=1}^{n} \gamma^{r_k - r_1} a_k u^{r_k}$$

$$\geq \gamma^{r_1} \sum_{k=1}^{n} a_k u^{r_k} = \gamma^{r_1} g(u).$$

Consequently, (7.10) holds for every $r \in (0, r_1]$. It should be noted that con-dition (7.13) does not necessarily hold for this choice of g.

7.2   Proof of Theorem 7.1

This section consists of the proof of Theorem 7.1. First, we will prove it in the ball, then in the annulus.

7.2.1   The case when $\Omega = B_R(x_0)$

For every $\varepsilon \in (0, R)$, we set

$$\varrho_\varepsilon := \frac{R}{R - \varepsilon} > 1$$

and consider the function

$$\bar{L}_\varepsilon(x) := L^{\min}_{[\lambda, B_R(x_0)]}(x_0 + \varrho_\varepsilon(x - x_0)), \qquad 0 \leq |x - x_0| \leq R - \varepsilon.$$

By definition, $\bar{L}_\varepsilon = \infty$ on $\partial B_{R-\varepsilon}(x_0)$. Moreover, for every $x \in B_{R-\varepsilon}(x_0)$, we have that

$$
\begin{aligned}
-\Delta \bar{L}_\varepsilon(x) &= -\varrho_\varepsilon^2 \Delta L^{\min}_{[\lambda, B_R(x_0)]}(x_0 + \varrho_\varepsilon(x - x_0)) \\
&= \varrho_\varepsilon^2 \lambda \bar{L}_\varepsilon(x) - \varrho_\varepsilon^2 \mathfrak{a}(R - \varrho_\varepsilon|x - x_0|)g(\bar{L}_\varepsilon(x))\bar{L}_\varepsilon(x) \\
&\geq \lambda \bar{L}_\varepsilon(x) - \varrho_\varepsilon^2 \mathfrak{a}(R - |x - x_0|)g(\bar{L}_\varepsilon(x))\bar{L}_\varepsilon(x) \\
&= \lambda \bar{L}_\varepsilon(x) - \varrho_\varepsilon^2 \mathfrak{a}(d(x))g(\bar{L}_\varepsilon(x))\bar{L}_\varepsilon(x),
\end{aligned}
$$

because $\varrho_\varepsilon > 1$, $\lambda \geq 0$, and, owing to (7.9),

$$\mathfrak{a}(R - \varrho_\varepsilon|x - x_0|) \leq \mathfrak{a}(R - |x - x_0|).$$

Let $\alpha > 0$ be satisfying (7.12) and consider the function

$$\hat{L}_\varepsilon := \varrho_\varepsilon^\alpha \bar{L}_\varepsilon \qquad \text{in } B_{R-\varepsilon}(x_0).$$

Then,

$$\hat{L}_\varepsilon = \infty \qquad \text{on } \partial B_{R-\varepsilon}(x_0)$$

and, for every $x \in B_{R-\varepsilon}(x_0)$,

$$-\Delta \hat{L}_\varepsilon(x) \geq \lambda \hat{L}_\varepsilon(x) - \varrho_\varepsilon^2 \mathfrak{a}(d(x))g(\varrho_\varepsilon^{-\alpha} \hat{L}_\varepsilon(x))\hat{L}_\varepsilon(x).$$

Thus, according to (7.12), we find that

$$-\Delta \hat{L}_\varepsilon(x) \geq \lambda \hat{L}_\varepsilon(x) - \mathfrak{a}(d(x))g(\hat{L}_\varepsilon(x))\hat{L}_\varepsilon(x)$$

for all $x \in B_{R-\varepsilon}(x_0)$. Therefore, $\hat{L}_\varepsilon$ provides us with a supersolution of the singular problem

$$\begin{cases} -\Delta L = \lambda L - \mathfrak{a}(d(x))g(L)L & \text{in } B_{R-\varepsilon}(x_0), \\ L = \infty & \text{on } \partial B_{R-\varepsilon}(x_0). \end{cases}$$

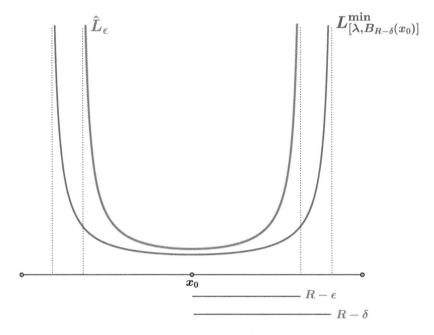

FIGURE 7.1: The functions $\hat{L}_\varepsilon$ and $L_{[\lambda, B_{R-\delta}(x_0)]}^{\min}$.

By Theorem 3.4, we already know that

$$L^{\max}_{[\lambda, B_R(x_0)]} = \lim_{\delta \downarrow 0} L^{\min}_{[\lambda, B_{R-\delta}(x_0)]}.$$

Moreover, as in a neighborhood of $\partial B_{R-\varepsilon}(x_0)$,

$$L^{\min}_{[\lambda, B_{R-\delta}(x_0)]} \leq \hat{L}_\varepsilon \qquad \text{for each } \delta < \varepsilon, \tag{7.17}$$

by Theorem 1.7, the estimate (7.17) holds in the entire $B_{R-\varepsilon}(x_0)$, as illustrated in Figure 7.1. Therefore, letting $\delta \downarrow 0$ in (7.17) yields

$$L^{\max}_{[\lambda, B_R(x_0)]}(x) \leq \hat{L}_\varepsilon(x) = \varrho_\varepsilon^\alpha L^{\min}_{[\lambda, B_R(x_0)]}(x_0 + \varrho_\varepsilon(x - x_0))$$

for all $\varepsilon \in (0, R)$ and $x \in B_{R-\varepsilon}(x_0)$. So, letting $\varepsilon \downarrow 0$ we find that

$$L^{\max}_{[\lambda, B_R(x_0)]}(x) \leq L^{\min}_{[\lambda, B_R(x_0)]}(x), \qquad |x - x_0| < R,$$

which ends the proof of the theorem in this case. □

7.2.2 The case when $\Omega = A_{R_1, R_2}(x_0)$

Then, setting

$$R_m := \frac{R_1 + R_2}{2}, \qquad r := |x - x_0|,$$

we have that

$$d(x) := \text{dist}(x, \partial\Omega) = \begin{cases} R_2 - r & \text{if } R_m \leq r \leq R_2, \\ r - R_1 & \text{if } R_1 \leq r \leq R_m. \end{cases}$$

Moreover, since $L^{\min}_{[\lambda, \Omega]}$ and $L^{\max}_{[\lambda, \Omega]}$ are radially symmetric, necessarily

$$L^{\min}_{[\lambda, \Omega]}(x) = \psi_{\min}(r), \qquad L^{\max}_{[\lambda, \Omega]}(x) = \psi_{\max}(r), \qquad x \in \Omega = A_{R_1, R_2}(x_0),$$

where $\psi_{\min}(r)$ and $\psi_{\max}(r)$ are the reflections about $r = R_m$ of the minimal and the maximal positive solutions, respectively, of the singular one-dimensional problem

$$\begin{cases} -\psi'' - \frac{N-1}{r}\psi' = \lambda\psi - \mathfrak{a}(R_2 - r)g(\psi)\psi, & R_m < r < R_2, \\ \psi'(R_m) = 0, \quad \psi(R_2) = \infty. \end{cases} \tag{7.18}$$

Next, we will show that any positive solution ψ of (7.18) satisfies

$$\psi'(r) \geq 0 \qquad \text{for all } r \in [R_m, R_2). \tag{7.19}$$

This is clear if $\lambda \leq 0$, because, in such case, multiplying the differential equation by r^{N-1} and integrating in (R_m, r) shows that, for every $r \in (R_m, R_2)$,

$$r^{N-1}\psi'(r) = \int_{R_m}^r s^{N-1}[\mathfrak{a}(R_2 - s)g(\psi(s)) - \lambda]\psi(s)\, ds > 0.$$

When $\lambda > 0$, our proof of (7.19) will proceed by contradiction. Suppose $\lambda > 0$ and (7.18) possesses a positive solution, ψ, for which there exists $\tilde{r} \in (R_m, R_2)$ such that

$$\psi'(\tilde{r}) < 0.$$

Then, since

$$\psi'(R_m) = 0, \qquad \lim_{r \uparrow R_2} \psi(r) = \infty,$$

there exist r_0 and r_1 such that

$$R_m \leq r_0 < \tilde{r} < r_1 < R_2$$

and

$$\begin{cases} \psi'(r_0) = \psi'(r_1) = 0, & \psi'(r) \leq 0 \quad \text{if} \quad r \in (r_0, r_1), \\ \psi''(r_0) \leq 0, & \psi''(r_1) \geq 0. \end{cases} \qquad (7.20)$$

Subsequently, we will consider the function H defined by

$$H(\xi) := \lambda \xi - \mathfrak{a}(R_2 - r_0)g(\xi)\xi, \qquad \xi > 0.$$

By (7.2), the value

$$\xi_0 := g^{-1}\left(\lambda/\mathfrak{a}(R_2 - r_0)\right)$$

provides us with the unique positive zero of $H(\xi)$. Actually, $H(\xi) > 0$ if $\xi \in (0, \xi_0)$, $H(\xi_0) = 0$, and $H(\xi) < 0$ if $\xi > \xi_0$.

Suppose $\psi(r_0) > \xi_0$. Then, due to (7.18) and (7.20), we find that

$$0 \leq -\psi''(r_0) = -\psi''(r_0) - \frac{N-1}{r_0}\psi'(r_0) = H(\psi(r_0)) < 0,$$

which is impossible. Thus,

$$\psi(r_0) \leq \xi_0, \qquad (7.21)$$

and hence, it follows from (7.18) and (7.20) that

$$0 \geq -\psi''(r_1) = -\psi''(r_1) - \frac{N-1}{r_1}\psi'(r_1) = \lambda\psi(r_1) - \mathfrak{a}(R_2 - r_1)g(\psi(r_1))\psi(r_1)$$

$$= \lambda\psi(r_1) - \mathfrak{a}(R_2 - r_0)g(\psi(r_1))\psi(r_1) + [\mathfrak{a}(R_2 - r_0) - \mathfrak{a}(R_2 - r_1)]g(\psi(r_1))\psi(r_1)$$

$$= H(\psi(r_1)) + [\mathfrak{a}(R_2 - r_0) - \mathfrak{a}(R_2 - r_1)]g(\psi(r_1))\psi(r_1).$$

As $\psi' \leq 0$, $\psi' \neq 0$, in (r_0, r_1), necessarily $\psi(r_1) < \psi(r_0)$. So, by (7.21), we find that $\psi(r_1) < \xi_0$, which implies

$$H(\psi(r_1)) > 0.$$

On the other hand, $r_0 < r_1$ implies $R_2 - r_0 > R_2 - r_1$ and hence, due to (7.9),

$$\mathfrak{a}(R_2 - r_0) \geq \mathfrak{a}(R_2 - r_1).$$

Therefore,

$$0 \geq H(\psi(r_1)) + [\mathfrak{a}(R_2 - r_0) - \mathfrak{a}(R_2 - r_1)]g(\psi(r_1))\psi(r_1) > 0,$$

which is impossible. Consequently, (7.19) holds.

Subsequently, we set

$$\eta_\varepsilon := \frac{R_2 - R_m}{R_2 - R_m - \varepsilon} > 1, \qquad \varepsilon \in (0, R_2 - R_m),$$

and consider the function $\bar{\psi}_\varepsilon$ defined by

$$\bar{\psi}_\varepsilon(r) := \psi_{\min}(\eta_\varepsilon(r - R_m) + R_m), \qquad R_m \leq r < R_2 - \varepsilon.$$

By definition, we have

$$\bar{\psi}'_\varepsilon(R_m) = \eta_\varepsilon \psi'_{\min}(R_m) = 0$$

and

$$\lim_{r \uparrow R_2 - \varepsilon} \bar{\psi}_\varepsilon(r) = \lim_{\varrho \uparrow R_2} \psi_{\min}(\varrho) = \infty.$$

Moreover, setting

$$\varrho := \eta_\varepsilon(r - R_m) + R_m, \qquad R_m \leq \varrho < R_2,$$

from (7.18) it becomes apparent that, for every, $r \in (R_m, R_2 - \varepsilon)$,

$$-\bar{\psi}''_\varepsilon(r) - \frac{N-1}{r}\bar{\psi}'_\varepsilon(r) = -\eta_\varepsilon^2 \psi''_{\min}(\varrho) - \frac{N-1}{r}\eta_\varepsilon \psi'_{\min}(\varrho)$$

$$= -\eta_\varepsilon^2 \psi''_{\min}(\varrho) - \frac{N-1}{\varrho}\frac{\varrho}{r}\eta_\varepsilon \psi'_{\min}(\varrho).$$

Thus, taking into account that

$$\lambda \geq 0, \quad \psi'_{\min} \geq 0, \quad \eta_\varepsilon > 1, \quad \frac{\varrho}{r} = \frac{R_m(1 - \eta_\varepsilon) + \eta_\varepsilon r}{r} \leq \eta_\varepsilon,$$

we find that

$$-\bar{\psi}''_\varepsilon(r) - \frac{N-1}{r}\bar{\psi}'_\varepsilon(r) = -\eta_\varepsilon^2 \psi''_{\min}(\varrho) - \frac{N-1}{\varrho}\frac{\varrho}{r}\eta_\varepsilon \psi'_{\min}(\varrho)$$

$$\geq \eta_\varepsilon^2 \left[-\psi''_{\min}(\varrho) - \frac{N-1}{\varrho}\psi'_{\min}(\varrho) \right]$$

$$= \eta_\varepsilon^2 \left[\lambda\psi_{\min}(\varrho) - \mathfrak{a}(R_2 - \varrho)g(\psi_{\min}(\varrho))\psi_{\min}(\varrho) \right]$$

$$\geq \lambda\psi_{\min}(\varrho) - \eta_\varepsilon^2 \mathfrak{a}(R_2 - \varrho)g(\psi_{\min}(\varrho))\psi_{\min}(\varrho).$$

On the other hand,

$$\bar{\psi}_\varepsilon(r) := \psi_{\min}(\varrho), \qquad \mathfrak{a}(R_2 - r) \ge \mathfrak{a}(R_2 - \varrho),$$

because $r \le \varrho$. Consequently, for every $r \in (R_m, R_2 - \varepsilon)$,

$$-\bar{\psi}''_\varepsilon(r) - \frac{N-1}{r}\bar{\psi}'_\varepsilon(r) \ge \lambda\bar{\psi}_\varepsilon(r) - \eta_\varepsilon^2\mathfrak{a}(R_2 - r)g(\bar{\psi}_\varepsilon(r))\bar{\psi}_\varepsilon(r).$$

Let $\alpha > 0$ be satisfying condition (7.12) and consider the function

$$\hat{\psi}_\varepsilon(r) := \eta_\varepsilon^\alpha\bar{\psi}_\varepsilon(r), \qquad R_m < r < R_2 - \varepsilon.$$

Then,

$$\hat{\psi}'_\varepsilon(R_m) = 0, \qquad \lim_{r\uparrow R_2-\varepsilon}\hat{\psi}_\varepsilon(r) = \infty,$$

and, for every $r \in (R_m, R_2 - \varepsilon)$,

$$-\hat{\psi}''_\varepsilon(r) - \frac{N-1}{r}\hat{\psi}'_\varepsilon(r) \ge \lambda\hat{\psi}_\varepsilon(r) - \eta_\varepsilon^2\mathfrak{a}(R_2 - r)g(\eta_\varepsilon^{-\alpha}\hat{\psi}_\varepsilon(r))\hat{\psi}_\varepsilon(r).$$

Consequently, thanks to (7.12), it becomes apparent that

$$-\hat{\psi}''_\varepsilon(r) - \frac{N-1}{r}\hat{\psi}'_\varepsilon(r) \ge \lambda\hat{\psi}_\varepsilon(r) - \mathfrak{a}(R_2-r)g(\hat{\psi}_\varepsilon(r))\hat{\psi}_\varepsilon(r), \quad R_m < r < R_2-\varepsilon.$$

Therefore, $\hat{\psi}_\varepsilon$ provides us with a supersolution of the singular problem

$$\begin{cases} -\psi'' - \frac{N-1}{r}\psi' = \lambda\psi - \mathfrak{a}(R_2 - r)g(\psi)\psi, & R_m < r < R_2 - \varepsilon, \\ \psi'(R_m) = 0, \quad \psi(R_2 - \varepsilon) = \infty. \end{cases}$$

According to Theorem 1.7,

$$\eta_\varepsilon^\alpha\bar{\psi}_\varepsilon(r) = \hat{\psi}_\varepsilon(r) \ge \psi_{\max}(r), \qquad R_m < r < R_2 - \varepsilon,$$

for sufficiently small $\varepsilon > 0$. Consequently, letting $\varepsilon \downarrow 0$ yields

$$\psi_{\min} \ge \psi_{\max} \qquad \text{in } (R_m, R_2)$$

which concludes the proof of the theorem. □

7.3 Exact blow-up rates on the boundary

The main result of this section provides us with the blow-up rate of the positive large solution of the singular problem

$$\begin{cases} -\Delta u = \lambda u - \mathfrak{a}(d(x))g(u)u & \text{in } \Omega, \\ u = \infty & \text{on } \partial\Omega, \end{cases} \tag{7.22}$$

where $\lambda \ge 0$ and Ω satisfies (7.3). Precisely, it can be stated as follows.

Theorem 7.3 *Suppose* (7.3), $\lambda \geq 0$, $\mathfrak{a} \in C[0, \infty)$ *satisfies* (7.9), $g \in C[0, \infty) \cap$ $C^1(0, \infty)$ *satisfies* (7.2), (7.10) *and* (7.16), *and*

$$\lim_{t \downarrow 0} \frac{\mathfrak{b}(t)\mathfrak{b}''(t)}{[\mathfrak{b}'(t)]^2} = I_0 > 0, \tag{7.23}$$

for some $R > 0$, *where* $\mathfrak{b}$ *stands for the function*

$$\mathfrak{b}(t) := \int_t^R \left(\int_0^y \mathfrak{a}^{\frac{1}{p+1}} \right)^{-\frac{p+1}{p-1}} dy, \qquad t \in (0, R]. \tag{7.24}$$

Then,

$$\lim_{d(x) \downarrow 0} \frac{L_{[\lambda,\Omega]}(x)}{\mathfrak{b}(d(x))} = I_0^{\frac{-p}{p-1}} \left(\frac{p+1}{p-1} \right)^{\frac{p+1}{p-1}} H^{\frac{-1}{p-1}}, \tag{7.25}$$

where $L_{[\lambda,\Omega]}$ *stands for the unique positive solution of* (7.22).

Owing to Lemma 6.4, we already know that $\mathfrak{b} \in C^2(0, R]$,

$$\mathfrak{b}(t) > 0, \qquad \mathfrak{b}'(t) < 0, \qquad \mathfrak{b}''(t) > 0, \qquad \forall\ t \in (0, R), \tag{7.26}$$

$$\lim_{t \downarrow 0} \mathfrak{b}(t) = \infty, \qquad \lim_{t \downarrow 0} \mathfrak{b}'(t) = -\infty, \tag{7.27}$$

and

$$\lim_{t \downarrow 0} \frac{-\mathfrak{b}'(t)}{\mathfrak{b}(t)} = \lim_{t \downarrow 0} \frac{\mathfrak{b}''(t)}{-\mathfrak{b}'(t)} = \lim_{t \downarrow 0} \frac{\mathfrak{b}''(t)}{\mathfrak{b}(t)} = \infty. \tag{7.28}$$

As in Section 6.2, in terms of the quotient function

$$\mathfrak{q}_{\mathfrak{b}}(t) := \frac{\mathfrak{b}(t)}{\mathfrak{b}'(t)} = \frac{1}{(\ln \mathfrak{b})'(t)}, \qquad t \in (0, R],$$

it is easily seen that (7.23) can be equivalently expressed in the form

$$\mathfrak{q}_{\mathfrak{b}} \in C^1[0, R] \qquad \text{with} \qquad \mathfrak{q}_{\mathfrak{b}}'(0) < 1.$$

According to Propositions 6.7 and 6.9, if there exists $\varepsilon > 0$ such that $\mathfrak{a} \in$ $C^3(0, \varepsilon]$, $\mathfrak{a}'(x) > 0$ and $(\ln \mathfrak{a})''(x) < 0$ for all $x \in (0, \varepsilon)$, and, in addition, $\frac{\mathfrak{a}\mathfrak{a}''}{(\mathfrak{a}')^2}$ is non-oscillating in $(0, \varepsilon)$, then (7.23) holds for all $p \geq 1$.

Proof of Theorem 7.3: According to Theorem 6.1, the singular problem

$$\begin{cases} u''(t) = \mathfrak{a}(t)g(u(t))u(t), & t > 0, \\ u(0) = \infty, \quad u(\infty) = 0, \end{cases} \tag{7.29}$$

possesses a unique positive solution, $\ell(t)$, $t > 0$. Moreover,

$$\ell(t) > 0, \qquad \ell'(t) < 0, \qquad \ell''(t) > 0, \qquad t > 0, \tag{7.30}$$

$$\lim_{t\downarrow 0} \ell'(t) = -\infty, \quad \lim_{t\uparrow\infty} \ell'(t) = 0, \tag{7.31}$$

and, due to Theorems 6.11 and 6.12,

$$\lim_{t\downarrow 0} \frac{-\ell'(t)}{\ell(t)} = \lim_{t\downarrow 0} \frac{\ell''(t)}{-\ell'(t)} = \lim_{t\downarrow 0} \frac{\ell''(t)}{\ell(t)} = \infty \tag{7.32}$$

and

$$\lim_{t\downarrow 0} \frac{\ell(t)}{\mathfrak{b}(t)} = I_0^{\frac{-p}{p-1}} \left(\frac{p+1}{p-1}\right)^{\frac{p+1}{p-1}} H^{\frac{-1}{p-1}}. \tag{7.33}$$

Subsequently, we will distinguish two different situations according to whether $\Omega = B_R(x_0)$, or $\Omega = A_{R_1,R_2}(x_0)$. Suppose

$$\Omega = B_R(x_0).$$

Then,

$$L_{[\lambda,\Omega]}(x) = \psi(r), \quad r = |x - x_0|,$$

where ψ is the unique positive solution of

$$\begin{cases} -\psi'' - \frac{N-1}{r}\psi' = \lambda\psi - \mathfrak{a}(R-r)\psi g(\psi) & \text{in } (0,R), \\ \psi'(0) = 0, \quad \psi(R) = \infty. \end{cases} \tag{7.34}$$

Firstly, we will show that for sufficiently small $\varepsilon > 0$ there exists a constant $A_\varepsilon > 0$ such that, for every $A > A_\varepsilon$, the function

$$\bar{\psi}_{\varepsilon,A}(r) := A + (1+\varepsilon)\left(\frac{r}{R}\right)^2 \ell(R-r), \quad 0 \le r < R, \tag{7.35}$$

provides us with a positive supersolution of (7.34). Indeed, by construction,

$$\bar{\psi}'_{\varepsilon,A}(0) = 0 \quad \text{and} \quad \lim_{r\uparrow R} \bar{\psi}_{\varepsilon,A}(r) = \infty.$$

Thus, $\bar{\psi}_{\varepsilon,A}$ is a supersolution of (7.34) if, and only if,

$$-2N\frac{1+\varepsilon}{R^2}\ell(R-r) + (N+3)\frac{r(1+\varepsilon)}{R^2}\ell'(R-r) - (1+\varepsilon)\left(\frac{r}{R}\right)^2 \ell''(R-r)$$

$$\ge \lambda\ell(R-r)\left[\frac{A}{\ell(R-r)} + (1+\varepsilon)\left(\frac{r}{R}\right)^2\right]$$

$$- \mathfrak{a}(R-r)\ell(R-r)g(\ell(R-r))\left[\frac{A}{\ell(R-r)} + (1+\varepsilon)\left(\frac{r}{R}\right)^2\right]\frac{g(\bar{\psi}_{\varepsilon,A}(r))}{g(\ell(R-r))}.$$

Dividing this inequality by $\ell''(R-r)$ and taking into account that

$$\ell'' = \mathfrak{a}g(\ell)\ell > 0,$$

it becomes apparent that $\bar{\psi}_{\varepsilon,A}$ is a supersolution of (7.34) if, and only if,

$$-2N\frac{1+\varepsilon}{R^2}\frac{\ell(R-r)}{\ell''(R-r)} + (N+3)\frac{r(1+\varepsilon)}{R^2}\frac{\ell'(R-r)}{\ell''(R-r)} - (1+\varepsilon)\left(\frac{r}{R}\right)^2$$
$$\geq \lambda\frac{\ell(R-r)}{\ell''(R-r)}\left[\frac{A}{\ell(R-r)} + (1+\varepsilon)\left(\frac{r}{R}\right)^2\right] \qquad (7.36)$$
$$-\left[\frac{A}{\ell(R-r)} + (1+\varepsilon)\left(\frac{r}{R}\right)^2\right]\frac{g(\bar{\psi}_{\varepsilon,A}(r))}{g(\ell(R-r))}.$$

As

$$\lim_{r\uparrow R}\ell(R-r) = \lim_{r\uparrow R}\bar{\psi}_{\varepsilon,A}(r) = \infty,$$

it follows from (7.16) and (7.35) that

$$\lim_{r\uparrow R}\frac{g(\bar{\psi}_{\varepsilon,A}(r))}{g(\ell(R-r))} = \lim_{r\uparrow R}\left(\frac{\bar{\psi}_{\varepsilon,A}(r)}{\ell(R-r)}\right)^{p-1} = (1+\varepsilon)^{p-1}. \qquad (7.37)$$

By (7.32) and (7.37) it becomes apparent that, at $r = R$, (7.36) reduces to

$$-(1+\varepsilon) \geq -(1+\varepsilon)^p,$$

which is satisfied because $1 + \varepsilon > 1$ and $p > 1$. By continuity, there exists $\delta = \delta(\varepsilon) > 0$ such that (7.36) holds for all $r \in [R-\delta, R)$. By enlarging the constant A, if necessary, we can assume that (7.36) also holds in the entire interval $[0, R]$, because ℓ and ℓ'' are positive and bounded away from zero on compact intervals of $(0, \infty)$, $\bar{\psi}_{\varepsilon,A} \geq A$ and we are assuming (7.2) and (7.16). Indeed, since g is increasing, we have that

$$g(\bar{\psi}_{\varepsilon,A}) \geq g(A) \sim HA^{p-1} \qquad \text{as } A \uparrow \infty.$$

Therefore, $\bar{\psi}_{\varepsilon,A}$ is a supersolution of (7.34) for sufficiently large $A > 0$.

Next, we will show that, for sufficiently small $0 < \varepsilon < 1$, there exists a negative constant, $C < 0$, such that

$$\underline{\psi}_{\varepsilon,C}(r) := \max\left\{0, C + (1-\varepsilon)\left(\frac{r}{R}\right)^2\ell(R-r)\right\}, \qquad 0 \leq r < R,$$

is a non-negative subsolution of (7.34). Indeed, by reversing the inequality in (7.36), it is apparent that $\underline{\psi}_{\varepsilon,C}$ is a subsolution of (7.34) if in the region

$$C + (1-\varepsilon)\left(\frac{r}{R}\right)^2\ell(R-r) \geq 0$$

the following inequality is satisfied

$$-2N\frac{1-\varepsilon}{R^2}\frac{\ell(R-r)}{\ell''(R-r)} + (N+3)\frac{r(1-\varepsilon)}{R^2}\frac{\ell'(R-r)}{\ell''(R-r)} - (1-\varepsilon)\left(\frac{r}{R}\right)^2$$
$$\leq \lambda\frac{\ell(R-r)}{\ell''(R-r)}\left[\frac{C}{\ell(R-r)} + (1-\varepsilon)\left(\frac{r}{R}\right)^2\right] \qquad (7.38)$$
$$-\left[\frac{C}{\ell(R-r)} + (1-\varepsilon)\left(\frac{r}{R}\right)^2\right]\frac{g(\underline{\psi}_{\varepsilon,C}(r))}{g(\ell(R-r))}.$$

Subsequently, we introduce the function α defined by

$$\alpha(r) := (1 - \varepsilon) \left(\frac{r}{R}\right)^2 \ell(R - r), \qquad r \in [0, R).$$

By construction,

$$\alpha(0) = 0, \qquad \lim_{r \uparrow R} \alpha(r) = \infty.$$

Moreover, α is increasing, because $\ell > 0$ and $\ell' < 0$. Thus, for every $C < 0$ there exists $z = z(C) \in (0, R)$ such that

$$C + \alpha(r) = C + (1 - \varepsilon) \left(\frac{r}{R}\right)^2 \ell(R - r) < 0 \qquad \text{if } r \in [0, z(C)),$$

whereas

$$C + \alpha(r) = C + (1 - \varepsilon) \left(\frac{r}{R}\right)^2 \ell(R - r) \geq 0 \qquad \text{if } r \in [z(C), R). \tag{7.39}$$

Note that the mapping $C \mapsto z(C)$ is decreasing and

$$\lim_{C \downarrow -\infty} z(C) = R, \qquad \lim_{C \uparrow 0} z(C) = 0. \tag{7.40}$$

Owing to (7.16), (7.29) and (7.39), it follows that

$$\lim_{r \uparrow R} \frac{g(\underline{\psi}_{\varepsilon,C}(r))}{g(\ell(R - r))} = \lim_{r \uparrow R} \left(\frac{\underline{\psi}_{\varepsilon,C}(r)}{\ell(R - r)}\right)^{p-1} = (1 - \varepsilon)^{p-1}. \tag{7.41}$$

Thus, thanks to (7.32) and (7.41), the inequality (7.38) at $r = R$ reduces to

$$-(1 - \varepsilon) \leq -(1 - \varepsilon)^p,$$

which holds true because $1 - \varepsilon \in (0, 1)$ and $p > 1$. By continuity, there exists $\delta = \delta(\varepsilon) > 0$ such that (7.38) also holds in $[R - \delta, R)$. Moreover, according to (7.40), there exists $C < 0$ such that

$$z(C) = R - \delta.$$

For this choice of C, it is apparent that $\underline{\psi}_{\varepsilon,C}$ provides us with a non-negative subsolution of (7.34).

Lastly, by construction, it becomes apparent that

$$\lim_{r \uparrow R} \frac{\overline{\psi}_{\varepsilon,A}(r)}{\ell(R - r)} = 1 + \varepsilon, \qquad \lim_{r \uparrow R} \frac{\underline{\psi}_{\varepsilon,C}(r)}{\ell(R - r)} = 1 - \varepsilon, \tag{7.42}$$

and

$$\underline{\psi}_{\varepsilon,C} \leq \overline{\psi}_{\varepsilon,A} \qquad \text{in } [0, R).$$

Thus, by the uniqueness of $L_{[\lambda,\Omega]}$ as a solution of the singular problem (7.6), it follows from Theorem 1.4 that

$$\underline{\psi}_{\varepsilon,C}(|x - x_0|) \leq L_{[\lambda,\Omega]}(x) \leq \bar{\psi}_{\varepsilon,A}(|x - x_0|), \qquad |x - x_0| < R,$$

for all $0 < \varepsilon < 1$. Consequently, by (7.42), one can infer that

$$1 - \varepsilon \leq \liminf_{d(x)\downarrow 0} \frac{L_{[\lambda,\Omega]}(x)}{\ell(d(x))} \leq \limsup_{d(x)\downarrow 0} \frac{L_{[\lambda,\Omega]}(x)}{\ell(d(x))} \leq 1 + \varepsilon \qquad (7.43)$$

for all $\varepsilon \in (0,1)$. Therefore, we can conclude that

$$\lim_{d(x)\downarrow 0} \frac{L_{[\lambda,\Omega]}(x)}{\ell(d(x))} = 1. \qquad (7.44)$$

Clearly, (7.25) follows readily by combining (7.44) with (7.33). This ends the proof of the theorem when Ω is a ball.

Subsequently, we assume

$$\Omega = A_{R_1,R_2}(x_0).$$

Then, setting

$$\delta(r) := \min\{R_2 - r, r - R_1\},$$

we have that

$$L_{[\lambda,\Omega]}(x) := \psi(r), \qquad r = |x - x_0|,$$

where ψ is the unique positive solution of

$$\begin{cases} -\psi'' - \frac{N-1}{r}\psi' = \lambda\psi - \mathfrak{a}(\delta(r))\psi g(\psi) & \text{in } (R_1, R_2), \\ \psi(R_1) = \psi(R_2) = \infty. \end{cases} \qquad (7.45)$$

Subsequently, we set

$$R_m := \frac{R_1 + R_2}{2}, \qquad \theta(r) := \begin{cases} \left(\frac{r - R_m}{R_1 - R_m}\right)^2 \ell(r - R_1) & \text{if } R_1 < r < R_m, \\ \left(\frac{r - R_m}{R_2 - R_m}\right)^2 \ell(R_2 - r) & \text{if } R_m \leq r < R_2. \end{cases}$$

Adapting the previous argument, for sufficiently small $\varepsilon > 0$ there exists a constant $A_\varepsilon > 0$ such that, for every $A > A_\varepsilon$, the function

$$\bar{\psi}_{\varepsilon,A}(r) := A + (1 + \varepsilon)\theta(r), \qquad r \in (R_1, R_2),$$

is a positive supersolution of (7.45). Similarly, there exists $C < 0$ such that

$$\underline{\psi}_{\varepsilon,C}(r) := \max\{0, C + (1 - \varepsilon)\theta(r)\}$$

is a non-negative subsolution of (7.45). Moreover, by construction,

$$\underline{\psi}_{\varepsilon,C} \leq \bar{\psi}_{\varepsilon,A}.$$

Thus, by Theorem 1.4, (7.45) possesses a positive solution, ψ_ε, such that

$$\underline{\psi}_{\varepsilon,C} \leq \psi_\varepsilon \leq \bar{\psi}_{\varepsilon,A}$$

for sufficiently large A. Consequently, since $L_{[\lambda,\Omega]}$ is the unique positive solution of (7.6), it is apparent that

$$\underline{\psi}_{\varepsilon,C} \leq \psi \leq \bar{\psi}_{\varepsilon,A}.$$

Therefore,

$$1 - \varepsilon \leq \lim_{r\uparrow R_2} \frac{\psi(r)}{\ell(R_2 - r)} = \lim_{r\downarrow R_1} \frac{\psi(r)}{\ell(r - R_1)} \leq 1 + \varepsilon. \tag{7.46}$$

As (7.46) holds for sufficiently small $\varepsilon > 0$, (7.44) holds.

Finally, combining (7.44) with (7.33) ends the proof of the theorem. $\qquad\square$

Remark 7.4 Let $m \geq 1$ and $K \subset \mathbb{R}^m$ be a compact subset. If the weight function $\mathfrak{a} = \mathfrak{a}_\mu$ is assumed to vary in a continuous way with respect to the parameter $\mu \in K$, in the sense that the map $K \to \mathcal{C}(\bar{\Omega})$, $\mu \to \mathfrak{a}_\mu$, is continuous, then the limit (7.25) holds uniformly in $\mu \in K$.

7.4   Simple application in population dynamics

Let $\Omega \subset \mathbb{R}^N$ be an arbitrary domain with smooth boundary, $\partial\Omega$, such that $\bar{B}_R(x_0) \subset \Omega$ and consider the function

$$a(x) := \begin{cases} -\mathfrak{a}(d(x)), & x \in \bar{B}_R(x_0), \\ 0, & x \in \bar{\Omega} \setminus \bar{B}_R(x_0), \end{cases}$$

as well as the associated parabolic model

$$\begin{cases} \frac{\partial u}{\partial t} - d\Delta u = \lambda u + a(x)g(u)u, & (x,t) \in \Omega \times (0,\infty), \\ u = 0, & (x,t) \in \partial\Omega \times (0,\infty), \\ u(x,0) = u_0(x), & x \in \Omega, \end{cases} \tag{7.47}$$

where $\mathfrak{a} \in \mathcal{C}[0,\infty)$ satisfies (7.9), g satisfies (7.2), (KO) and (7.10), $d > 0$ and $u_0 \in \mathcal{C}(\bar{\Omega})$ satisfies $u_0 > 0$, i.e., $u_0 \geq 0$ and $u_0 \neq 0$.

In the context of population dynamics, (7.47) models the evolution of a single species u dispersing randomly in the territory Ω, which consists of two regions. In $B_R(x_0)$, $a < 0$ and u grows according to a generalized logistic law, while in

$$\Omega_0 := \Omega \setminus \bar{B}_R(x_0),$$

$a = 0$ and u exhibits a genuine Malthusian growth. As usual, the parameter λ is the intrinsic growth rate of the population u and u_0 is the initial population distribution. As we are working under homogeneous Dirichlet conditions, the inhabiting area Ω is assumed to be entirely surrounded by completely hostile regions.

Setting

$$\sigma_0 := \lambda_1[-\Delta, \Omega], \qquad \sigma_1 := \lambda_1[-\Delta, B_R(x_0)],$$

and combining Theorem 5.2 with Theorem 7.1, the next result holds. Note that $\lambda > 0$ if $\lambda \geq d\sigma_1$. Consequently, in such case we are within the range of applicability of Theorem 7.1.

Theorem 7.5 *Suppose* $\mathfrak{a} \in \mathcal{C}[0, \infty)$ *and* g *satisfy* (7.9), *and* (7.2), (KO) *and* (7.10), *respectively. Let* $u(x, t)$ *denote the unique solution of* (7.47). *The following properties hold:*

(a) *If* $\lambda \leq d\sigma_0$, *then*

$$\lim_{t\uparrow\infty} u(\cdot, t) = 0 \qquad in \quad \mathcal{C}(\bar{\Omega}).$$

(b) *If* $d\sigma_0 < \lambda < d\sigma_1$, *then*

$$\lim_{t\uparrow\infty} u(\cdot, t) = \theta_{[\lambda,\Omega]} \qquad in \quad \mathcal{C}(\bar{\Omega}),$$

where $\theta_{[\lambda,\Omega]}$ *is the unique positive steady-state solution of* (7.47).

(c) *If* $\lambda \geq d\sigma_1$, *then*

$$\lim_{t\uparrow\infty} u(\cdot, t) = \infty \qquad in \quad \bar{\Omega}_0 \setminus \partial\Omega,$$

while

$$\lim_{t\uparrow\infty} u(\cdot, t) = L_{[\lambda, B_R(x_0)]} \qquad in \quad B_R(x_0),$$

where $L_{[\lambda, B_R(x_0)]}$ *stands for the unique positive solution of the singular problem*

$$\begin{cases} -d\Delta u = \lambda u - \mathfrak{a}(d(x))g(u)u & in \ B_R(x_0), \\ u = \infty & on \ \partial B_R(x_0). \end{cases}$$

7.5   Comments on Chapter 7

Theorem 7.1 goes back to J. López-Gómez [159, 160, 161] for the special case when

$$g(u) = u^{p-1}, \qquad u \geq 0, \tag{7.48}$$

for some $p > 1$, and to J. López-Gómez [162] for the general case when g satisfies (7.2), (KO) and (7.10). Lemma 7.2 goes back to S. Cano-Casanova and J. López-Gómez [39], as well as Theorem 7.3, which is the main result of [39]. It extends some previous results of the author for the special case (7.48) [159, 160, 161].

Theorem 7.1 is an extremely sharp result because it establishes the uniqueness of the positive large solution from the *generalized convexity* of $g(u)$, imposed through condition (7.10), and the monotonicity of the weight function $\mathfrak{a}$, by means of a tricky use of the strong maximum principle. Consequently, the knowledge of the exact blow-up rates of the large positive solutions on $\partial\Omega$ is far from necessary for the uniqueness; however it is imperative in most of the specialized literature. The proof of Theorem 7.1 can be adapted to get more general uniqueness results valid for general cooperative systems based on J. López-Gómez and L. Maire [164]. The tremendous advantage of our approach relies on the fact that it does not invoke the nature of the decay rate of $\mathfrak{a}(x)$ as $\operatorname{dist}(x, \partial\Omega) \downarrow 0$, nor the type of growth of $g(u)$ as $u \uparrow \infty$, though those are indeed imperative for ascertaining the exact blow-up rate of the positive large solution on $\partial\Omega$.

When S. Cano-Casanova and J. López-Gómez [39] was published, some other results of very different nature, with completely different techniques, had been found by Y. Du and Q. Huang [80], J. García-Melián, R. Letelier and J. C. Sabina de Lis [101], F. C. Cirstea and V. Radulescu [55, 56], T. Ouyang and Z. Xie [205, 206], and J. García-Melián [96, 97]. Other uniqueness results in balls, even for degenerate weights, had been given by M. Chuaqui et al. [53], [52]. Some of these results were based on ideas and technical devices taken from [154, 161]. But, adopting the spirit of Y. Du and Q. Huang [80] and J. García-Melián, R. Letelier and J. C. Sabina de Lis [101], to find out the exact blow-up rates of the positive large solutions on the boundary, most of these papers required $\mathfrak{a}(x)$ to decay like a fixed power of

$$d(x) := \operatorname{dist}(x, \partial\Omega),$$

as $d(x) \downarrow 0$, which is unnecessary to get uniqueness, at least in the radially symmetric case, as established by Theorem 7.1.

By some very classical results of C. Loewner and L. Nirenberg [138], V. A. Kondratiev and V. A. Nikishin [125], C. Bandle and M. Marcus [17], A. C. Lazer and P. J. McKenna [129], [131], M. Marcus and L. Véron [188], and L. Véron [230], when $\mathfrak{a}(0) > 0$, the monotonicity condition (7.9) is not necessary for the validity of Theorem 7.1. In such case, under condition (7.48), a very classical result is the existence of a constant $C > 0$ such that

$$\lim_{d(x)\downarrow 0} CL(x)d^{-\frac{2}{p-1}}(x) = 1 \qquad (7.49)$$

for any positive solution $L(x)$ of (7.6). Consequently, according to the results of Chapter 8, (7.6) possesses a unique positive solution.

These classical results were substantially sharpened by Y. Du and Q. Huang [80] to cover the more general case when

$$C_0 := \lim_{t \downarrow 0} \frac{\mathfrak{a}(t)}{t^\gamma} > 0 \qquad (7.50)$$

for some constant $\gamma \geq 0$. By establishing that any positive solution $L(x)$ of (7.6) satisfies

$$\lim_{d(x) \downarrow 0} \frac{L(x)}{[\alpha(\alpha+1)/C_0]^{\frac{1}{p-1}} d^{-\alpha}(x)} = 1 \qquad \text{with} \qquad \alpha := \frac{\gamma+2}{p-1}, \qquad (7.51)$$

Y. Du and Q. Huang [80] found the uniqueness of the positive solution of (7.6) for a general class of smooth domains Ω, not necessarily radially symmetric. Some time later, J. García-Melián, R. Letelier and J. C. Sabina de Lis [101] substantially sharpened (7.51) by establishing that if

$$\mathfrak{a}(x) = C_0 d^\gamma(x)[1 + C_1 d(x) + o(d(x))] \qquad \text{as } d(x) \downarrow 0,$$

then, for every $\xi \in \partial\Omega$,

$$L(x) = [\alpha(\alpha+1)/C_0]^{\frac{1}{p-1}} d^{-\alpha}(x)[1 + B(\xi)d(x) + o(d(x))], \quad x \to \xi, \quad (7.52)$$

where

$$B(\xi) := \frac{(N-1)H(\xi) - (\alpha+1)C_1}{\gamma+p+3}$$

and $H(\xi)$ is the mean curvature of $\partial\Omega$ at ξ. These results were inspired by some previous findings of G. Díaz and R. Letelier [76] and M. A. del Pino and R. Letelier [71], which extended the pioneering results of C. Bandle and M. Marcus [18] for radially symmetric problems. Essentially, all these results established that the more curved towards the exterior a domain is around a given point of its boundary, the higher the blow-up rate of that point of the solution is.

Incidentally, although [80] and [101] were written independently, [80] was received by the *SIAM Journal on Mathematical Analysis* on February 22, 1999, whereas [101] was received by the *Proceedings of the American Mathematical Society* on April 17, 2000, after publication of [80]. It should be noted that [71] was received by the editors on December 14, 1999, and not published until 2002.

Also in a general domain Ω with smooth boundary and not necessarily radially symmetric, M. Chuaqui et al. [53], [52] characterized the existence and the uniqueness of the positive large solution of

$$\Delta u = \mathfrak{a}(x)u^p$$

when $\mathfrak{a}(x)$ satisfies

$$C_1 d^\gamma(x) \leq \mathfrak{a}(x) \leq C_2 d^\gamma(x), \qquad x \in \Omega,$$

for some $\gamma \leq 0$. Essentially, it was established that the equation cannot admit a positive large solution in Ω if $\gamma \leq -2$, while it admits a unique large positive solution if $\gamma \in (-2, 0)$. Some further refinements of these results were given by J. García-Melián [97].

Almost simultaneously, F. C. Cirstea and V. Radulescu [55, 56, 57] generalized some of the results of [80] and [101] by covering the more general situation when $\mathfrak{a} \in \mathcal{C}^1[0, \infty)$ satisfies $\mathfrak{a}(0) = 0$ and

$$\lim_{t \downarrow 0} \frac{\int_0^t \sqrt{\mathfrak{a}}}{\sqrt{\mathfrak{a}(t)}} = 0, \qquad C_1 := \lim_{t \downarrow 0} \frac{d}{dt} \frac{\int_0^t \sqrt{\mathfrak{a}}}{\sqrt{\mathfrak{a}(t)}} \in [0, 1]. \tag{7.53}$$

Note that (7.53) holds, with $C_1 = \frac{2}{\gamma+2} \in (0, 1]$, under condition (7.50). Moreover, setting

$$Q(t) := \frac{\int_0^t \sqrt{\mathfrak{a}}}{\sqrt{\mathfrak{a}(t)}}, \qquad t \geq 0,$$

(7.53) also holds provided

$$\mathfrak{a}' \geq 0, \qquad Q \in \mathcal{C}^1[0, R], \qquad Q(0) = 0. \tag{7.54}$$

Indeed, $Q(0) = 0$ entails the validity of the first limit of (7.53), and the second one can be inferred from the identity

$$0 \leq \frac{\mathfrak{a}'(t)}{\mathfrak{a}(t)} \frac{\int_0^t \sqrt{\mathfrak{a}}}{\sqrt{\mathfrak{a}(t)}} = 2\left(1 - \frac{d}{dt} \frac{\int_0^t \sqrt{\mathfrak{a}}}{\sqrt{\mathfrak{a}(t)}}\right), \qquad t > 0, \tag{7.55}$$

because, taking into account that $Q \in \mathcal{C}^1[0, R]$, the next limit is well defined:

$$C_1 := Q'(0) = \lim_{t \downarrow 0} \frac{d}{dt} \frac{\int_0^t \sqrt{\mathfrak{a}}}{\sqrt{\mathfrak{a}(t)}}.$$

Moreover, by (7.55), $C_1 \leq 1$. Finally, since $Q(0) = 0$ and $Q(t) \geq 0$ for all $t > 0$, necessarily

$$C_1 = Q'(0) \geq 0,$$

which ends the proof of (7.53).

More recently, T. Ouyang and Z. Xie [205] demonstrated the uniqueness in the special case $\Omega = B_R(x_0)$ by imposing

$$\mathfrak{Q} \in \mathcal{C}^1[0, R] \quad \text{and} \quad \mathfrak{Q}(0) = 0, \tag{7.56}$$

where the quotient function $\mathfrak{Q}$ is defined by

$$\mathfrak{Q}(t) := \frac{\int_0^t \mathfrak{a}}{\mathfrak{a}(t)}, \qquad t \geq 0. \tag{7.57}$$

Condition (7.56) is rather reminiscent of (7.54), though it might be stronger than (7.54), because, thanks to the Hölder inequality,

$$Q(t) = \frac{\int_0^t \sqrt{\mathfrak{a}}}{\sqrt{\mathfrak{a}(t)}} \leq \frac{\left(\int_0^t \mathfrak{a}\right)^{\frac{1}{2}} t^{\frac{1}{2}}}{\sqrt{\mathfrak{a}(t)}} = \left(\frac{\int_0^t \mathfrak{a}}{\mathfrak{a}(t)}\right)^{\frac{1}{2}} t^{\frac{1}{2}} = \sqrt{t \mathfrak{Q}(t)},$$

and hence, $Q(0) = 0$ if $\mathfrak{Q}(0) = 0$. For every $t > 0$,

$$\mathfrak{Q}'(t) = 1 - \frac{\mathfrak{a}'(t)\int_0^t \mathfrak{a}}{\mathfrak{a}^2(t)} = 1 - \frac{I''(t)I(t)}{[I'(t)]^2},$$

where

$$I(t) = \int_0^t \mathfrak{a}(s)\,ds, \qquad t \geq 0,$$

and it becomes apparent how condition (7.56) is closely related to (7.23).

Nevertheless, except in [159, 160, 161], to prove the uniqueness, the usual strategy adopted in most of the available references consists of establishing that all large solutions have the same blow-up rate on the boundary to infer the uniqueness from this property, as will be discussed in Chapter 8.

Theorem 7.1 does not impose any special requirement on $\mathfrak{a}$, like (7.50), (7.53), (7.54), (7.56), or (7.23), beyond its monotonicity. Condition (7.9) provides us with a uniqueness theorem for which the knowledge of the exact blow-up rates of the large solutions on $\partial\Omega$ is not needed. We conjecture that, for every $g(u)$ satisfying (7.2), (KO) and (7.10), and any $\mathfrak{a} \in \mathcal{C}[0, \infty)$ such that $\mathfrak{a}(0) = 0$ and $\mathfrak{a}(t) \geq \mathfrak{a}(s) > 0$ for sufficiently small $t \geq s > 0$, the problem (7.6) has a unique solution. Our conjecture relies on Theorem 7.1 and on the fact that there always exist $\mathfrak{a}_1, \mathfrak{a}_2 \in \mathcal{C}[0, \infty)$ non-decreasing with

$$\mathfrak{a}_1(0) = \mathfrak{a}_2(0) = 0, \qquad \mathfrak{a}_1 \leq \mathfrak{a} \leq \mathfrak{a}_2, \qquad \mathfrak{a}_1(t) = \mathfrak{a}(t) = \mathfrak{a}_2(t) \quad \text{if} \quad t \sim 0.$$

According to Theorem 7.1, for each $i \in \{1, 2\}$, the problem

$$\begin{cases} -\Delta u = \lambda u - \mathfrak{a}_i(d(x))g(u)u & \text{in } \Omega, \\ u = \infty & \text{on } \partial\Omega, \end{cases}$$

has a unique positive solution if $\lambda \geq 0$. Let denote it by L_i. By comparison,

$$L_2 \leq L \leq L_1$$

for every solution L of (7.6). As the blow-up rates of the large solutions should only depend on the values of $\mathfrak{a}$ around $\partial\Omega$ and $\mathfrak{a} = \mathfrak{a}_1 = \mathfrak{a}_2$ in a neighborhood of $\partial\Omega$, necessarily L_1 and L_2, and hence L, have identical blow-up rates on $\partial\Omega$.

The discussion carried out in the final section of Chapter 9 reveals that imposing $\lambda \geq 0$ might be imperative for the validity of most of the results of this chapter. Indeed, at least in the context of superlinear indefinite problems, the nature of the problem changes dramatically when λ changes sign.

Chapter 8

General uniqueness results

This chapter establishes the uniqueness of the positive solution of the singular boundary value problem

$$\begin{cases} -d\Delta u = \lambda u + a(x)g(u)u & \text{in} \quad \Omega_j, \\ u = 0 & \text{on} \quad \partial\Omega_j \cap \partial\Omega, \\ u = \infty & \text{on} \quad \partial\Omega_j \setminus \partial\Omega, \end{cases} \qquad (8.1)$$

under the following assumptions:

(i) $d > 0$, $\lambda \geq 0$, and $a(x)$ satisfies Hypothesis (Ha).

(ii) $g \in \mathcal{C}[0, \infty) \cap \mathcal{C}^1(0, \infty)$ satisfies (7.2), (7.10) and (7.16).

(iii) For each $1 \leq j \leq q_0$,

$$\Omega_j := \Omega \setminus \left(\bar{\Omega}_{0,1} \cup \cdots \cup \bar{\Omega}_{0,j}\right), \qquad (8.2)$$

where we are denoting

$$\Omega_{0,j} = \bigcup_{i=1}^{m_j} \Omega_{0,j}^i, \qquad 1 \leq j \leq q_0.$$

It should be remembered that according to Hypothesis (Ha), $\Omega_{0,j}^i$, $1 \leq j \leq q_0$, $1 \leq i \leq m_j$, are the components of

$$\Omega_0 = \text{int } a^{-1}(0).$$

More generally, given any subdomain $D \subset \Omega$ such that ∂D consists of entire components of $\partial\Omega$ and $\partial\Omega_0$, we will establish the existence of at most one positive solution of

$$\begin{cases} -d\Delta u = \lambda u + a(x)g(u)u & \text{in} \quad D, \\ \mathfrak{B}u = 0 & \text{on} \quad \Gamma_{\mathfrak{B}}, \\ u = \infty & \text{on} \quad \Gamma_\infty, \end{cases} \qquad (8.3)$$

where Γ_∞ is an open and closed subset of ∂D,

$$\Gamma_{\mathfrak{B}} := \partial D \setminus \Gamma_\infty,$$

and $\mathfrak{B}$ is a general mixed boundary operator on $\Gamma_{\mathfrak{B}}$ like (1.49). More precisely, for a given open and closed subset of $\Gamma_{\mathfrak{B}}$, $\Gamma_{\mathfrak{B}}^0$,

$$\Gamma_{\mathfrak{B}}^1 := \Gamma_{\mathfrak{B}} \setminus \Gamma_{\mathfrak{B}}^0$$

and $\xi \in C^1(\bar{D})$, we will take

$$\mathfrak{B}\xi := \begin{cases} \xi, & \text{on } \Gamma_{\mathfrak{B}}^0, \\ \partial_\nu \xi + b(x)\xi, & \text{on } \Gamma_{\mathfrak{B}}^1, \end{cases}$$

where $b \in C^{1+\nu}(\Gamma_{\mathfrak{B}}^1)$ and $\nu \in C^1(\Gamma_{\mathfrak{B}}^1; \mathbb{R}^N)$ is an outward pointing nowhere tangent vector field. Consequently, $\mathfrak{B}$ is the Dirichlet boundary operator on $\Gamma_{\mathfrak{B}}^0$ and the Neumann or a first order regular oblique derivative boundary operator on $\Gamma_{\mathfrak{B}}^1$. It should be noted that either $\Gamma_{\mathfrak{B}}^0$, or $\Gamma_{\mathfrak{B}}^1$ or both simultaneously may be empty.

Dividing by d the differential equation and setting

$$\mu := \frac{\lambda}{d}, \qquad \mathfrak{a} := -\frac{a}{d},$$

(8.3) can be equivalently written in the form

$$\begin{cases} -\Delta u = \mu u - \mathfrak{a}(x)g(u)u & \text{in } D, \\ \mathfrak{B}u = 0 & \text{on } \Gamma_{\mathfrak{B}}, \\ u = \infty & \text{on } \Gamma_\infty, \end{cases} \qquad (8.4)$$

which is the problem considered in this chapter.

Based on the uniqueness results of Chapter 7, the main result of this chapter establishes the uniqueness of a positive solution of (8.4) through the *boundary normal sections* of $\mathfrak{a}(x)$ on ∂D, which are going to be introduced in Section 8.1.

The organization of the chapter is the following: Section 8.1 introduces the concept of boundary normal section, Section 8.2 states the main theorem of this chapter, which provides us with the exact blow-up rates of the positive solutions of (8.4) for a wide class of non-oscillatory weight functions $\mathfrak{a}(x)$, Section 8.3 gives the proof of the main theorem and, finally, Section 8.4 uses it to establish the uniqueness of the positive solution of (8.4) if it exists.

8.1 Boundary normal section of $\mathfrak{a}$

Throughout this chapter,

$$\begin{aligned} n : \partial D &\longrightarrow \mathbb{R}^N \\ x &\longmapsto n_x \end{aligned}$$

stands for the outward normal vector field to ∂D. As ∂D is of class $\mathcal{C}^{2+\nu}$, n is well defined. Actually, thanks to Theorem 1.9 of López-Gómez [163], ∂D satisfies the uniform interior sphere property in the strong sense on ∂D. Similarly, it also satisfies an exterior sphere property in the strong sense on ∂D. These properties are extremely useful in this chapter. According to Definition 1.2 of [163], D is said to satisfy the uniform interior sphere property in the strong sense on ∂D if there is $R > 0$ such that for every $z \in D$ with

$$\mathrm{dist}\,(z, \partial D) \leq R$$

there exists a point $\pi(z) \in \partial D$ such that

$$\mathrm{dist}\,(z, \partial \Omega) = |z - \pi(z)|, \qquad B_R\left(\pi(z) + R\frac{z - \pi(z)}{|z - \pi(z)|}\right) \subset D.$$

Shortening $R > 0$, if necessary, and setting

$$z_x := x - Rn_x, \qquad x \in \partial D,$$

we have that, for every $x \in \partial D$,

$$|x - z_x| = R, \qquad B_R(z_x) \subset D, \qquad \bar{B}_R(z_x) \cap \partial D = \{x\}. \tag{8.5}$$

Moreover, one can also get

$$\bar{B}_R(x + Rn_x) \cap \bar{D} = \{x\} \qquad \text{for all } x \in \partial D. \tag{8.6}$$

Throughout this chapter, we will assume that R has been chosen to satisfy (8.5) and (8.6).

The next concept plays an important role in the general theory developed in this chapter.

Definition 8.1 (Boundary normal sections) *For each $x \in \partial D$, the function $\mathfrak{a}_x \in \mathcal{C}[0, R]$ defined by*

$$\mathfrak{a}_x(t) := \mathfrak{a}\,(x - tn_x), \qquad t \in [0, R],$$

is said to be the boundary normal section of $\mathfrak{a}$ at $x \in \partial D$.

Roughly speaking, the boundary normal section of $\mathfrak{a}$ at $x \in \partial D$ is the local restriction of $\mathfrak{a}$ along the normal line to D through x. Subsequently, for every $x \in \Gamma_\infty$ and $\eta \in (0, R]$, we will denote by $\mathfrak{a}_{x,\eta} \in \mathcal{C}[0, \infty)$ the function

$$\mathfrak{a}_{x,\eta}(t) := \begin{cases} \mathfrak{a}_x(t) & \text{if } t \in [0, \eta), \\ \mathfrak{a}_x(\eta) & \text{if } t \geq \eta, \end{cases} \tag{8.7}$$

which is one of the multiple (global) prolongations of the (local) boundary normal section.

8.2 Boundary blow-up rate of the large solutions

The main result of this chapter can be stated as follows.

Theorem 8.2 *Suppose $\mu \geq 0$, $\mathfrak{a} \in \mathcal{C}(\bar{D})$, $\mathfrak{a} \geq 0$, $\mathfrak{a}(x) > 0$ for all $x \in D$ near Γ_∞, $\mathfrak{a} \in \mathcal{C}^1((\Gamma_\infty + B_\eta) \cap \bar{D})$ for some $\eta \in (0, R]$, and the boundary normal sections $\mathfrak{a}_x$, $x \in \Gamma_\infty$, satisfy $\mathfrak{a}_x \in \mathcal{C}^3(0, \eta]$,*

$$\mathfrak{a}_x(0) = 0, \quad \mathfrak{a}'_x(t) > 0 \quad \wedge \quad (\log \mathfrak{a}_x)''(t) < 0 \quad \forall \ t \in (0, \eta), \tag{8.8}$$

$(\mathfrak{a}_x/\mathfrak{a}'_x)'$ is non-oscillating, as discussed by Definition 6.6, and

$$\lim_{\substack{\Gamma_\infty \ni y \to x \\ t \downarrow 0}} \frac{\mathfrak{a}_y(t)}{\mathfrak{a}_x(t)} = 1 \quad \text{uniformly in } x \in \Gamma_\infty. \tag{8.9}$$

Suppose, in addition, that $g \in \mathcal{C}[0, \infty) \cap \mathcal{C}^1(0, \infty)$ satisfies (7.2), (7.10) and (7.16). Then, any positive solution, L, of (8.4) satisfies

$$\lim_{t \downarrow 0} \frac{L(x - tn_x)}{\mathfrak{b}_x(t)} = I_{0,x}^{-\frac{p}{p-1}} \left(\frac{p+1}{p-1}\right)^{\frac{p+1}{p-1}} H^{\frac{-1}{p-1}} \tag{8.10}$$

for all $x \in \Gamma_\infty$, where

$$\mathfrak{b}_x(t) := \int_t^R \left(\int_0^s \mathfrak{a}_x^{\frac{1}{p+1}}\right)^{-\frac{p+1}{p-1}} ds, \quad I_{0,x} := \lim_{t \downarrow 0} \frac{\mathfrak{b}_x(t)\mathfrak{b}''_x(t)}{[\mathfrak{b}'_x(t)]^2}. \tag{8.11}$$

As we have already discussed after the statement of Proposition 6.7, the local logarithmic concavity imposed in (8.8) is rather natural, since $\mathfrak{a}_x(0) = 0$ implies

$$\lim_{t \downarrow 0} \ln \mathfrak{a}_x(t) = -\infty.$$

The additional regularity $\mathfrak{a}_x \in \mathcal{C}^3(0, \eta]$ together with (8.8) and the non-oscillating behavior of $(\mathfrak{a}_x/\mathfrak{a}'_x)'$ in $(0, \eta]$ is required to apply Proposition 6.9 since, according to it, the functions

$$Q_{\mathfrak{a}_x}(t) := \frac{\mathfrak{a}'_x(t)}{\mathfrak{a}_x(t)} \cdot \frac{\int_0^t \mathfrak{a}_x^{\frac{1}{p+1}}}{\mathfrak{a}_x^{\frac{1}{p+1}}(t)}, \quad t \in (0, \eta], \tag{8.12}$$

are non-oscillating for all $p > 1$. Thus, owing to Proposition 6.7, for every $p > 1$ and $x \in \Gamma_\infty$ the limit $I_{0,x}$ in (8.11) is well defined and satisfies $I_{0,x} \in [1, \infty)$.

Naturally, for every $p > 1$ the choice

$$g(u) := u^{p-1}, \quad u \geq 0,$$

satisfies all the requirements of Theorem 8.2.

8.3 Proof of Theorem 8.2

Fix $\varepsilon > 0$. Due to (8.9), there exist $\varrho = \varrho(\varepsilon) \in (0, \eta)$ and $\delta = \delta(\varepsilon) > 0$ such that, for every $x, y \in \Gamma_\infty$ with dist $(x, y) \leq \varrho$,

$$1 - \varepsilon \leq \frac{\mathfrak{a}_y(t)}{\mathfrak{a}_x(t)} = \frac{\mathfrak{a}\,(y - tn_y)}{\mathfrak{a}\,(x - tn_x)} \leq 1 + \varepsilon \quad \text{for all } t \in [0, \delta]. \tag{8.13}$$

Subsequently, we set

$$\mathcal{Q} := \{y - tn_y \;:\; (y, t) \in \Gamma_\infty \times [0, \delta]\} = \{z \in \bar{D} \;:\; \text{dist}\,(z, \Gamma_\infty) \leq \delta\}.$$

Since Γ_∞ is of class $\mathcal{C}^2$, ϱ and δ can be shortened if necessary so that, for every $z \in \mathcal{Q}$, there exists a unique

$$(\pi(z), t(z)) \in \Gamma_\infty \times [0, \delta]$$

such that

$$\begin{cases} z = \pi(z) - t(z)n_{\pi(z)}, \\[2mm] t(z) = |z - \pi(z)| = \text{dist}\,(z, \partial D) = \text{dist}\,(z, \Gamma_\infty). \end{cases} \tag{8.14}$$

Throughout the rest of the proof, ϱ and δ are chosen to satisfy these requirements.

According to the discussion at the beginning of Section 8.1, there exists $R > 0$ such that

$$R < \min\left\{\frac{\varrho}{2}, \frac{\delta}{4}\right\}$$

and

$$B_R(x - Rn_x) \subset \text{int } \mathcal{Q}, \qquad \bar{B}_R(x - Rn_x) \cap \partial D = \{x\},$$

for all $x \in \Gamma_\infty$. For such choice of R, there exists $\alpha_0 > 0$ such that, for every $x \in \Gamma_\infty$ and $a \in (0, \alpha_0]$,

$$\bar{B}_R(x - (R + \alpha)n_x) \subset \text{int } \mathcal{Q}.$$

Thus,

$$\mathcal{B}_x := \bigcup_{0 \leq \alpha \leq \alpha_0} \bar{B}_R(x - (R + \alpha)n_x) \subset \mathcal{Q}.$$

By construction, it follows from (8.13) and (8.14) that for every $x \in \Gamma_\infty$, $\alpha \in [0, \alpha_0]$ and $z \in \bar{B}_R(x - (R + \alpha)n_x)$,

$$\begin{aligned} \mathfrak{a}(z) = \mathfrak{a}(\pi(z) - t(z)n_{\pi(z)}) &\geq (1 - \varepsilon)\mathfrak{a}(x - t(z)n_x) \\ &= (1 - \varepsilon)\mathfrak{a}_x(t(z)) = (1 - \varepsilon)\mathfrak{a}_x(\text{dist}\,(z, \partial D)) \\ &\geq (1 - \varepsilon)\mathfrak{a}_x\left(\text{dist}\,(z, \partial B_R(x - (R + \alpha)n_x))\right). \end{aligned}$$

Moreover, since $z \in \bar{B}_R(x - (R + \alpha)n_x)$, we have that

$$\text{dist}\,(z, \partial B_R(x - (R + \alpha)n_x)) = R - |z - x + (R + \alpha)n_x|$$

and hence,

$$\mathfrak{a}(z) \geq (1 - \varepsilon)\mathfrak{a}_x(R - r_\alpha), \qquad r_\alpha \equiv |z - x + (R + \alpha)n_x|, \qquad (8.15)$$

for all $x \in \Gamma_\infty$, $\alpha \in [0, \alpha_0]$ and $z \in \bar{B}_R(x - (R + \alpha)n_x)$.

Let L be a positive solution of (8.4). By (8.15), L provides us with a positive (bounded) subsolution of the semilinear singular boundary value problem

$$\begin{cases} -\Delta u = \mu u - (1-\varepsilon)\mathfrak{a}_x(R - r_\alpha)g(u)u & \text{in } B_R(x - (R + \alpha)n_x), \\ u = \infty & \text{on } \partial B_R(x - (R + \alpha)n_x), \end{cases} \qquad (8.16)$$

for all $x \in \Gamma_\infty$ and $\alpha \in [0, \alpha_0]$. Thanks to (8.8), Theorem 7.1 guarantees that, for each $x \in \Gamma_\infty$ and $\alpha \in [0, \alpha_0]$, (8.16) possesses a unique positive solution, $\bar{L}_{\varepsilon,x,\alpha}$. Moreover, by uniqueness,

$$\bar{L}_{\varepsilon,x,\alpha} = \bar{L}_{\varepsilon,x,0}(\cdot + \alpha n_x) \quad \text{for all } (x, \alpha) \in \Gamma_\infty \times [0, \alpha_0]. \qquad (8.17)$$

On the other hand, according to Theorem 3.4, we already know that

$$\bar{L}_{\varepsilon,x,\alpha} = \lim_{M \uparrow \infty} \bar{u}_{[\varepsilon,x,\alpha,M]},$$

where $\bar{u}_{[\varepsilon,x,\alpha,M]}$ stands for the unique positive solution of

$$\begin{cases} -\Delta u = \mu u - (1-\varepsilon)\mathfrak{a}_x(R - r_\alpha)g(u)u & \text{in } B_R(x - (R + \alpha)n_x), \\ u = M & \text{on } \partial B_R(x - (R + \alpha)n_x). \end{cases}$$

Moreover, thanks to, e.g., Theorem 1.7, the map

$$M \mapsto \bar{u}_{[\varepsilon,x,\alpha,M]}$$

is increasing. Hence, setting

$$M := \max_{\partial B_R(x - (R + \alpha)n_x)} L, \qquad x \in \Gamma_\infty, \ \alpha \in (0, \alpha_0],$$

it follows from Theorem 1.7 that

$$L \leq \bar{u}_{[\varepsilon,x,\alpha,M]} \leq \bar{L}_{\varepsilon,x,\alpha} \quad \text{in } B_R(x - (R + \alpha)n_x)$$

for all $x \in \Gamma_\infty$ and $\alpha \in (0, \alpha_0]$. Thus, letting $\alpha \downarrow 0$ yields

$$L \leq \bar{L}_{\varepsilon,x,0} \quad \text{in } B_R(x - Rn_x) \qquad (8.18)$$

for all $x \in \Gamma_\infty$ and sufficiently small $\varepsilon > 0$.

On the other hand, if $\varepsilon > \vartheta$ we have that

$$-(1 - \varepsilon)\mathfrak{a}_x \geq -(1 - \vartheta)\mathfrak{a}_x$$

for all $x \in \Gamma_\infty$ and hence, $\bar{u}_{[\vartheta,x,0,M]}$ provides us with a subsolution of

$$\begin{cases} -\Delta u = \mu u - (1-\varepsilon)\mathfrak{a}_x(R-r_0)g(u)u & \text{in } B_R(x - Rn_x), \\ u = M & \text{on } \partial B_R(x - Rn_x). \end{cases}$$

Consequently, thanks to Theorem 1.7, we find that

$$\bar{u}_{[\vartheta,x,0,M]} \leq \bar{u}_{[\varepsilon,x,0,M]} \qquad \text{for all } M > 0.$$

So, letting $M \to \infty$, from Theorem 3.4 it becomes apparent that

$$\bar{L}_{\vartheta,x,0} \leq \bar{L}_{\varepsilon,x,0}$$

for all $x \in \Gamma_\infty$ and $\vartheta < \varepsilon$. By a rather standard compactness argument, this estimate shows that

$$\bar{L}_x := \lim_{\varepsilon \downarrow 0} \bar{L}_{\varepsilon,x,0}$$

is well defined. Actually, it provides us with the unique positive solution of the singular problem

$$\begin{cases} -\Delta u = \mu u - \mathfrak{a}_x(R-r_0)g(u)u & \text{in } B_R(x - Rn_x), \\ u = \infty & \text{on } \partial B_R(x - Rn_x), \end{cases} \tag{8.19}$$

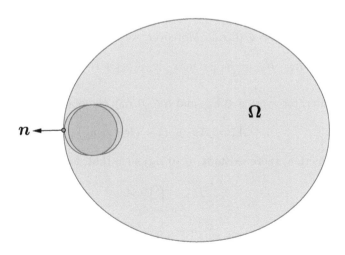

FIGURE 8.1: The balls where the supersolutions $\bar{L}_{\varepsilon,x,\alpha}$ are supported.

for all $x \in \Gamma_\infty$. It should be noted that in the special case when, for some $p > 1$,

$$g(u) = Hu^{p-1}$$

then, by a simple re-scaling argument,

$$\bar{L}_{\varepsilon,x,0} = (1-\varepsilon)^{-\frac{1}{p-1}} \bar{L}_x.$$

Now, letting $\varepsilon \downarrow 0$ in (8.18) shows that

$$L \leq \bar{L}_x \quad \text{in } B_R(x - Rn_x) \quad \text{for all } x \in \Gamma_\infty. \tag{8.20}$$

By Theorem 7.3, we already know that

$$\lim_{t \downarrow 0} \frac{\bar{L}_x(x - tn_x)}{\mathfrak{b}_x(t)} = I_{0,x}^{\frac{-p}{p-1}} \left(\frac{p+1}{p-1}\right)^{\frac{p+1}{p-1}} H^{\frac{-1}{p-1}}. \tag{8.21}$$

Therefore, according to (8.20) and (8.21) we find that

$$\lim_{t \downarrow 0} \frac{L(x - tn_x)}{\mathfrak{b}_x(t)} \leq I_{0,x}^{\frac{-p}{p-1}} \left(\frac{p+1}{p-1}\right)^{\frac{p+1}{p-1}} H^{\frac{-1}{p-1}} \tag{8.22}$$

for all $x \in \Gamma_\infty$.

Subsequently, we suppose that $R > 0$ and $\alpha_0 > 0$ have been shortened, if necessary, so that

$$\bar{B}_R(x + Rn_x) \cap \bar{D} = \{x\} \quad \text{and} \quad \bar{B}_R(x + (R+\alpha)n_x) \subset \mathbb{R}^N \setminus \bar{D}$$

for all $x \in \Gamma_\infty$ and $\alpha \in (0, \alpha_0]$. Moreover, we will set

$$R_1 := R, \quad R_m := R_1 + \max_{z_1, z_2 \in \bar{D}} |z_1 - z_2| + 1, \quad R_2 := 2R_m - R_1,$$

and consider, for every $x \in \Gamma_\infty$ and $\alpha \in [0, \alpha_0]$, the open annulus

$$A_\alpha := A_{R_1, R_2}\left(x + (R+\alpha)n_x\right).$$

By construction, there exists $\alpha_1 \in (0, \alpha_0)$ such that, for every $x \in \Gamma_\infty$,

$$D \subset \bigcap_{0 \leq \alpha \leq \alpha_1} A_\alpha.$$

Next, we pick a sufficiently small $\xi > 0$ such that, for every $x \in \Gamma_\infty$,

$$\bar{A}_{R_1, R_1+2\xi}(x + Rn_x) \cap \bar{D} \subset \mathcal{Q},$$

and consider the constant

$$C := \|\mathfrak{a}\|_{\mathcal{C}(\bar{D})} + 1.$$

By (8.13), for every
$$z \in \bar{A}_{R_1, R_1 + \xi}(x + Rn_x) \cap \bar{D}$$

the next estimate holds
$$\mathfrak{a}(z) = \mathfrak{a}(\pi(z) - t(z)n_{\pi(z)}) \leq (1 + \varepsilon)\mathfrak{a}(x - t(z)n_x),$$

because, since $R < \frac{\varrho}{2}$,
$$|x - \pi(z)| \leq R_1 + \xi < \varrho.$$

Thus,
$$\begin{aligned}
\mathfrak{a}(z) &\leq (1 + \varepsilon)\mathfrak{a}_x(t(z)) = (1 + \varepsilon)\mathfrak{a}_x(\text{dist}\,(z, \Gamma_\infty)) \\
&\leq (1 + \varepsilon)\mathfrak{a}_x(\text{dist}\,(z, x)) \leq (1 + \varepsilon)\mathfrak{a}_x(\text{dist}\,(z, \partial B_R(x + Rn_x)) \\
&\leq (1 + \varepsilon)\mathfrak{a}_x(\text{dist}\,(z, \partial B_R(x + (R + \alpha)n_x))
\end{aligned}$$

for all $\alpha \in [0, \alpha_1]$. Hence,
$$\mathfrak{a}(z) \leq (1 + \varepsilon)\mathfrak{a}_x(\text{dist}\,(z, \partial A_\alpha)) \tag{8.23}$$

for all
$$(z, \alpha) \in \left[\bar{A}_{R_1, R_1 + \xi}(x + Rn_x) \cap \bar{D}\right] \times [0, \alpha_1]. \tag{8.24}$$

Subsequently, for sufficiently large $n \geq 1$ with
$$\xi < \xi\left(1 + \frac{1}{n}\right) < R < \frac{\varrho}{2} < \eta,$$

we will consider the function $\tilde{\mathfrak{a}}_x$ defined by
$$\tilde{\mathfrak{a}}_x(t) := \begin{cases} \mathfrak{a}_x(t) & \text{if } 0 \leq t \leq \xi, \\ n\left(\frac{t}{\xi} - 1\right)[C - \mathfrak{a}_x(\xi)] + \mathfrak{a}_x(\xi) & \text{if } \xi < t \leq \xi\left(1 + \frac{1}{n}\right), \\ C & \text{if } \xi\left(1 + \frac{1}{n}\right) < t \leq R_m. \end{cases}$$

It is a non-decreasing continuous extension of $\mathfrak{a}_x$ from the interval $[0, \xi]$ to the interval $[0, R_m]$ such that, for sufficiently large $n \geq 1$,
$$\mathfrak{a}(z) \leq (1 + \varepsilon)\tilde{\mathfrak{a}}_x(\text{dist}\,(z, \partial A_\alpha)) \quad \text{for all } (z, \alpha) \in D \times [0, \alpha_1]. \tag{8.25}$$

Indeed, due to (8.23), (8.25) holds in the range (8.24). In particular, if $\text{dist}\,(z, \partial A_\alpha) \leq \xi$ for some $\alpha \in [0, \alpha_1]$, then (8.24) holds and hence,
$$\mathfrak{a}(z) \leq (1 + \varepsilon)\mathfrak{a}_x(\text{dist}\,(z, \partial A_\alpha)) = (1 + \varepsilon)\tilde{\mathfrak{a}}_x(\text{dist}\,(z, \partial A_\alpha)).$$

Similarly, if
$$\text{dist}\,(z, \partial A_\alpha) > \xi\left(1 + \frac{1}{n}\right)$$

for some $\alpha \in [0, \alpha_1]$, then

$$\mathfrak{a}(z) \leq C < (1+\varepsilon)C = (1+\varepsilon)\tilde{\mathfrak{a}}_x(\text{dist}\,(z, \partial A_\alpha))$$

and hence, (8.25) also holds. Lastly, suppose that, for some $\alpha \in [0, \alpha_1]$,

$$\xi < \text{dist}\,(z, \partial A_\alpha) \leq \xi\left(1 + \frac{1}{n}\right).$$

The set of these points z lies in the compact set

$$\mathcal{K} := \left\{ z \in \bar{D} \ : \ \text{dist}\,(z, \partial A_\alpha) \leq \frac{\varrho}{2} \right\}$$

where, owing to (8.8), $\mathfrak{a}$ is a function of class $\mathcal{C}^1$, because $\frac{\varrho}{2} < \eta$. In particular, $\nabla \mathfrak{a}$ is bounded in $\mathcal{K}$ and, consequently, since

$$\tilde{\mathfrak{a}}'_x(t) = \frac{n}{\xi}[C - \mathfrak{a}_x(\xi)] \geq \frac{n}{\xi} \qquad \text{if} \ \ \xi < t \leq \xi\left(1 + \frac{1}{n}\right),$$

it becomes apparent that (8.25) holds for sufficiently large $n \geq 1$.

Due to (8.25), L is a supersolution of the singular problem

$$\begin{cases} -\Delta u = \mu u - (1+\varepsilon)\tilde{\mathfrak{a}}_x(\text{dist}\,(z, \partial A_\alpha))g(u)u & \text{in } A_\alpha, \\ u = \infty & \text{on } \partial A_\alpha, \end{cases} \tag{8.26}$$

for each $\alpha \in [0, \alpha_1]$. Thanks to (8.8), it follows from Theorem 7.1 that (8.26) possesses a unique positive solution, $\underline{L}_{\varepsilon, x, \alpha}$. By the uniqueness,

$$\underline{L}_{\varepsilon, x, \alpha} = \underline{L}_{\varepsilon, x, 0}(\cdot - \alpha n_x), \qquad \alpha \in [0, \alpha_1].$$

Actually, for each $\alpha \in (0, \alpha_1]$, the restriction $\underline{L}_{\varepsilon, x, \alpha}|_{\bar{D}}$ is a bounded solution of

$$-\Delta u = \mu u - (1+\varepsilon)\tilde{\mathfrak{a}}_x(\text{dist}\,(z, \partial A_\alpha))g(u)u$$

in D and, therefore, thanks to Theorem 1.7,

$$\underline{L}_{\varepsilon, x, \alpha} = \underline{L}_{\varepsilon, x, 0}(\cdot - \alpha n_x) \leq L \qquad \text{in } D.$$

On the other hand, by Theorem 3.4, we already know that

$$\underline{L}_{\varepsilon, x, \alpha} = \lim_{M \uparrow \infty} \underline{u}_{[\varepsilon, x, \alpha, M]},$$

where $\underline{u}_{[\varepsilon, x, \alpha, M]}$ stands for the unique positive solution of

$$\begin{cases} -\Delta u = \mu u - (1+\varepsilon)\tilde{\mathfrak{a}}_x(\text{dist}\,(z, \partial A_\alpha))g(u)u & \text{in } A_\alpha, \\ u = M & \text{on } \partial A_\alpha. \end{cases}$$

Moreover, the map

$$M \mapsto \underline{u}_{[\varepsilon, x, \alpha, M]}$$

is increasing. Hence,

$$\underline{u}_{[\varepsilon,x,\alpha,M]} \le \underline{L}_{\varepsilon,x,\alpha} \le L \quad \text{in} \quad D$$

for all $x \in \Gamma_\infty$ and $\alpha \in (0, \alpha_0]$. Thus, letting $\alpha \downarrow 0$ yields

$$\underline{u}_{[\varepsilon,x,0,M]} \le \underline{L}_{\varepsilon,x,0} \le L \quad \text{in} \quad D \tag{8.27}$$

for all $x \in \Gamma_\infty$ and sufficiently small $\varepsilon > 0$.

Lastly, note that if $\varepsilon > \vartheta$, then we have that

$$-(1+\vartheta)\tilde{a}_x \ge -(1+\varepsilon)\tilde{a}_x$$

for all $x \in \Gamma_\infty$. So, $\underline{u}_{[\vartheta,x,0,M]}$ provides us with a supersolution of

$$\begin{cases} -\Delta u = \mu u - (1+\varepsilon)\tilde{a}_x(\text{dist}\,(z, \partial A_0))g(u)u & \text{in} \ A_0, \\ u = M & \text{on} \ \partial A_0. \end{cases}$$

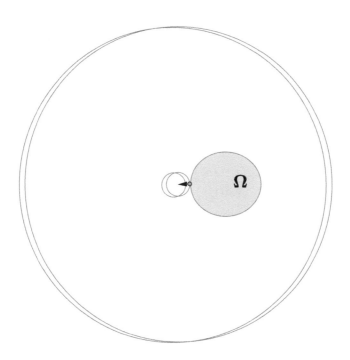

FIGURE 8.2: The annuli where the subsolutions are supported.

Consequently, thanks to Theorem 1.7, we find that

$$\underline{u}_{[\varepsilon,x,0,M]} \leq \underline{u}_{[\vartheta,x,0,M]} \qquad \text{for all } M > 0.$$

So, letting $M \to \infty$ shows that

$$\underline{L}_{\varepsilon,x,0} \leq \underline{L}_{\vartheta,x,0} \leq L \quad \text{in } D \tag{8.28}$$

for all $x \in \Gamma_\infty$ and $\vartheta < \varepsilon$. Therefore, by a rather standard compactness argument, it follows from (8.28) that

$$\underline{L}_x := \lim_{\varepsilon \downarrow 0} \underline{L}_{\varepsilon,x,0}$$

is well defined and actually provides us with the unique positive solution of the singular problem

$$\begin{cases} -\Delta u = \mu u - \tilde{a}_x(\text{dist}\,(z, \partial A_0))g(u)u & \text{in } A_0, \\ u = \infty & \text{on } \partial A_0, \end{cases} \tag{8.29}$$

for all $x \in \Gamma_\infty$. Moreover, according to (8.28), we find that

$$\underline{L}_x \leq L \qquad \text{in } D \tag{8.30}$$

for all $x \in \Gamma_\infty$.

It should be noted that in the special case when $g(u) = Hu^{p-1}$ for some $p > 1$, then

$$\underline{L}_{\varepsilon,x,0} = (1+\varepsilon)^{\frac{-1}{p-1}} \underline{L}_x$$

for all $\varepsilon > 0$.

By Theorem 7.3 we already know that

$$\lim_{t \downarrow 0} \frac{\underline{L}_x(x - tn_x)}{b_x(t)} = I_{0,x}^{\frac{-p}{p-1}} \left(\frac{p+1}{p-1} \right)^{\frac{p+1}{p-1}} H^{\frac{-1}{p-1}}. \tag{8.31}$$

Thus, from (8.30) and (8.31) we can infer that

$$\lim_{t \downarrow 0} \frac{L(x - tn_x)}{b_x(t)} \geq I_{0,x}^{\frac{-p}{p-1}} \left(\frac{p+1}{p-1} \right)^{\frac{p+1}{p-1}} H^{\frac{-1}{p-1}}$$

for all $x \in \Gamma_\infty$. Therefore, by (8.22),

$$\lim_{t \downarrow 0} \frac{L(x - tn_x)}{b_x(t)} = I_{0,x}^{\frac{-p}{p-1}} \left(\frac{p+1}{p-1} \right)^{\frac{p+1}{p-1}} H^{\frac{-1}{p-1}}$$

for all $x \in \Gamma_\infty$, which ends the proof of Theorem 8.2. $\square$

Actually, as a consequence from the proof of Theorem 8.2 and of Remark 7.4 the next result holds.

Corollary 8.3 *Under the same assumptions of Theorem 8.2, for any pair of positive solutions, L_1 and L_2, of (8.4), the function*

$$q(z) := \frac{L_1(z)}{L_2(z)}, \qquad z \in D,$$

satisfies

$$\lim_{d(z)\downarrow 0} q(z) = 1, \qquad d(z) := \mathrm{dist}\,(z, \Gamma_\infty), \qquad z \in D.$$

Proof: According to (8.20) and (8.30), we have that

$$\frac{\underline{L}_{\pi(z)}(z)}{\overline{L}_{\pi(z)}(z)} \le \frac{L_1(z)}{L_2(z)} \le \frac{\overline{L}_{\pi(z)}(z)}{\underline{L}_{\pi(z)}(z)} \tag{8.32}$$

for all $z \in D$ with $d(z) \le R$. Moreover, setting

$$J(x) := I_{0,x}^{\frac{-p}{p-1}} \left(\frac{p+1}{p-1}\right)^{\frac{p+1}{p-1}} H^{\frac{-1}{p-1}}, \qquad x \in \Gamma_\infty,$$

we find from (8.8) and Remark 7.4 that

$$\lim_{d(z)\downarrow 0} \frac{\overline{L}_{\pi(z)}(z)}{\mathfrak{b}_{\pi(z)}(d(z))J(\pi(z))} = \lim_{d(z)\downarrow 0} \frac{\underline{L}_{\pi(z)}(z)}{\mathfrak{b}_{\pi(z)}(d(z))J(\pi(z))} = 1.$$

Thus,

$$\lim_{d(z)\downarrow 0} \frac{\overline{L}_{\pi(z)}(z)}{\underline{L}_{\pi(z)}(z)} = \lim_{d(z)\downarrow 0} \frac{\overline{L}_{\pi(z)}(z)}{\mathfrak{b}_{\pi(z)}(d(z))J(\pi(z))} \lim_{d(z)\downarrow 0} \frac{\mathfrak{b}_{\pi(z)}(d(z))J(\pi(z))}{\underline{L}_{\pi(z)}(z)} = 1.$$

Therefore, also

$$\lim_{d(z)\downarrow 0} \frac{\underline{L}_{\pi(z)}(z)}{\overline{L}_{\pi(z)}(z)} = 1$$

and letting $d(z) \to 0$ in (8.32) ends the proof. $\qquad\square$

8.4 Special case when $\mathfrak{a}(x) > 0$ for some $x \in \partial\Omega$

Although we imposed $\mathfrak{a}(x) = 0$ in the statement of Theorem 8.2, this condition is not necessary for its validity. Indeed, if $\mathfrak{a}(x) > 0$ for some $x \in \partial\Omega$, then, in a neighborhood of x, $\mathfrak{a}_x$ lies between two positive constants, e.g.,

$$\mathfrak{a}(x) - \eta < \mathfrak{a}_x(t) < \mathfrak{a}(x) + \eta \quad \text{for all} \quad t \in (0, \varepsilon],$$

and the localization method of the proof of Theorem 8.2 can be easily adapted to show the validity of (8.10) with

$$\mathfrak{b}_x(t) := \int_t^R \left(\int_0^s C_0^{\frac{1}{p+1}} \right)^{-\frac{p+1}{p-1}} ds, \qquad I_{0,x} := \lim_{t \downarrow 0} \frac{\mathfrak{b}_x(t)\mathfrak{b}_x''(t)}{[\mathfrak{b}_x'(t)]^2},$$

where

$$C_0 = \mathfrak{a}_x(0) = \mathfrak{a}(x).$$

Since $\mathfrak{a}(x) > 0$, the non-oscillation conditions on $\mathfrak{a}$ and its derivatives are not necessary for the validity of (8.10), of course. In such case, a direct calculation shows that

$$\mathfrak{b}_x(t) = C_0^{-\frac{1}{p-1}} \frac{p-1}{2} \left(t^{-\frac{2}{p-1}} - R^{-\frac{2}{p-1}} \right)$$

for sufficiently small $t > 0$. Hence,

$$I_{0,x} := \lim_{t \downarrow 0} \frac{\mathfrak{b}_x(t)\mathfrak{b}_x''(t)}{[\mathfrak{b}_x'(t)]^2} = \frac{p+1}{2}$$

and consequently, according to (8.10),

$$\lim_{t \downarrow 0} \frac{L(x - tn_x)}{C_0^{-\frac{1}{p-1}} \frac{p-1}{2} \left(t^{-\frac{2}{p-1}} - R^{-\frac{2}{p-1}} \right)} = \left(\frac{p+1}{2} \right)^{-\frac{p}{p-1}} \left(\frac{p+1}{p-1} \right)^{\frac{p+1}{p-1}} H^{\frac{-1}{p-1}}.$$

Therefore, rearranging terms and simplifying yields

$$\lim_{t \downarrow 0} \frac{L(x - tn_x)}{t^{-\frac{2}{p-1}}} = \mathfrak{a}^{-\frac{1}{p-1}}(x) \left(\frac{2}{p-1} \frac{p+1}{p-1} \right)^{\frac{1}{p-1}} H^{\frac{-1}{p-1}},$$

which provides us with (7.51) if $H = 1$.

8.5   Uniqueness of the large solution

As an immediate consequence of Corollary 8.3, the following uniqueness result holds.

Theorem 8.4 *Under the same assumptions of Theorem 8.2, (8.4) admits, at most, a unique positive solution, L.*

Proof: Suppose L_1 and L_2 are two positive solutions of (8.4). Then, according to Corollary 8.3, the auxiliary function

$$q(z) := \begin{cases} L_1(z)/L_2(z) & z \in D, \\ 1 & z \in \Gamma_\infty, \end{cases}$$

is continuous in
$$K_R := \{z \in \bar{D} \; : \; \text{dist}\,(z, \Gamma_\infty) \le R\}$$
for sufficiently small $R > 0$. As K_R is compact, $q(z)$ is uniformly continuous in K_R. Thus, for every $\varepsilon > 0$ there exists $\delta > 0$ such that
$$|q(z) - q(\pi(z))| = |q(z) - 1| < \varepsilon \qquad \text{if } d(z) = |z - \pi(z)| \le \delta.$$
Therefore,
$$(1 - \varepsilon)L_2 \le L_1 \le (1 + \varepsilon)L_2 \quad \text{in } D_\delta := \{z \in D \; : \; \text{dist}\,(z, \Gamma_\infty) \le \delta\}.$$

Next, we will consider the boundary value problem
$$\begin{cases} -\Delta u = \mu u - \mathfrak{a}(x)g(u)u & \text{in } D \setminus D_\delta, \\ \mathfrak{B}u = 0 & \text{on } \Gamma_\mathfrak{B}, \\ u = L_1 & \text{on } \partial(D \setminus D_\delta) \setminus \Gamma_\mathfrak{B}. \end{cases} \tag{8.33}$$

By adapting Theorem 1.7, through Theorem 7.1 of [163], it is easily seen that L_1 is the unique solution of (8.33).

On the other hand, since g is increasing, we find that
$$\begin{aligned} -\Delta((1 - \varepsilon)L_2) &= \mu(1 - \varepsilon)L_2 - \mathfrak{a}(x)g(L_2)(1 - \varepsilon)L_2 \\ &\le \mu(1 - \varepsilon)L_2 - \mathfrak{a}(x)g((1 - \varepsilon)L_2)(1 - \varepsilon)L_2 \end{aligned}$$
in $D \setminus D_\delta$. Moreover,
$$\mathfrak{B}((1 - \varepsilon)L_1) = (1 - \varepsilon)\mathfrak{B}L_1 = 0 \qquad \text{on } \Gamma_\mathfrak{B},$$
and
$$(1 - \varepsilon)L_2 \le L_1 \qquad \text{on } \partial(D \setminus D_\delta) \setminus \Gamma_\mathfrak{B}.$$
Thus, $(1 - \varepsilon)L_2$ is a subsolution of (8.33). Similarly, $(1 + \varepsilon)L_2$ provides us with a supersolution of (8.33). Therefore, thanks again to Theorem 7.1 of [163] and adapting the argument of the proof of Theorem 1.7, it becomes apparent that
$$(1 - \varepsilon)L_2 \le L_1 \le (1 + \varepsilon)L_2 \qquad \text{in } D \setminus D_\delta.$$
Consequently,
$$(1 - \varepsilon)L_2 \le L_1 \le (1 + \varepsilon)L_2 \qquad \text{in } D$$
and letting $\varepsilon \downarrow 0$ yields $L_1 = L_2$, which ends the proof. $\qquad \square$

The numerical example of Section 5.3 in the domain $\Omega_1 = B_{0.3}$ satisfies all the assumptions of Theorem 8.2 and hence, the large solution $L_{[\lambda, B_{0.3}]}$ is unique for all $\lambda < \sigma_2$. Indeed, for the choice of the weight function in Section 5.3, we have that, for sufficiently small $\varepsilon > 0$ and every $x \in \partial B_{0.3}$,
$$\mathfrak{a}_x(t) = \sin(5\pi(0.8 + t)) \qquad \text{for all } t \in [0, \varepsilon).$$

Since

$$\mathfrak{a}'_x(t) = 5\pi \cos(5\pi(0.8+t)), \qquad \mathfrak{a}''_x(t) = -(5\pi)^2 \sin(5\pi(0.8+t)),$$

it is apparent that

$$\mathfrak{a}_x(t) > 0, \quad \mathfrak{a}'_x(t) > 0, \quad \mathfrak{a}''_x(t) < 0,$$

for all $t \in (0, \varepsilon]$. Hence, (8.8) holds. Naturally, as $\mathfrak{a}_x$ is independent of x, (8.9) also holds. Finally, since

$$\left(\frac{\mathfrak{a}_x(t)}{\mathfrak{a}'_x(t)}\right)' = \frac{[\mathfrak{a}'_x(t)]^2 - \mathfrak{a}''_x(t)\mathfrak{a}_x(t)}{[\mathfrak{a}'_x(t)]^2} = \sec^2(5\pi(0.8+t))$$

we find that

$$\left(\frac{\mathfrak{a}_x(t)}{\mathfrak{a}'_x(t)}\right)'' = 2\sec^2(5\pi(0.8+t))\tan(5\pi(0.8+t)) > 0$$

for sufficiently small $t > 0$. Consequently, $(\mathfrak{a}_x/\mathfrak{a}'_x)'$ is indeed non-oscillating and all the assumptions of Theorem 8.2 are fulfilled.

8.6   Comments on Chapter 8

The results of this chapter go back to J. López-Gómez [161]. The localization method used in the proof of Theorem 8.2 to determine the exact blow-up rate on the boundary of the large solutions goes back to J. López-Gómez [154]. Since then, it has been used very often in the specialized literature.

Three years after [161] was published, J. García-Melián [98], inspired by S. Dumont et al. [81], found some sharp uniqueness results for the simplest prototype model

$$\begin{cases} \Delta u = \mathfrak{a}(x)f(u) & \text{in } \Omega, \\ u = \infty & \text{on } \partial\Omega, \end{cases} \tag{8.34}$$

for a rather general class of functions $f(u)$, but, once again, he had to impose

$$C_1 d^\gamma(x) \le \mathfrak{a}(x) \le C_2 d^\gamma(x) \tag{8.35}$$

for $x \in \Omega$ near $\partial\Omega$, with $\gamma \ge 0$ and C_1, C_2 positive constants. To make the general results of this chapter valid for wide classes of weight functions $a(x)$ not satisfying (8.35), one must pay the price of imposing (7.16), i.e., $f(u) \sim u^p$, with $p > 1$, for sufficiently large u.

In the very special case when $\mathfrak{a} = 1$, a necessary and sufficient condition for the existence of a positive solution of (8.34) when f is increasing is the classical Keller–Osserman condition

$$\int_u^\infty \frac{dx}{\sqrt{\int_0^x f}} < \infty \quad \text{for some } u > 0 \tag{8.36}$$

(see J. B. Keller [123] and R. Osserman [202]). However, the uniqueness of the large solution for general weight functions $\mathfrak{a}(x)$ is a very trickly problem, not completely understood yet, even for (8.34), which is a very simplified version of the general model dealt with in this book.

Although the uniqueness results of Chapter 7 are based on the maximum principle, at this stage of the book it is apparent that the blow-up rates of the positive solutions of (8.34) are closely related to the problem of the uniqueness. Indeed, according to Theorem 6.8 of Y. Du [79] and Theorem 1.6 of S. Dumont at al. [81], it is well known that, in general, the boundary behavior of the positive solutions $L(x)$ of (8.34) is given by

$$\lim_{d(x)\downarrow 0} \frac{\Psi(L(x))}{d(x)} = 1, \qquad d(x) := \text{dist}\,(x, \partial\Omega), \tag{8.37}$$

where

$$\Psi(t) := \frac{1}{\sqrt{2}} \int_t^\infty \frac{dx}{\sqrt{\int_0^x f}}. \tag{8.38}$$

But, in order to get the exact blow-up rate of the positive solutions $L(x)$ of (8.34), one needs to impose some additional restrictions on the nonlinearity $f(u)$, for instance,

$$\liminf_{r\uparrow\infty} \frac{\Psi(ru)}{\Psi(u)} > 1 \qquad \text{for all } r \in (0, 1). \tag{8.39}$$

Indeed, in such case, it is well known that

$$\lim_{d(x)\downarrow 0} \frac{L(x)}{\Phi(d(x))} = 1 \tag{8.40}$$

where $\Phi := \Psi^{-1}$ (e.g., Corollary 6.9 of Y. Du [79]). It turns out that under condition (8.40)

$$\lim_{d(x)\downarrow 0} \frac{L_1(x)}{L_2(x)} = 1$$

for any pair, L_1 and L_2, of positive solutions of (8.34). Consequently, imposing the additional monotonicity assumption that

$$f(u) = g(u)u, \qquad u > 0$$

with $g(u)$ increasing, the uniqueness of the positive solution of (8.34) holds, which is a classical result of C. Bandle and M. Marcus [17]. Note that (8.39) holds under condition

$$H := \lim_{u \uparrow \infty} \frac{f(u)}{u^p} > 0$$

for some $p > 1$. The reader is sent to J. García-Melián [98] for a complete discussion from a classical perspective. This book pays attention to a much more general spatially heterogeneous prototype model, where the nonlinearity is far from being monotonic and most of the previous results fail.

The localization method used in the proof of Theorem 8.2 has been widely used in the literature to find a number of refinements of some of the results covered in the last three chapters. For example, T. Ouyang and Z. Xie [205, 206] adapted it to get their own results.

The literature on the uniqueness of large solutions for the equations covered in this book and other related equations with gradient terms, singular coefficients or even $p(x)$-Laplace operators, is really huge. The interested reader might wish to have a look at Z. Zhang [246, 247, 248, 249, 250, 251, 252], Q. Zhang, X. Liu and Z. Qiu [243], Z. Xie [238], P. Feng [86], M. Wang and L. Wei [231], C. Mu, S. Huang, Q. Tian and L. Liu [196], H. Li, P. Y. H. Pang and M. Wang [135, 136], M. M. Boureanu [30], Q. Zhang, Y. Wang and Z. Qiu [244], Y. Wang and M. Wang [233], L. Wei [234], S. Huang and Q. Tian [114], S. Huang, Q. Tian, S. Zhang, J. Xi and Z. Fan [117], Y. Chen and M. Wang [45, 46, 47, 48], Q. Zhang [242], S. Huang, Q. Tian, S. Zhang and J. Xi [116], M. Marras and G. Porru [191], S. Huang, W. T. Li, Q. Tian and C. Mu [118], L. Wei and J. Zhu [236], S. Alarcón, J. García-Melián and A. Quaas [5], S. Alarcón, G. Díaz, R. Letelier and J. M. Rey [3], S. Alarcón, G. Díaz and J. M. Rey [4], Z. Xie and C. Zhao [239], C. Liu and Z. Yang [137], C. Aneda and G. Porru [15], S. Huang, Q. Tian and Y. Mi [115], S. Nakamori and K. Takimoto [198], Y. Chen, P. Y. H. Pang and M. Wang [49], N. Belhaj Rhouma, A. Drissi and W. Sayeb [22], Q. Zhang and C. Zhao [245], Y. Liang, Q. Zhang and C. Zhao [133], Y. Chen, Y. Zhu and R. Hao [51], L. Wei and M. Wang [235], T. Shibata [221], X. Ji and J. Bao [121], Z. Zhang, Y. Ma, L. Mi and X. Li [253], Y. Chen and Y. Zhu [50], H. Feng and C. Zhong, [85], W. Lei and M. Wang [132], Z. Zhang and L. Mi [254], H. Yang and Y. Chang [241], L. Mi and B. Liu [194], D. Repovs [218], J. Bao, X. Ji and H. Li [19], L. Chen, Y. Chen and D. Luo [44], and W. Wang, H. Gong and S. Zheng [232], among many others.

Some related results by the author and his coworkers on the porous media equation and variable exponents are found in M. Delgado, J. López-Gómez and A. Suárez [73, 74, 75], J. López-Gómez and A. Suárez [182], and J. López-Gómez [156].

Part III

Metasolutions do arise everywhere

Chapter 9

A paradigmatic superlinear indefinite problem

This chapter studies the parabolic problem

$$
\begin{cases}
\frac{\partial u}{\partial t} - d\Delta u = \lambda u + a(x)|u|^p u & \text{in} \quad \Omega \times (0, \infty), \\
u = 0 & \text{on} \quad \partial\Omega \times (0, \infty), \\
u(\cdot, 0) = u_0 > 0 & \text{in} \quad \Omega,
\end{cases}
\tag{9.1}
$$

where $d > 0$, $p \geq 1$, $\lambda \in \mathbb{R}$, Ω is a bounded domain of $\mathbb{R}^N$ with $N \geq 1$ of class $\mathcal{C}^{2+\nu}$, and $a \in \mathcal{C}^\nu(\bar{\Omega})$ is arbitrary. Since in this chapter we are not requiring $a(x)$ to satisfy $a \leq 0$, as we have done in the previous ones, (9.1) is said to be a parabolic problem of *superlinear indefinite type*. Naturally, we are most interested in the general case when $a(x)$ changes sign in Ω. Precisely, throughout this chapter we assume that

Hypothesis (HA)

There are x_+, $x_- \in \Omega$ such that $a(x_+) > 0$ and $a(x_-) < 0$, the open sets

$$\Omega_+ := \{x \in \Omega : a(x) > 0\}, \quad \Omega_- := \{x \in \Omega : a(x) < 0\}, \quad \Omega_0 := \operatorname{int} a^{-1}(0),$$

are of class $C^{2+\nu}$, and the negative part of $a(x)$,

$$a_-(x) = \min\{0, a(x)\}, \qquad x \in \Omega,$$

satisfies Hypothesis (Ha) *if Ω_0 is non-empty.*

As they are of class $C^{2+\nu}$, the open sets Ω_+, Ω_- and Ω_0 at most have finitely many components. Figure 9.1 shows an admissible nodal configuration of $a(x)$ with Ω_0 non-empty. The problem (9.1) is *superlinear* because $a > 0$ somewhere, though of *indefinite type* because simultaneously it is of *sublinear type* in Ω_-. Indeed, for every $u > 0$,

$$\lambda u + a(x)|u|^p u \begin{cases} < \lambda u & \text{if } a(x) < 0, \\[2mm] = \lambda u & \text{if } a(x) = 0, \\[2mm] > \lambda u & \text{if } a(x) > 0. \end{cases}$$

Naturally, these problems might be also called *sublinear of indefinite type*.
    The positive part of the weight function a, a_+, is defined by

$$a_+(x) := \max\{0, a(x)\} = a(x) - a_-(x), \qquad x \in \Omega.$$

Obviously,

$$a_+ \geq 0, \qquad a_- \leq 0, \qquad a = a_+ + a_-,$$

and the relative sizes of a_+ and a_- measure the sublinear or superlinear character of the parabolic problem (9.1). Indeed, for sufficiently small a_- the model (9.1) should be utterly superlinear, while for small a_+ the model should exhibit a sublinear behavior, in a sense to be discussed later.
    Closely related to problem (9.1) arises the *associated sublinear problem*

$$\begin{cases} \frac{\partial u}{\partial t} - d\Delta u = \lambda u + a_-(x)|u|^p u & \text{in } \Omega \times (0, \infty), \\ u = 0 & \text{on } \partial\Omega \times (0, \infty), \\ u(\cdot, 0) = u_0 > 0 & \text{in } \Omega, \end{cases} \qquad (9.2)$$

whose dynamics has been already analyzed in Part I of this book. As the solutions of (9.2) are subsolutions of (9.1), if we denote by $u(x, t; \lambda, a, u_0)$ and $u(x, t; \lambda, a_-, u_0)$ the solutions of (9.1) and (9.2), respectively, then, thanks to the parabolic maximum principle, we find that

$$u(x, t; \lambda, a_-, u_0) < u(x, t; \lambda, a, u_0) \qquad \text{for all } (x, t) \in \Omega \times (0, T)$$

for all time time T for which $u(x,t;\lambda,a,u_0)$ is well defined. In particular, the metasolutions of (9.2) push up the solutions of (9.1). Consequently, these metasolutions should play a significant role not only for describing the dynamics of (9.2) but also for ascertaining the global dynamics of (9.1). By the superlinear character of (9.1) in Ω_+, the local solutions of (9.1), those defined for $t \in [0,T)$, might not be globally defined for all time $t > 0$, as they can blow up in a finite time. Indeed, it is well known that (9.1) possesses a unique classical solution, $u := u(x,t;\lambda,a,u_0)$, which is defined in a maximal existence interval of the form $[0,T_{\max})$ for some $T_{\max} \le +\infty$. Moreover, u blows up in $L^\infty(\Omega)$ at time $T_b := T_{\max}$ if $T_{\max} < +\infty$, i.e.,

$$\lim_{t\uparrow T_b} \|u(\cdot,t;\lambda,a,u_0)\|_{\mathcal{C}(\bar\Omega)} = \infty$$

(see, e.g., D. Henry [112] or P. Quittner and P. Souplet [213]). Nevertheless, in many circumstances, the dynamics of (9.1) are regulated by its non-negative steady states, which are the non-negative solutions of

$$\begin{cases} -d\Delta u = \lambda u + a(x)|u|^p u & \text{in} \quad \Omega, \\ u = 0 & \text{on} \quad \partial\Omega. \end{cases} \tag{9.3}$$

Because λ in this chapter is viewed as a continuation parameter, the solutions of (9.3) are regarded as solution couples (λ, u). According to [163, Ch. 5],

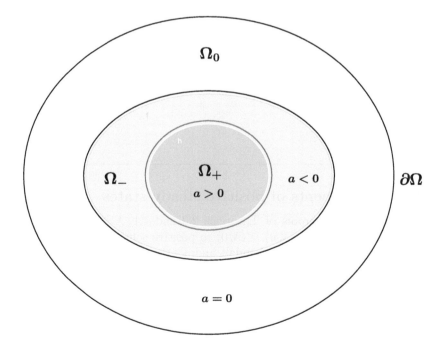

FIGURE 9.1: An admissible nodal configuration of $a(x)$.

any weak solution of (9.3) is a classical solution in $\mathcal{C}^{2+\nu}(\bar{\Omega})$. Obviously, (9.3) admits two types of non-negative solutions: the trivial one, $u = 0$, and the positive solutions, $u \geq 0$, $u \neq 0$. Arguing as in Lemma 1.6, it is apparent that any positive solution of (9.3), u, satisfies $u \gg 0$. These solutions are supersolutions of the underlying sublinear elliptic problem

$$\begin{cases} -d\Delta u = \lambda u + a_-(x)|u|^p u & \text{in } \Omega, \\ u = 0 & \text{on } \partial\Omega, \end{cases} \tag{9.4}$$

whose non-negative solutions are the steady states of (9.2). By Theorem 4.1, (9.4) admits a positive solution if, and only if, $d\sigma_0 < \lambda < d\sigma_1$, where

$$\sigma_1 := \lambda_1[-\Delta, \Omega \setminus \bar{\Omega}_-].$$

For the special nodal configuration represented in Figure 9.1, $\Omega \setminus \Omega_-$ consists of two components, Ω_0 and Ω_+. Hence,

$$\lambda_1[-\Delta, \Omega \setminus \bar{\Omega}_-] := \min\{\lambda_1[-\Delta, \Omega_0], \lambda_1[-\Delta, \Omega_+]\}. \tag{9.5}$$

From the point of view of population dynamics, (9.1) is a generalized diffusive logistic equation that incorporates some facilitation effects between the individuals of the species u, measured by a_+, in the patch Ω_+. It seems rather reasonable that, according to the amount of resources in each patch of Ω, the interaction between the individuals of the species might change from either patch. Naturally, in patches with shortcomings of resources, the individuals will be forced to compete for them, whereas in abundance the individuals will facilitate the others. So, Ω_- models a region with harsh environmental conditions, while Ω_+ models a region where cooperative synergy between the individuals occurs. In Ω_0 the individuals are allowed to grow exponentially. Adopting this perspective, the main goal of this chapter is to ascertain the combined effects of intra-specific competition and facilitation between the individuals of a single species.

9.1   Components of positive steady states

Arguing as in the proofs of Theorems 2.3 and 4.1, $\lambda = d\sigma_0$ is the unique bifurcation value from $(\lambda, u) = (\lambda, 0)$ to positive solutions of (9.3). Actually, by the theorem of M. G. Crandall and P. H. Rabinowitz [59], a curve of positive solutions of (9.3) emanates from $(\lambda, u) = (\lambda, 0)$ at $\lambda = d\sigma_0$. More precisely, if $\varphi \gg 0$ stands for the unique principal eigenfunction associated with $\sigma_0 := \lambda_1[-\Delta, \Omega]$ such that

$$\int_\Omega \varphi^2 = 1,$$

then the following result holds.

Proposition 9.1 *There exist $s_0 > 0$ and two (unique) maps of class C^1*

$$\mu : (-s_0, s_0) \to \mathbb{R}, \qquad v : (-s_0, s_0) \to C^{2+\nu}(\bar{\Omega}),$$

such that

$$\mu(0) = 0, \quad v(0) = 0, \quad \int_\Omega v(s)\varphi \, dx = 0 \quad \text{for all } s \in (-s_0, s_0),$$

and the couple

$$(\lambda(s), u(s)) := (d\sigma_0 + \mu(s), s(\varphi + v(s))) \tag{9.6}$$

is a solution of (9.3) for every $s \in (-s_0, s_0)$. Moreover, for sufficiently small s_0, these are the unique non-trivial solutions of (9.3) in a neighborhood of $(\lambda, u) = (d\sigma_0, 0) \in \mathbb{R} \times C^{2+\nu}(\bar{\Omega})$, and

$$\lim_{s \to 0} \frac{\mu(s)}{|s|^p} = -\int_\Omega a(x)\varphi^{p+2}(x) \, dx. \tag{9.7}$$

Proof: It suffices to show (9.7), as the remaining assertions can be easily derived from the main theorem of [59], as in Part I. Substituting (9.6) into (9.3), dividing by s, using the definition of φ and rearranging terms yields

$$(-d\Delta - d\sigma_0)v(s) = \mu(s)(\varphi + v(s)) + |s|^p a(\varphi + v(s))^{p+1} \quad \text{for all } s \in (-s_0, s_0).$$

Thus, multiplying this identity by φ, integrating in Ω and applying the formula of integration by parts, we find that

$$\mu(s) \int_\Omega \varphi(\varphi + v(s)) \, dx = -|s|^p \int_\Omega a\varphi(\varphi + v(s))^{p+1} \, dx, \qquad s \in (-s_0, s_0).$$

Dividing this relation by $|s|^p$ and letting $s \to 0$ provides (9.7), which ends the proof. $\square$

As $v(0) = 0$ and $\varphi \gg 0$, for sufficiently small $s_0 > 0$ the couple

$$(\lambda(s), u(s)) := (d\sigma_0 + \mu(s), s(\varphi + v(s)))$$

is a positive solution of (9.3) for all $s \in (0, s_0)$. Thus, by setting

$$\mathfrak{D} := -\int_\Omega a(x)\varphi^{p+2}(x) \, dx,$$

the bifurcation to positive solutions is supercritical if $\mathfrak{D} > 0$, i.e., $\lambda(s) > d\sigma_0$ for small $s > 0$, while it is subcritical if $\mathfrak{D} < 0$, i.e., $\lambda(s) < d\sigma_0$ for small $s > 0$.

According to [163, Th. 7.1.4] the component of positive solutions of (9.3), $\mathfrak{C}^+$, bifurcating from $(\lambda, u) = (\lambda, 0)$ at $\lambda = d\sigma_0$ is unbounded in $\mathbb{R} \times C^{2+\nu}(\bar{\Omega})$. A component is a closed and connected subset of the set of positive solutions that it is maximal for the inclusion. The next result establishes that $\mathfrak{C}^+$ bends backward globally if $\mathfrak{D} \le 0$.

Proposition 9.2 *If* (9.3) *admits a positive solution,* (λ, u), *with* $\lambda \geq d\sigma_0$, *then* $\mathfrak{D} > 0$. *Therefore, it cannot admit a positive solution if* $\lambda \geq d\sigma_0$ *and* $\mathfrak{D} \leq 0$. *In particular,*

$$\mathfrak{C}^+ \subset (-\infty, d\sigma_0) \times \mathcal{C}^{2+\nu}(\bar{\Omega}) \quad \text{if} \quad \mathfrak{D} \leq 0.$$

The proof of this proposition is based on the next result, which is a very classical property going back to M. Picone [209].

Lemma 9.3 *Let* $u, v \in \mathcal{C}^2(\Omega)$ *be two arbitrary functions such that*

$$u = v = 0 \quad \text{on} \quad \partial\Omega \quad \text{and} \quad \frac{v}{u} \in \mathcal{C}^1(\Omega) \cap \mathcal{C}(\bar{\Omega}).$$

Then, for every $f \in \mathcal{C}^1[0, \infty)$ *the next identity holds*

$$\int_\Omega f\left(\frac{v}{u}\right)(v\Delta u - u\Delta v) = \int_\Omega f'\left(\frac{v}{u}\right) u^2 \left|\nabla \frac{v}{u}\right|^2. \tag{9.8}$$

Proof: The following chain of identities holds

$$f\left(\frac{v}{u}\right)(v\Delta u - u\Delta v) = f\left(\frac{v}{u}\right)(v \operatorname{div} \nabla u - u \operatorname{div} \nabla v) = f\left(\frac{v}{u}\right) \operatorname{div}(v\nabla u - u\nabla v)$$

$$= \operatorname{div}\left[f\left(\frac{v}{u}\right)(v\nabla u - u\nabla v)\right] - f'\left(\frac{v}{u}\right)\left\langle \nabla \frac{v}{u}, v\nabla u - u\nabla v\right\rangle$$

$$= \operatorname{div}\left[f\left(\frac{v}{u}\right)(v\nabla u - u\nabla v)\right] + f'\left(\frac{v}{u}\right) u^2 \left\langle \nabla \frac{v}{u}, \nabla \frac{v}{u}\right\rangle.$$

Consequently, by integrating in Ω, applying the divergence theorem, and taking into account that $u = v = 0$ on $\partial\Omega$, (9.8) holds. $\quad\square$

Proof of Proposition 9.2: Let (λ, u) be a positive solution of (9.3). Then, since $\varphi \gg 0$ and $u \gg 0$, we have that $\frac{\varphi}{u} \in \mathcal{C}^1(\bar{\Omega})$ and

$$d\left(\frac{\varphi}{u}\right)^{p+1}(\varphi\Delta u - u\Delta\varphi) = -\left(\frac{\varphi}{u}\right)^{p+2} u^2(\lambda - d\sigma_0) - a(x)\varphi^{p+2}. \tag{9.9}$$

On the other hand, by (9.8),

$$\int_\Omega \left(\frac{\varphi}{u}\right)^{p+1}(\varphi\Delta u - u\Delta\varphi) = \int_\Omega (p+1)\left(\frac{\varphi}{u}\right)^p u^2 \left|\nabla \frac{\varphi}{u}\right|^2 > 0,$$

since u cannot be a multiple of φ. Therefore, integrating (9.9) in Ω yields

$$-(\lambda - d\sigma_0)\int_\Omega \left(\frac{\varphi}{u}\right)^{p+2} u^2 > \int_\Omega a(x)\varphi^{p+2}(x)\,dx.$$

Suppose (9.3) admits a positive solution, (λ, u), with $\lambda \geq d\sigma_0$. Then, $\int_\Omega a\varphi^{p+1}\,dx < 0$ and therefore, $\mathfrak{D} > 0$. This ends the proof. $\quad\square$

According to the linearized stability principle, the local stability of $u = 0$ as a solution of (9.1) is given by the sign of the eigenvalues of the problem

$$\begin{cases} (-d\Delta - \lambda)\psi = \tau\psi & \text{in } \Omega, \\ \psi = 0 & \text{on } \partial\Omega. \end{cases} \qquad (9.10)$$

Actually, the sign of the principal eigenvalue of (9.10) provides us with the local qualitative character of $(\lambda, u) = (\lambda, 0)$. Indeed, if $d\sigma_0 - \lambda > 0$, i.e., $\lambda < d\sigma_0$, then the trivial solution is linearly asymptotically stable, while if $d\sigma_0 - \lambda < 0$, i.e., $\lambda > d\sigma_0$, then it is linearly unstable. At $\lambda = d\sigma_0$, the value of λ where the local character of 0 changes, it is said that zero is neutrally stable and is attractive depending on the global nature of the nonlinearity. Thus, by the exchange stability principle of M. G. Crandall and P. H. Rabinowitz [60], the bifurcated solution $(\lambda(s), u(s))$, $s > 0$, is linearly stable for sufficiently small $s > 0$ if $\mathfrak{D} > 0$, while it is linearly unstable with a one-dimensional unstable manifold, if $\mathfrak{D} < 0$, i.e., if it bifurcates to the left of $d\sigma_0$.

The next result provides us with an optimal necessary condition for the existence of a positive solution of (9.3).

Proposition 9.4 *Suppose* (9.3) *possesses a positive solution,* (λ, u). *Then,*

$$\lambda < \lambda_\omega := \lambda_\omega(a) \equiv \lambda_1[-d\Delta, \Omega \setminus \bar{\Omega}_-]. \qquad (9.11)$$

Proof: Arguing as in the proof of Lemma 1.6, it follows that

$$\lambda = \lambda_1[-d\Delta - a(x)u^p(x), \Omega]. \qquad (9.12)$$

Thus, by the monotonicity properties of the principal eigenvalue,

$$\lambda < \lambda_1[-d\Delta - a(x)u^p(x), \Omega \setminus \bar{\Omega}_-] \le \lambda_1[-d\Delta, \Omega \setminus \bar{\Omega}_-],$$

which ends the proof. $\square$

According to (9.11), the larger Ω_- the smaller $\Omega \setminus \bar{\Omega}_-$ and hence, the larger $\lambda_\omega(a)$. Actually, thanks to Faber–Krahn inequality, we already know that

$$\lim_{|\Omega \setminus \bar{\Omega}_-| \downarrow 0} \lambda_\omega(a) = \infty,$$

in complete agreement with Theorem 2.2, because the associated sublinear elliptic problem (9.4) can be regarded as a limiting problem from (9.3) as Ω_- approximates Ω. It should be noted that, in such case, $\Omega_0 = \emptyset$.

Conversely, we have that $\lambda_\omega(a) > d\sigma_0$ and, actually, $\lambda_\omega(a)$ can be taken as close as we wish to $d\sigma_0$ by choosing Ω_- sufficiently small. Therefore, by playing with the weight function $a(x)$, the threshold $\lambda_\omega(a)$ can indeed take any value within the interval $(d\sigma_0, \infty)$.

According to Theorem 4.1 and (9.5), the associated sublinear problem (9.4) admits a positive solution, $\theta_{[\lambda, a_-]}$, if and only if

$$d\sigma_0 < \lambda < \lambda_\omega(a) = \lambda_1[-d\Delta, \Omega \setminus \Omega_-].$$

Thus, since $\varepsilon\varphi$ provides us with a positive subsolution of (9.4) for sufficiently small ε and any $\lambda > d\sigma_0$, and the positive solutions of (9.3) are supersolutions of (9.4), it becomes apparent that (9.3) cannot admit a positive solution for $\lambda \geq \lambda_w(a)$, as has been already established by Proposition 9.4 with another argument. The next theorem establishes that if the superlinear indefinite problem (9.3) is a perturbation of the associated sublinear model (9.4), in the sense that

$$(a_+)_M \equiv \max_{\bar{\Omega}} a_+ = \max_{\bar{\Omega}_+} a_+$$

is sufficiently small, then (9.3) admits a positive solution for λ arbitrarily close to $\lambda_w(a)$. From this perspective, the universal upper bound $\lambda_w(a)$ for the set of values of λ for which (9.3) has a positive solution, J_λ, is optimal.

Theorem 9.5 *For every $\varepsilon > 0$ there exists $\delta > 0$ such that (9.3) possesses a positive solution for each $\lambda \in (d\sigma_0, \lambda_w(a) - \varepsilon]$ provided $(a_+)_M < \delta$.*

Proof: Consider the operator $\mathfrak{G} : \mathbb{R} \times C_0(\bar{\Omega}) \times \mathbb{R} \to C_0(\bar{\Omega})$ defined by

$$\mathfrak{G}(\lambda, u, \delta) = u - (-d\Delta)^{-1} \left[\lambda u + (a_- + \delta a_+) |u|^p u\right],$$

which is of class $\mathcal{C}^2$, since $p \geq 1$. Note that the positive zeroes of

$$\mathfrak{F}(\lambda, u) := \mathfrak{G}(\lambda, u, 0)$$

are the positive solutions of the associated sublinear problem (9.4). By Theorem 4.1, (9.4) possesses a positive solution, $\theta_{[\lambda, a_-]}$, if, and only if, $d\sigma_0 < \lambda < \lambda_w(a)$. Moreover, going back to the proof of Theorem 4.1, the positive solution $(\lambda, \theta_{[\lambda, a_-]})$ is non-degenerate in the sense that

$$D_u\mathfrak{F}(\lambda, \theta_{[\lambda, a_-]}) : C_0(\bar{\Omega}) \longrightarrow C_0(\bar{\Omega})$$

is a linear topological isomorphism. Given $\varepsilon > 0$, set

$$\lambda_0 := \lambda_w(a) - \varepsilon.$$

Then, applying the implicit function theorem to the operator $\mathfrak{G}$ at

$$(\lambda, u, \delta) = (\lambda_0, \theta_{[\lambda_0, a_-]}, 0)$$

shows that there exist $\delta_0 > 0$ and a (unique) function of class $\mathcal{C}^2$,

$$u : (-\delta_0, \delta_0) \to C_0(\bar{\Omega})$$

such that

$$u(0) = \theta_{[\lambda_0, a_-]} \quad \text{and} \quad \mathfrak{G}(\lambda_0, u(\delta), \delta) = 0 \quad \text{for all} \quad \delta \in (-\delta_0, \delta_0).$$

By elliptic regularity, $(\lambda_0, u(\delta))$ is a positive solution of (9.3) if $\delta < \delta_0$. Finally, let $\lambda \in (d\sigma_0, \lambda_0)$ be. It is very easy to see that $(\lambda, \eta\varphi)$ provides us with a subsolution of (9.3) for sufficiently small $\eta > 0$. Moreover, since $\lambda < \lambda_0$, $(\lambda_0, u(\delta))$ is a supersolution of (9.3). As $\eta\varphi < u(\delta)$ for sufficiently small η, it follows from Theorem 1.4 that (9.3) admits a positive solution for each $\lambda \in (d\sigma_0, \lambda_0]$. Note that Theorem 1.4 is valid for general $a \in C^\nu(\bar{\Omega})$. This ends the proof. □

But, nevertheless, although the previous results suggest that (9.3) and (9.4) can share very close solutions for large intervals of λ, there is nothing to do between the global behavior of the components $\mathfrak{C}^+$ for either model. Even for arbitrarily small a_+, the components behave in a rather different way. Actually, though $\mathfrak{C}^+$ is the unique component of positive solutions of the sublinear problem (9.4), the superlinear indefinite problem (9.3) might exhibit an arbitrarily large number of components. As a consequence, the global dynamics of (9.1) and (9.2) are qualitatively different, even for an arbitrarily small a_+ supported in an arbitrarily small compact subset of Ω. Indeed, suppose $\Omega = (0, 1)$, $d = 1$, $p = 4$ and

$$
a(x) = \begin{cases} \frac{1}{2}\sin(3\pi x) & \text{if } x \in \left[0, \frac{1}{3}\right] \cup \left[\frac{2}{3}, 1\right], \\[2mm] \sin(3\pi x), & \text{if } x \in \left(\frac{1}{3}, \frac{2}{3}\right). \end{cases} \tag{9.13}
$$

In this example,

$$
\Omega_0 = \emptyset, \qquad \Omega_- = \left(\frac{1}{3}, \frac{2}{3}\right), \qquad \Omega_+ = \left(0, \frac{2}{3}\right) \cup \left(\frac{2}{3}, 1\right),
$$

and

$$
\sigma_0 = \lambda_1[-\Delta, \Omega] = \pi^2, \qquad \lambda_\omega = \lambda_1[-\Delta, \Omega \setminus \bar{\Omega}_-] = (3\pi)^2.
$$

Moreover, $\varphi(x) = \sqrt{2}\sin(\pi x)$. Hence,

$$
\mathfrak{D} := -\int_0^1 a(x)\sin^6(\pi x)\,dx \simeq 1.3816 > 0.
$$

Consequently, the component $\mathfrak{C}^+$ bifurcates from $u = 0$ at $\lambda = \pi^2$ towards the right of π^2. The pictures on the first row of Figure 9.2 show $\mathfrak{C}^+$ at two different scales. They plot the value of the parameter λ versus the L^2-norm of the non-negative solutions. The remaining four plots show the profiles of a series of positive solutions along $\mathfrak{C}^+$. They were computed by coupling a pure spectral method with collocation by R. Gómez-Reñasco and appeared in J. López-Gómez [149, Fig. 7.3]. In all those bifurcation diagrams, continuous lines are filled in by stable solutions and dashed lines by unstable ones. The left plot on the first row shows the bifurcation diagram for $0 \leq \lambda \leq 20$, while the second one shows it for $-10^3 \leq \lambda \leq 10^2$. Naturally, the details shown in the first plot are lost in the second one as an effect of the different scales.

The component $\mathfrak{C}^+$ emanates supercritically from $u = 0$ at $\lambda = \pi^2$ and has a subcritical turning point at $\lambda \sim 17.1615$, where it turns backward. The solutions on $\mathfrak{C}^+$ are linearly asymptotically stable and increase with λ until λ reaches the turning point. Then, they become unstable for all further val-

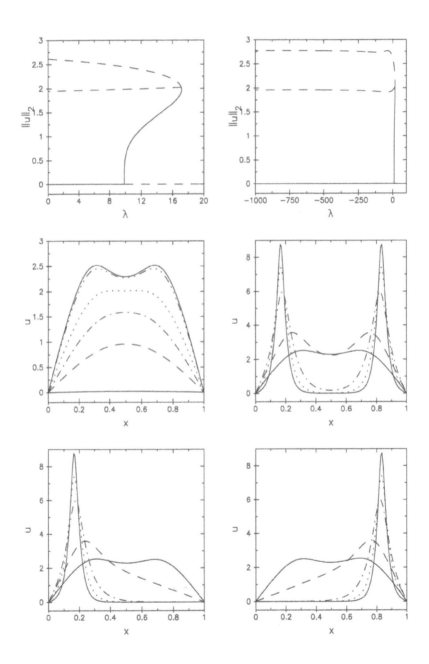

FIGURE 9.2: Bifurcation diagram and solution plots for the choice (9.13).

ues of the parameter. The unstable manifolds of the positive solutions along the upper half-branch are one-dimensional from the turning point until another critical value, $\lambda \simeq 17.1142$, where they become two-dimensional. So, a secondary bifurcation occurs at $\lambda \simeq 17.1142$. The secondary bifurcation is subcritical and each of the solutions on the secondary branches possesses a one-dimensional unstable manifold, though the bifurcation diagrams on the first row of Figure 9.2 can only show one of these secondary branches based on the symmetry of the problem. Indeed, as $a(x)$ is symmetric about 0.5, the solutions on the secondary branches are reflections about 0.5 of the solutions on the other. So, they have identical L^2-norms and cannot be differentiated in those plots.

The first figure of the second row of Figure 9.2 shows the plots of the solutions on the lower half-branch of $\mathfrak{C}^+$ for the values $\lambda = 9.8697$, 10.1620, 12.0291, 15.2227, 17.1614 and 17.1137, whereas the second one shows the plots of the solutions on the upper half of the primary branch for $\lambda = 17.1137$, 0.0, -200.0, -500.0 and -1000.0. These solutions exhibit a two-peak layer behavior as $\lambda \downarrow -\infty$ with the two peaks approximating the two local maxima of $a(x)$. Outside these local maxima, the solutions approximate zero as $\lambda \downarrow -\infty$.

The third row of Figure 9.2 shows a series of plots of the solutions on each of the two secondary branches, for $\lambda = 17.1101$, 0.0, -200.0, -500.0 and -1000.0. Each of the plots in the left picture is a reflection about 0.5 of the corresponding one on the right. These solutions exhibit a single peak layer behavior as $\lambda \downarrow -\infty$. According to the secondary branch where the solution lies, the peak is localized around one maxima or the other.

Now, instead of (9.13), we will make the choice

$$
a(x) = \begin{cases} \frac{1}{2}\sin(5\pi x), & x \in \left[0, \frac{1}{5}\right] \cup \left[\frac{2}{5}, \frac{3}{5}\right] \cup \left[\frac{4}{5}, 1\right], \\ \sin(5\pi x), & x \in \left(\frac{1}{5}, \frac{2}{5}\right) \cup \left(\frac{3}{5}, \frac{4}{5}\right). \end{cases} \tag{9.14}
$$

In this case,

$$
\Omega_0 = \emptyset, \quad \Omega_- = \left(\frac{1}{5}, \frac{2}{5}\right) \cup \left(\frac{3}{5}, \frac{4}{5}\right), \quad \Omega_+ = \left(0, \frac{1}{5}\right) \cup \left(\frac{2}{5}, \frac{3}{5}\right) \cup \left(\frac{4}{5}, 1\right).
$$

Moreover, $\varphi(x) = \sqrt{2}\sin(\pi x)$, $\sigma_0 = \pi^2$ and

$$
\lambda_\omega = \lambda_1[-\Delta, \Omega \setminus \bar{\Omega}_-] = (5\pi)^2, \quad \mathfrak{D} = -\int_0^1 a(x)\sin^6(\pi x)\,dx \simeq 0.6136 > 0.
$$

Thus, also in this case $\mathfrak{C}^+$ bifurcates supercritically from $u = 0$.

Figure 9.3 shows the bifurcation diagram computed for this special choice. The first row plots it for $\lambda \in (-20, 15)$, while the second row shows it for $\lambda \in (-200, 50)$. A dramatic change with respect to the previous case occurs, as for this choice of $a(x)$ we have computed four components of positive solutions, namely, the component $\mathfrak{C}^+$ that bifurcates from $u = 0$ at $\lambda = \pi^2$, plus three additional *global subcritical folding* type components, $\mathfrak{F}_1$, $\mathfrak{F}_2$ and $\mathfrak{F}_3$. As the solutions of $\mathfrak{F}_2$ are reflections about 0.5 of the solutions of $\mathfrak{F}_3$, they share the L^2-norm and hence, cannot be differentiated in the bifurcation diagrams. This explains why we decided to plot the component $\mathfrak{F}_2$ in the left diagrams of Figure 9.3, together with $\mathfrak{F}_1$, while $\mathfrak{F}_3$ has been plotted on the right.

Subsequently, we explain briefly the scheme adopted to compute these components. The component $\mathfrak{C}^+$ was constructed by bifurcation from $u = 0$

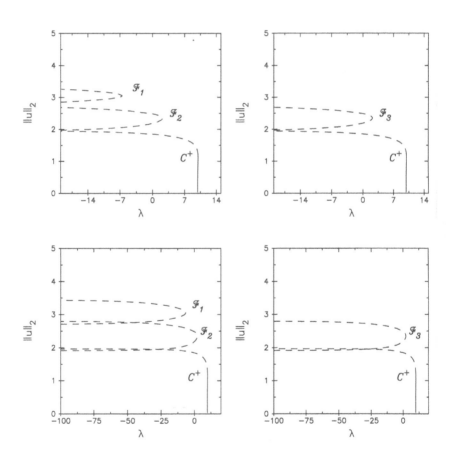

FIGURE 9.3: Bifurcation diagram for the choice (9.14).

at $\lambda = \pi^2$ and global continuation. It exhibits a subcritical turning point at $\lambda_t \simeq 9.9719$ where it turns backward. As for the choice (9.13), the solutions on $\mathfrak{C}^+$ are linearly asymptotically stable and increase with λ until λ reaches λ_t where they become unstable, with a one-dimensional unstable manifold for the entire interval where we computed them. Figure 9.4 plots a series of solution plots along $\mathfrak{C}^+$ for $\lambda = 9.8697$ and 9.9544 on the lower half-branch and $\lambda = 0.5414 \times 10^{-2}$, -100.0025 and -200.0033 on the upper one. It should be noted that the positive solution $\lambda = 9.8697$ is very small, because λ is very close to π^2. All these solutions have a single peak around the central maximum of $a(x)$, 0.5; the smaller λ, the more emphasized the peak.

Our numerical experiments for the choice (9.14) suggest that (4.3) should have, at least,

$$\sum_{j=1}^{3} \binom{3}{j} = 2^3 - 1 = 7$$

positive solutions for sufficiently small $\lambda < 0$: three among them with a single peak around each of the three local maxima of $a(x)$, another three with two peaks and one further solution with three peaks. The solutions on $\mathfrak{C}^+$ provide us with the solutions having a single peak around the central maximum. To compute the remaining solutions we adopted the following methodology. First,

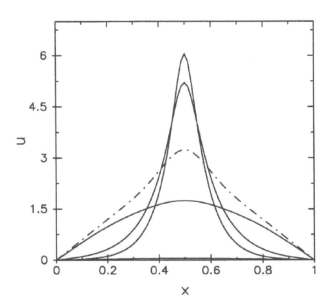

FIGURE 9.4: Plots of a series of solutions on $\mathfrak{C}^+$ for the choice (9.14).

we solved (9.3) with $\Omega = (0,1)$ and $p = 4$ for the special choices

$$
a_\varepsilon(x) := \begin{cases} \frac{1}{2}\sin(5\pi x), & x \in \left[0, \frac{1}{5}\right] \cup \left[\frac{4}{5}, 1\right], \\ \epsilon \sin(5\pi x), & x \in \left[\frac{2}{5}, \frac{3}{5}\right], \\ \sin(5\pi x), & x \in \left(\frac{1}{5}, \frac{2}{5}\right) \cup \left(\frac{3}{5}, \frac{4}{5}\right), \end{cases} \tag{9.15}
$$

where $\epsilon \in [0, 0.5]$ is regarded as a secondary real parameter. When $\epsilon = 0.5$, (9.15) equals (9.14), while for $\epsilon = 0$ the nodal behavior of $a_0(x)$ is reminiscent of the nodal behavior of $a(x)$ for the choice (9.13). Naturally, it is expected the bifurcation diagram for the choice $a_0(x)$ will be similar to the one plotted in Figure 9.2. Indeed, the component $\mathfrak{C}^+$ for the choice $a = a_0$, like the one shown in Figure 9.2, bifurcates supercritically from $u = 0$ at $\lambda = \pi^2$ and exhibits a subcritical turning point at $\lambda_t \simeq 45.4077$ where it turns backward. Then, along the upper half-branch, it exhibits a pitchfork subcritical bifurcation at $\lambda_b \sim 45.4077$ where the unstable manifolds of the solutions become two-dimensional. Thus, for every $\lambda \in (-\infty, \lambda_b)$ the problem (4.3) possesses at least three positive solutions: two of them with a single peak around each of the local maxima 0.1 and 0.9 and the third one with two peaks at these these maxima. The second step consisted in fixing a value of λ far away from the bifurcation point λ_b, e.g., $\lambda = -110.0575$, and using ϵ instead of λ as the main continuation parameter to compute the solution from $\varepsilon = 0$ up to $\epsilon = 0.5$. We began computing the perturbation of the solution with two peaks. After we computed it, we switched off the path-following parameter from ε to λ to compute the entire component $\mathfrak{F}_1$, which exhibits a subcritical turning point at $\lambda_{t,1} = -6.5421$.

Each of the solutions on the upper half-branch of $\mathfrak{F}_1$ has a three-dimensional unstable manifold and three peaks, each around 0.1, 0.5 and 0.9, while the solutions along its lower half-branch have bi-dimensional unstable manifolds and two peaks around 0.1 and 0.9. Figure 9.6(a) shows a series of solution plots along the upper half-branch of $\mathfrak{F}_1$, for $\lambda = -6.5461$, -99.9985 and -200.0005, as well as the plots of a series of solutions along its lower half-branch, for $\lambda = -6.5425$, -100.0580 and -200.0555.

To compute the component $\mathfrak{F}_2$ we proceeded as in the previous case, but now perturbing the solution of (9.3) with a single peak around 0.9 from $\varepsilon = 0$ up to $\varepsilon = 0.5$ for $\lambda = -110.0006$. Then, we used λ instead of ε as the main continuation parameter to construct $\mathfrak{F}_2$, which exhibits its turning point at $\lambda_{t,2} = 2.1657$. The solutions on the upper half-branch of this component have two-dimensional unstable manifolds and exhibit two peaks around 0.5 and 0.9, while the solutions on its lower half-branch have one-dimensional unstable manifolds and one single peak around 0.9. Figure 9.6(b) shows the plots of some of the solutions along $\mathfrak{F}_2$ for the values $\lambda = 2.1656$, 0.7686×10^{-1}, -100.0068 and -200.0062, on its upper half-branch, and for $\lambda = 2.1656$, -0.2747×10^{-1}, -100.0007 and -200.0027 on the lower one.

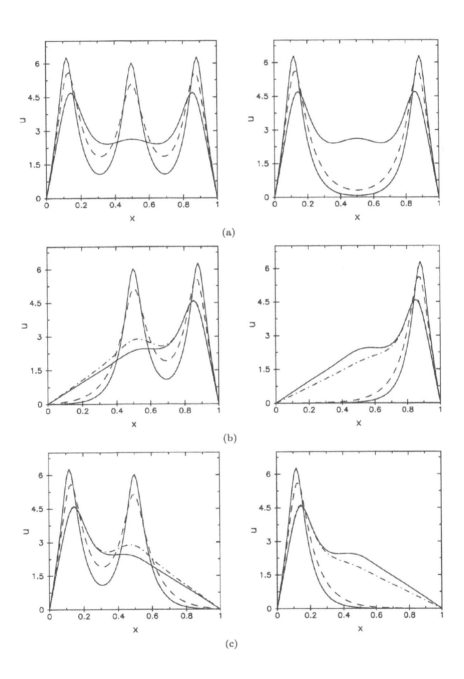

FIGURE 9.5: Plots of a series of positive solutions of (9.3) for the choice (9.14) on the components $\mathfrak{F}_1$, (a), $\mathfrak{F}_2$, (b), and $\mathfrak{F}_3$, (c).

Finally, to construct the fourth component, $\mathfrak{F}_3$, we adopted the same strategy as before, but on this occasion starting at $\lambda = -109.9927$ from the positive solution with a single peak around 0.1. The solutions on its upper half-branch have two-dimensional unstable manifolds and exhibit two peaks around the local minima 0.1 and 0.5, while the solutions on its lower half-branch have one-dimensional unstable manifolds and possess a single peak around 0.1, the turning point of this component placed at the same value of λ, $\lambda_{t,2} = 2.1657$ as the turning point of $\mathfrak{F}_2$. Figure 9.6(c) shows the plots of a series of solutions along $\mathfrak{F}_3$.

All those numerical experiments strongly suggest that we can have an arbitrarily large number of solutions by simply choosing $a(x)$ sufficiently wavy. As a matter of fact, varying appropriately $a(x)$ in (9.3) might provide us with very intricate bifurcation diagrams even in one spatial dimension.

Our numerical experiments support the idea that varying coefficients in a one-dimensional model has a similar effect as varying support domains in higher dimensional problems. In the context of higher dimensional superlinear problems, it is well documented that breaking down the convexity of the domain can provoke multiplicity of positive solutions even when $a(x)$ is a positive constant. More precisely, joining two balls with the same radius by a thin narrow strip provokes a similar effect as varying the coefficient $a(x)$ in model (9.3), E. N. Dancer [64]. Naturally, we expect that varying the coefficients in higher dimensional problems will increase the complexity of the bifurcation diagrams as much as we wish, since in these problems one can play not only with the shape of the support domain but also with the nodal behavior of the weight function $a(x)$.

9.2   Local structure at stable positive solutions

Given a positive solution (λ_0, u_0) of (9.3), its stability as a steady state of (9.1) is determined by the spectrum of the linearized operator (2.16), which in our present situation becomes

$$\mathfrak{L}(\lambda_0, u_0) := -d\Delta - \lambda_0 - (p+1)a(x)u_0^p$$

subject to homogeneous Dirichlet boundary conditions on $\partial\Omega$. Indeed, when (λ_0, u_0) is hyperbolic, the dimension of its unstable manifold equals the sum of the algebraic multiplicities of the negative eigenvalues of $\mathfrak{L}(\lambda_0, u_0)$. As $\mathfrak{L}(\lambda_0, u_0)$ is self-adjoint, their eigenvalues are real and semi-simple. So, their algebraic and geometric multiplicities coincide, equaling 1 if $N = 1$.

More generally, for any $(\lambda, u) \in \mathbb{R} \times C^2(\Omega)$ with $u \geq 0$ in Ω, we will denote

$$\mathfrak{L}(\lambda, u) := -d\Delta - \lambda - (p+1)a(x)u^p.$$

Subsequently, (λ_0, u_0) is said to be

- *Linearly asymptotically stable if* $\lambda_1[\mathcal{L}(\lambda_0, u_0), \Omega] > 0$

- *Linearly unstable if* $\lambda_1[\mathcal{L}(\lambda_0, u_0), \Omega] < 0$

- *Neutrally stable if* $\lambda_1[\mathcal{L}(\lambda_0, u_0), \Omega] = 0$

By the linearized stability principle, (λ_0, u_0) is exponentially asymptotically stable if it is linearly asymptotically stable and unstable if it is linearly unstable.

The next two results provide us with the structure of the solution set of (9.3) around any linearly stable positive solution.

Proposition 9.6 *Let* (λ_0, u_0) *be a linearly asymptotically stable positive solution of* (9.3). *Then, there exist* $\varepsilon > 0$ *and a map of class* $\mathcal{C}^2$

$$u : (\lambda_0 - \varepsilon, \lambda_0 + \varepsilon) \to \mathcal{C}_0(\bar{\Omega})$$

with $u(\lambda_0) = u_0$ *such that:*

- $(\lambda, u(\lambda))$ *is a positive solution of* (9.3) *for all* $\lambda \in (\lambda_0 - \varepsilon, \lambda_0 + \varepsilon)$.

- *The map*
$$
\begin{array}{ccc}
(\lambda_0 - \epsilon, \lambda_0 + \epsilon) & \longrightarrow & \mathcal{C}_0(\bar{\Omega}) \\
\lambda & \mapsto & u(\lambda)
\end{array}
$$

  is increasing.

- *There exists a neighborhood* $\mathcal{N}_0$ *of* (λ_0, u_0) *in* $\mathbb{R} \times \mathcal{C}_0(\bar{\Omega})$ *such that* $(\lambda, u) = (\lambda, u(\lambda))$ *for some* $\lambda \in (\lambda_0 - \epsilon, \lambda_0 + \epsilon)$ *whenever* $(\lambda, u) \in \mathcal{N}_0$ *is a positive solution of* (9.3).

Proof: As in the proof of Theorem 2.3, the solutions of (9.3) are the zeros of the operator $\mathfrak{F} : \mathbb{R} \times \mathcal{C}_0(\bar{\Omega}) \to \mathcal{C}_0(\bar{\Omega})$ defined by

$$\mathfrak{F}(\lambda, u) = u - (-d\Delta)^{-1} \left(\lambda u + a|u|^p u \right). \tag{9.16}$$

Since $p \geq 1$, $\mathfrak{F}$ is of class $\mathcal{C}^2$. We are assuming that $\mathfrak{F}(\lambda_0, u_0) = 0$ and, since $\lambda_1[\mathcal{L}(\lambda_0, u_0), \Omega] > 0$, the linearized operator $D_u\mathfrak{F}(\lambda_0, u_0)$ is an isomorphism. Thus, the local existence and the uniqueness of a solution curve through (λ_0, u_0) are guaranteed by the implicit function theorem. Moreover, by implicit differentiation, it follows that

$$\mathcal{L}(\lambda, u(\lambda))u'(\lambda) = u(\lambda) > 0.$$

Hence, owing to Theorem 1.1, $u'(\lambda) \gg 0$. It should be pointed out that the linear asymptotic stability of (λ_0, u_0) also entails

$$\lambda_1[\mathcal{L}(\lambda, u(\lambda)), \Omega] > 0$$

for λ sufficiently close to λ_0, because the principal eigenvalue varies continuously with the potential. The proof is complete. $\quad\square$

Proposition 9.7 *Let (λ_0, u_0) be a neutrally stable positive solution of (9.3) and let $\psi_0 > 0$ be a principal eigenfunction associated with*

$$\lambda_1[\mathcal{L}(\lambda_0, u_0), \Omega] = 0.$$

Then, there exist $\epsilon > 0$ and a map of class C^2

$$(\lambda, u) : (-\varepsilon, \varepsilon) \to \mathbb{R} \times C_0(\bar{\Omega})$$

such that

- *$(\lambda(0), u(0)) = (\lambda_0, u_0)$ and $(\lambda(s), u(s))$ is a positive solution of (9.3) for all $s \in (-\epsilon, \epsilon)$. Moreover,*

$$v(s) := u(s) - u_0 - s\psi_0 = O(s^2), \qquad \lambda(s) = \lambda_0 + s^2\lambda_2 + o(s^2),$$

as $s \to 0$, with

$$\int_\Omega v(s)\psi_0 \, dx = 0 \quad \text{for all} \quad s \in (-\varepsilon, \varepsilon)$$

and

$$\lambda_2 = -\frac{p(p+1)\int_\Omega au_0^{p-1}\psi_0^3}{2 \int_\Omega u_0\psi_0} < 0. \tag{9.17}$$

- *There exists a neighborhood, $\mathcal{N}_0$, of (λ_0, u_0) in $\mathbb{R} \times C_0(\bar{\Omega})$ such that $(\lambda, u) = (\lambda(s), u(s))$ for some $s \in (-\epsilon, \epsilon)$ if $(\lambda, u) \in \mathcal{N}_0$ solves (9.3).*

- *For every $s \in (-\varepsilon_0, \varepsilon_0)$,*

$$\operatorname{sign} \lambda'(s) = \operatorname{sign} \lambda_1[\mathcal{L}(\lambda(s), u(s)), \Omega]. \tag{9.18}$$

Summarizing, the solution set of (9.3) in a neighborhood of (λ_0, u_0) has the structure of a quadratic subcritical turning point. Moreover, the solutions on the upper half branch of this turning point are linearly unstable, while those on the lower one are linearly asymptotically stable.

Proof: We already know that the solutions of (9.3) are the zeros of the nonlinear operator $\mathfrak{F} : \mathbb{R} \times C_0(\bar{\Omega}) \to C_0(\bar{\Omega})$ defined in (9.16). Since

$$D_u\mathfrak{F}(\lambda_0, u_0)u = u - (-d\Delta)^{-1}\left[\lambda_0 u + (p+1)a(x)u_0^p u\right],$$

$$\mathcal{L}(\lambda_0, u_0)\psi_0 := (-d\Delta - \lambda_0 - (p+1)a(x)u_0^p)\psi_0 = 0,$$

and zero is a simple eigenvalue of $\mathcal{L}(\lambda_0, u_0)$, it becomes apparent that

$$N[D_u\mathfrak{F}(\lambda_0, u_0)] = \operatorname{span}[\psi_0].$$

We claim that

$$D_\lambda\mathfrak{F}(\lambda_0, u_0) = -(-d\Delta)^{-1}u_0 \notin R[D_u\mathfrak{F}(\lambda_0, u_0)]. \tag{9.19}$$

Indeed, if there is $u \in \mathcal{C}_0(\bar{\Omega})$ such that

$$D_u \mathfrak{F}(\lambda_0, u_0)u = u - (-d\Delta)^{-1}[\lambda_0 u + (p+1)a(x)u_0^p u] = -(-d\Delta)^{-1}u_0,$$

by elliptic regularity we find that $u \in \mathcal{C}^{2+\nu}(\bar{\Omega})$ and

$$-d\Delta u = \lambda_0 u + (p+1)a(x)u_0^p u - u_0.$$

Thus, multiplying this equation by ψ_0 and integrating in Ω yields

$$\int_\Omega \psi_0 u_0 \, dx = 0,$$

which is impossible. Therefore, (9.19) holds and hence,

$$N[D_{(u,\lambda)}\mathfrak{F}(\lambda_0, u_0)] = \operatorname{span}[(\psi_0, 0)]. \tag{9.20}$$

Let X denote the L^2-orthogonal of ψ_0 in $\mathcal{C}_0(\bar{\Omega})$ and set $Z := X \times \mathbb{R}$. According to (9.20),

$$\mathcal{C}_0(\bar{\Omega}) \times \mathbb{R} = N[D_{(u,\lambda)}\mathfrak{F}(\lambda_0, u_0)] \oplus Z.$$

So, each element $(u, \lambda) \in \mathcal{C}_0(\bar{\Omega}) \times \mathbb{R}$ possesses a unique decomposition as

$$(u, \lambda) = s(\psi_0, 0) + z$$

for some $s \in \mathbb{R}$ and $z \in Z$. Indeed, one can make the choice

$$z = (u - s\psi_0, \lambda), \qquad s := \frac{\int_\Omega u_0 \psi_0}{\int_\Omega \psi_0^2}.$$

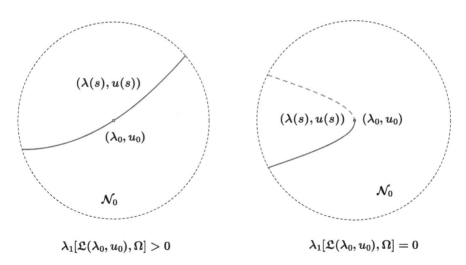

$$\lambda_1[\mathcal{L}(\lambda_0, u_0), \Omega] > 0 \qquad\qquad \lambda_1[\mathcal{L}(\lambda_0, u_0), \Omega] = 0$$

FIGURE 9.6: Local structure of the set of positive solutions of (9.3) around a linearly stable solution and a neutrally stable solution.

Consequently, $\mathfrak{F}(u, \lambda) = 0$ can be expressed, equivalently, as $\mathcal{H}(z, s) = 0$, where

$$\mathcal{H}(z, s) := \mathfrak{F}((u_0, \lambda_0) + s(\psi_0, 0) + z).$$

In particular, the solutions $(u, \lambda) \in \mathfrak{F}^{-1}(0)$ in a neighborhood of (u_0, λ_0) are in correspondence with the solutions $(z, s) \in \mathcal{H}^{-1}(0)$ in a neighborhood of $(0, 0)$. By construction,

$$\mathcal{H}(0, 0) = \mathfrak{F}(u_0, \lambda_0) = 0.$$

Moreover, the operator

$$D_z \mathcal{H}(0, 0) = D_{(u, \lambda)} \mathfrak{F}(u_0, \lambda_0)|_Z$$

is a topological isomorphism, because, since it is a compact perturbation of the identity map, it is Fredholm of index zero. Thus, by the implicit function theorem, there exist $\varepsilon > 0$ and a map of class $\mathcal{C}^2$, $z : (-\varepsilon, \varepsilon) \to Z$, such that $z(0) = 0$ and

$$\mathcal{H}(z(s), s) = 0 \quad \text{for all } s \in (-\varepsilon, \varepsilon).$$

Moreover, there exists a neighborhood, $\mathcal{N}$, of $(0, 0)$ in $Z \times \mathbb{R}$ such that $z = z(s)$ if $(z, s) \in \mathcal{N}$ with $\mathcal{H}(z, s) = 0$. Consequently, in a neighborhood of (u_0, λ_0) in $\mathcal{C}_0(\bar{\Omega}) \times \mathbb{R}$, $\mathcal{N}_0$, the solution set $\mathfrak{F}^{-1}(0)$ consists of a $\mathcal{C}^2$-curve, $(u(s), \lambda(s))$, such that

$$v(s) := u(s) - u_0 - s\psi_0 = O(s^2),$$
$$\lambda(s) = \lambda_0 + s\lambda'(0) + s^2\lambda_2 + o(s^2),$$

for all $s \in (-\varepsilon, \varepsilon)$. As for every $s \in (-\epsilon, \epsilon)$ we have that

$$-d\Delta u(s) = \lambda(s)u(s) + a(x)u_0^{p+1}\left(1 + s\frac{\psi_0}{u_0} + \frac{v(s)}{u_0}\right)^{p+1}; \qquad (9.21)$$

differentiating with respect to s at $s = 0$ and rearranging terms yields

$$0 = \mathfrak{L}(\lambda_0, u_0)\psi_0 = \lambda'(0)u_0$$

and hence, $\lambda'(0) = 0$. Now, differentiating (9.21) twice with respect to s, particularizing at $s = 0$ and rearranging terms yields

$$\mathfrak{L}(\lambda_0, u_0)v''(0) = \lambda''(0)u_0 + a(x)p(p+1)u_0^{p-1}\psi_0^2. \qquad (9.22)$$

Note that, for every $p > 0$, the function

$$u_0^{p-1}\psi_0^2 = u_0^p\psi_0\frac{\psi_0}{u_0}$$

is well defined, since $\psi_0 \gg 0$ and $u_0 \gg 0$. Multiplying (9.22) by ψ_0, integrating in Ω and applying the formula of integration by parts we find that

$$\lambda''(0) = -p(p+1)\frac{\int_\Omega au_0^{p-1}\psi_0^3}{\int_\Omega u_0\psi_0}.$$

Thus, to complete the proof of (9.17) it suffices to show that

$$\int_{\Omega} a u_0^{p-1} \psi_0^3 > 0. \tag{9.23}$$

Thanks to Lemma 9.3,

$$d \int_{\Omega} \left(\frac{\psi_0}{u_0} \right)^2 (\psi_0 \Delta u_0 - u_0 \Delta \psi_0) = 2 \int_{\Omega} \psi_0 u_0 \left| \nabla \frac{\psi_0}{u_0} \right|^2 > 0, \tag{9.24}$$

because ψ_0 cannot be a multiple of u_0. On the other hand,

$$
\begin{aligned}
d(\psi_0 \Delta u_0 - u_0 \Delta \psi_0) &= -\psi_0(-d\Delta u_0) + u_0(-d\Delta \psi_0) \\
&= -\psi_0(\lambda_0 u_0 + a u_0^{p+1}) + u_0[\lambda_0 + (p+1)a u_0^p]\psi_0 \\
&= p a u_0^{p+1} \psi_0.
\end{aligned}
$$

Therefore, (9.23) can be easily inferred from (9.24).

To establish (9.18) we argue as follows. Since

$$u(s) = (-d\Delta)^{-1} \left(\lambda(s)u(s) + a u^{p+1}(s) \right) \quad \text{for all } s \in (-\varepsilon, \varepsilon),$$

differentiating with respect to s yields

$$u'(s) = (-d\Delta)^{-1} \left(\lambda'(s)u(s) + \lambda(s)u'(s) + a(p+1)u^p(s)u'(s) \right)$$

for all $s \in (-\varepsilon, \varepsilon)$, where $' = \frac{d}{ds}$. Thus, by elliptic regularity, inverting $-d\Delta$ and rearranging terms, shows that

$$\mathfrak{L}(\lambda(s), u(s))u'(s) = \lambda'(s)u(s) \quad \text{for all } s \in (-\varepsilon, \varepsilon). \tag{9.25}$$

Since

$$u'(s) = \psi_0 + O(s) \gg 0 \quad \text{and} \quad u(s) = u_0 + O(s) \gg 0$$

for sufficiently small s, it follows from Theorem 1.1 that $\lambda'(s) > 0$ if, and only if, $\lambda_1[\mathfrak{L}(\lambda(s), u(s)), \Omega] > 0$. Indeed, by (9.25), $\lambda'(s) > 0$ implies

$$\mathfrak{L}(\lambda(s), u(s))u'(s) > 0$$

and hence, $u'(s) \gg 0$ provides with a strict positive supersolution of $\mathfrak{L}(\lambda(s), u(s))$ in Ω. Thus,

$$\lambda_1[\mathfrak{L}(\lambda(s), u(s)), \Omega] > 0.$$

Conversely, if $\lambda_1[\mathfrak{L}(\lambda(s), u(s)), \Omega] > 0$, then the resolvent of $\mathfrak{L}(\lambda(s), u(s))$ in Ω is strongly positive and hence, (9.25) implies that $\lambda'(s) > 0$. Similarly, if $\lambda'(s) = 0$ then,

$$\lambda_1[\mathfrak{L}(\lambda(s), u(s)), \Omega] = 0$$

and, conversely, if $\lambda_1[\mathcal{L}(\lambda(s), u(s)), \Omega] = 0$ and we denote by $\psi_s \gg 0$ any principal eigenfunction associated with $\lambda_1[\mathcal{L}(\lambda(s), u(s)), \Omega] = 0$, then multiplying (9.25) by ψ_s and integrating in Ω yields

$$\lambda'(s) \int_\Omega \psi_s u(s) \, dx = 0,$$

which implies $\lambda'(s) = 0$ and ends the proof of (9.18).

Finally, since $\lambda'(s) = 2s\lambda_2 + o(s)$, we find from (9.17) that $\lambda'(s) > 0$ if $s < 0$, while $\lambda'(s) < 0$ if $s > 0$. Therefore,

$$\lambda_1[\mathcal{L}(\lambda(s), u(s)), \Omega] \begin{cases} > 0 & \text{if} \quad s < 0, \\ < 0 & \text{if} \quad s > 0, \end{cases}$$

which ends the proof. □

Corollary 9.8 *Let (λ_0, u_0) be a neutrally stable positive solution of (9.3). Then, there exists $\epsilon > 0$ such that (9.3) possesses two positive solutions for each $\lambda \in (\lambda_0 - \epsilon, \lambda_0)$. Moreover, one of them is linearly asymptotically stable and some other linearly unstable. Furthermore, (9.3) cannot admit a positive solution if $\lambda > \lambda_0$ and*

$$|\lambda - \lambda_0| + \|u - u_0\|_{C(\bar{\Omega})} < \epsilon.$$

Figure 9.6 summarizes at a single glance the main findings of Propositions 9.6, 9.7 and Corollary 9.8.

9.3   Existence of stable positive solutions

The following result establishes that $u = 0$ is the unique linearly stable non-negative solution of (9.3) for all $\lambda \leq d\sigma_0$.

Theorem 9.9 *Let (λ_0, u_0) be a positive solution of (9.3) with $\lambda_0 \leq d\sigma_0$. Then, $\lambda_1[\mathcal{L}(\lambda_0, u_0), \Omega] < 0$. Consequently, it is unstable.*

Proof: If $a > 0$ the proof is very easy. Indeed, by (9.12), the monotonicity of the principal eigenvalue with respect to the potential yields

$$\lambda_1[\mathcal{L}(\lambda_0, u_0), \Omega] = \lambda_1[-d\Delta - \lambda_0 - (p+1)au_0^p, \Omega]$$
$$< \lambda_1[-d\Delta - \lambda_0 - au_0^p, \Omega] = 0,$$

since $a > 0$. Actually, in this case it also follows from (9.12) that

$$\lambda_0 = \lambda_1[-d\Delta - au_0^p, \Omega] < \lambda_1[-d\Delta, \Omega] = d\sigma_0.$$

Hence, (9.3) cannot admit a positive solution if $\lambda \geq d\sigma_0$.

The proof in the general case when a changes sign is less obvious. It proceeds by contradiction. Suppose (9.3) admits a positive solution (λ_0, u_0) with

$$\lambda_0 \leq d\sigma_0, \qquad \lambda_1[\mathcal{L}(\lambda_0, u_0), \Omega] \geq 0.$$

Then, thanks to Propositions 9.6 and 9.7, there exists a positive solution $(\tilde{\lambda}, \tilde{u})$ of (9.3) such that

$$\tilde{\lambda} \leq d\sigma_0, \qquad \lambda_1[\mathcal{L}(\tilde{\lambda}, \tilde{u})] > 0.$$

By Proposition 9.6, $(\tilde{\lambda}, \tilde{u})$ lies on a smooth curve $(\lambda, u(\lambda))$, $\lambda \simeq \tilde{\lambda}$, of positive solutions of (9.3) such that

$$\lambda_1[\mathcal{L}(\lambda, u(\lambda)), \Omega] > 0, \qquad \lambda \simeq \tilde{\lambda}.$$

By global continuation of the curve $(\lambda, u(\lambda))$ for $\lambda < \tilde{\lambda}$, some of the following options occur:

(O1) $u(\lambda) > 0$ and $\lambda_1[\mathcal{L}(\lambda, u(\lambda)), \Omega] > 0$ for all $\lambda \leq \tilde{\lambda}$.

(O2) There is $\lambda_b < \tilde{\lambda}$ such that

$$u(\lambda) > 0 \quad \text{and} \quad \lambda_1[\mathcal{L}(\lambda, u(\lambda)), \Omega] > 0 \qquad (9.26)$$

for all $\lambda \in (\lambda_b, \tilde{\lambda}]$, but $u(\lambda_b) = 0$.

(O3) There is $\lambda_t < \tilde{\lambda}$ such that (9.26) holds for all $\lambda \in (\lambda_t, \tilde{\lambda}]$, but

$$u(\lambda_t) > 0 \quad \text{and} \quad \lambda_1[\mathcal{L}(\lambda_t, u(\lambda_t)), \Omega] = 0.$$

The option (O2) cannot occur because

$$\lambda_b < \tilde{\lambda} \leq d\sigma_0$$

and $\lambda = d\sigma_0$ is the unique bifurcation value to positive solutions from $u = 0$. By Corollary 9.8, the option (O3) cannot occur either. Therefore, the first option occurs. Note that, according to Proposition 9.6,

$$u(\lambda) < u(\tilde{\lambda}) = \tilde{u} \quad \text{for all} \quad \lambda < \tilde{\lambda}.$$

Now, for every $\lambda < 0$, let $x_\lambda \in \Omega$ be such that

$$u(\lambda)(x_\lambda) = \|u(\lambda)\|_{\mathcal{C}(\bar{\Omega})}.$$

Since

$$0 \leq -d\Delta u(\lambda)(x_\lambda) = \lambda u(x_\lambda) + a(x_\lambda)[u(\lambda)]^{p+1}(x_\lambda),$$

it becomes apparent that

$$\lambda + a(x_\lambda)[u(\lambda)]^p(x_\lambda) \geq 0$$

and hence,

$$a(x_\lambda) > 0, \qquad \lim_{\lambda \downarrow -\infty} u(\lambda)(x_\lambda) = \infty,$$

which contradicts the estimate

$$u(\lambda)(x_\lambda) < \tilde{u}(x_\lambda) \leq \|\tilde{u}\|_{\mathcal{C}(\bar{\Omega})}$$

and ends the proof.   □

Corollary 9.10 *Suppose*

$$\mathfrak{D} := - \int_\Omega a(x)\varphi^{p+2}(x) \, dx \leq 0.$$

Then, all the positive solutions of (9.3) are linearly unstable.

Proof: Thanks to Proposition 9.2, (9.3) cannot admit a positive solution if $\lambda \geq d\sigma_0$. Theorem 9.9 ends the proof.   □

Theorem 9.11 *The problem (9.3) admits a linearly stable positive solution, (λ, u), if, and only if, $\mathfrak{D} > 0$. Moreover, $\lambda > d\sigma_0$.*

Proof: Thanks to Corollary 9.10, $\mathfrak{D} > 0$ if (9.3) has a linearly stable positive solution. Moreover, in such case, $\lambda > d\sigma_0$, by Theorem 9.9. Conversely, assume

$$\mathfrak{D} > 0.$$

Then, thanks to Proposition 9.1, the component $\mathfrak{C}^+$ bifurcates supercritically from $(\lambda, 0)$ and in a neighborhood of $(\lambda, u) = (d\sigma_0, 0)$ consists of the curve $(\lambda(s), u(s))$, $s > 0$, defined in (9.6). Moreover, since $\mathfrak{D} > 0$, it follows from (9.7) that $\lambda'(s) > 0$ for sufficiently small $s > 0$. Hence, differentiating (9.3) with respect to s yields

$$\mathfrak{L}(\lambda(s), u(s))u'(s) = \lambda'(s)u(s) > 0. \tag{9.27}$$

Let $\psi_s \gg 0$ be a principal eigenfunction associated with $\lambda_1[\mathfrak{L}(\lambda(s), u(s)), \Omega]$. Multiplying (9.27) by ψ_s, integrating in Ω and applying the formula of integration by parts gives

$$\lambda_1[\mathfrak{L}(\lambda(s), u(s)), \Omega] \int_\Omega \psi_s u'(s) = \lambda'(s) \int_\Omega \psi_s u(s) > 0. \tag{9.28}$$

On the other hand,

$$u'(s) = \varphi + O(s) > 0 \quad \text{as} \quad s \to 0.$$

Therefore, (9.28) implies

$$\lambda_1[\mathfrak{L}(\lambda(s), u(s)), \Omega] > 0.$$

This shows that in a neighborhood of the bifurcation point, $(\lambda, u) = (d\sigma_0, 0)$, the component $\mathfrak{C}^+$ consists of linearly asymptotically stable positive solutions. The proof is complete.   □

9.4   Uniqueness of the stable positive solution

The main result of this section is the following.

Theorem 9.12 *Suppose* $\mathfrak{D} > 0$ *and let* $\lambda_0 > d\sigma_0$ *be such that* (9.3) *possesses a linearly stable positive solution* (λ_0, u_0), *i.e.*,

$$\lambda_1[\mathfrak{L}(\lambda_0, u_0), \Omega] \geq 0.$$

Then, for every $\lambda \in (d\sigma_0, \lambda_0]$, *the problem* (9.3) *has a unique linearly stable positive solution,* $(\lambda, \theta_\lambda)$. *Moreover:*

(a) *For every* $\lambda \in (d\sigma_0, \lambda_0)$, *the positive solution* $(\lambda, \theta_\lambda)$ *is linearly asymptotically stable, i.e.* $\lambda_1[\mathfrak{L}(\lambda, \theta_\lambda), \Omega] > 0$.

(b) *The map*

$$(d\sigma_0, \lambda_0) \longmapsto \mathcal{C}_0(\bar{\Omega})$$
$$\lambda \mapsto \theta_\lambda$$

is of class $\mathcal{C}^2$ *and, for every* $x \in \Omega$, *the map* $\lambda \mapsto \theta_\lambda(x)$ *is increasing.*

(c) *The following relations hold*

$$\lim_{\lambda \downarrow d\sigma_0} \|\theta_\lambda\|_{\mathcal{C}(\bar{\Omega})} = 0, \qquad \lim_{\lambda \uparrow \lambda_0} \|\theta_\lambda - u_0\|_{\mathcal{C}(\bar{\Omega})} = 0. \qquad (9.29)$$

(d) *For every* $\lambda \in (d\sigma_0, \lambda_0]$, $(\lambda, \theta_\lambda)$ *is the minimal positive solution of* (9.3).

Therefore, the minimal positive solution of (9.3) *is the unique linearly stable positive steady state solution of* (9.1) *and, actually, it is linearly asymptotically stable if* $\lambda < \lambda_0$.

Theorem 9.12 establishes that, for every $\lambda \in (d\sigma_0, \lambda_0)$ the minimal positive solution is linearly asymptotically stable and that any other positive steady state of (9.1) must be linearly unstable. Moreover, u_0 must be the minimal positive solution of (9.3) at $\lambda = \lambda_0$.

Remark 9.13 Since $\mathfrak{D} > 0$, the existence of a linearly stable positive solution, (λ_0, u_0), with $\lambda_0 > d\sigma_0$ is guaranteed by Theorem 9.11. By Corollary 9.10, $\mathfrak{D} > 0$ is necessary for the existence of (λ_0, u_0).

In the proof of Theorem 9.12 we will use the following result.

Proposition 9.14 *Suppose* $\lambda > d\sigma_0$ *and* (9.3) *possesses a positive solution,* θ, *then admits a minimal positive solution,* $\theta_{\min}$. *Moreover, the minimal positive solution is linearly stable, i.e.,* $\lambda_1[\mathfrak{L}(\lambda, \theta_{\min}), \Omega] \geq 0$.

Proof: Assume $\lambda > d\sigma_0$ and θ represents a positive solution of (9.3). Then, $\underline{u} := \varepsilon\varphi$ is a subsolution of (9.3) for sufficiently small $\varepsilon > 0$. Moreover, by shortening $\varepsilon > 0$, we can assume that $\underline{u} = \varepsilon\varphi < \theta$. Then, the unique solution of (9.1) with $u_0 = \varepsilon\varphi$, $u(x, t; \lambda, a, \varepsilon\varphi)$, is bounded above by θ and it is increasing in time. In particular, it is globally defined in time and, according to M. Langlais and D. Phillips [128], the limit

$$\theta_\varepsilon = \lim_{t\uparrow\infty} u(\cdot, t; \lambda, a, \varepsilon\varphi)$$

is well defined and it provides us with a positive solution of (9.3). Moreover, $\theta_\varepsilon \leq \theta$ and $\theta_\varepsilon \leq \theta_\delta$ if $\varepsilon < \delta$, by construction. Naturally, the minimal positive solution of (9.3) can be defined through

$$\theta_{\min} := \min_{\varepsilon > 0} \theta_\varepsilon \gg 0.$$

Indeed, necessarily $\theta_{\min}$ is a non-negative solution of (9.3). Moreover, $\theta_{\min} \gg 0$ because we are assuming $\lambda > d\sigma_0$ and the unique bifurcation value to positive solutions from $u = 0$ is $\lambda = d\sigma_0$. Let θ be an arbitrary positive solution of (9.3). Then, the argument above establishes that $\theta_\varepsilon \leq \theta$ for sufficiently small $\varepsilon > 0$. Consequently, $\theta_{\min} \leq \theta$. So, $\theta_{\min}$ is indeed the minimal positive solution of (9.3). It remains to show that

$$\lambda_1[\mathcal{L}(\lambda, \theta_{\min}), \Omega] \geq 0. \tag{9.30}$$

On the contrary, suppose that

$$\lambda_1 := \lambda_1[\mathcal{L}(\lambda, \theta_{\min}), \Omega] < 0.$$

Let $\psi \gg 0$ be any positive eigenfunction associated with λ_1. Then,

$$\bar{u} := \theta_{\min} - \varepsilon\psi$$

is a supersolution of (9.3) for sufficiently small $\varepsilon > 0$. Indeed,

$$
\begin{aligned}
-d\Delta\bar{u} &= -d\Delta\theta_{\min} + \varepsilon d\Delta\psi \\
&= \lambda\theta_{\min} + a\theta_{\min}^{p+1} - \varepsilon\lambda\psi - \varepsilon(p+1)a\theta_{\min}^p\psi - \varepsilon\lambda_1\psi \\
&= \lambda\bar{u} + a\left(\theta_{\min}^{p+1} - \varepsilon(p+1)\theta_{\min}^p\psi\right) - \varepsilon\lambda_1\psi.
\end{aligned}
$$

On the other hand, setting

$$f(\varepsilon) := (\theta_{\min} - \varepsilon\psi)^{p+1},$$

it is apparent that

$$\bar{u}^{p+1} = f(\varepsilon) = f(0) + \varepsilon f'(0) + o(\varepsilon) = \theta_{\min}^{p+1} - \varepsilon(p+1)\psi\theta_{\min}^p + o(\varepsilon)$$

as $\varepsilon \downarrow 0$. Consequently,

$$-d\Delta\bar{u} = \lambda\bar{u} + a\bar{u}^{p+1} + o(\varepsilon) - \varepsilon\lambda_1\psi > \lambda\bar{u} + a\bar{u}^{p+1}$$

for sufficiently small $\varepsilon > 0$, because we are assuming that $\lambda_1 < 0$. Therefore, $\bar{u}$ is a positive supersolution of (9.1) for sufficiently small $\varepsilon > 0$. As (9.3) possesses arbitrarily small positive subsolutions, $\varepsilon\varphi$, the problem (9.3) has a positive solution between $\varepsilon\varphi$ and $\bar{u} < \theta_{\min}$, by Theorem 1.2. This contradicts the minimality of $\theta_{\min}$ and ends the proof. $\quad\square$

Proof of Theorem 9.12: Suppose $\mathfrak{D} > 0$ and let (λ_0, u_0) be a positive solution of (9.3) with $\lambda_0 > d\sigma_0$. Thanks to Propositions 9.6 and 9.7, there exist $\epsilon > 0$ and a map of class $\mathcal{C}^2$,

$$
\begin{array}{ccc}
(\lambda_0 - \epsilon, \lambda_0) & \longmapsto & \mathcal{C}_0(\bar{\Omega}) \\
\lambda & \mapsto & \theta_\lambda
\end{array}
$$

such that $(\lambda, \theta_\lambda)$ is a positive solution of (9.3) with

$$\lambda_1[\mathfrak{L}(\lambda, \theta_\lambda), \Omega] > 0 \quad \text{for all} \quad \lambda \in (\lambda_0 - \epsilon, \lambda_0).$$

Moreover,

$$\lim_{\lambda\uparrow\lambda_0} \|\theta_\lambda - u_0\|_{\mathcal{C}(\bar{\Omega})} = 0.$$

By global continuation of the curve $(\lambda, \theta_\lambda)$ towards the left of λ_0, one of the following options occurs:

(O1) $\theta_\lambda > 0$ and $\lambda_1[\mathfrak{L}(\lambda, \theta_\lambda), \Omega] > 0$ for all $\lambda < \lambda_0$.

(O2) There exists $\lambda_b < \lambda_0$ such that

$$\theta_\lambda > 0, \qquad \lambda_1[\mathfrak{L}(\lambda, \theta_\lambda), \Omega] > 0 \quad \text{for all} \quad \lambda \in (\lambda_b, \lambda_0]$$

and $\theta_{\lambda_b} = 0$.

(O3) There exists $\lambda_t < \lambda_0$ such that

$$\theta_\lambda > 0, \qquad \lambda_1[\mathfrak{L}(\lambda, \theta_\lambda), \Omega] > 0 \quad \text{for all} \quad \lambda \in (\lambda_t, \lambda_0],$$

$$\theta_{\lambda_t} > 0 \text{ and } \lambda_1[\mathfrak{L}(\lambda_t, \theta_{\lambda_t}), \Omega] = 0.$$

By Theorem 9.9, the option (O1) is excluded. According to Proposition 9.7, the option (O3) cannot occur either. Therefore, (O2) occurs. Moreover, since $d\sigma_0$ is the unique value of λ for which bifurcation to positive solutions from $u = 0$ occurs, necessarily

$$\lambda_b = d\sigma_0 \quad \text{and} \quad \lim_{\lambda\downarrow d\sigma_0} \|\theta_\lambda\|_{\mathcal{C}(\bar{\Omega})} = 0.$$

Subsequently, we set

$$\theta_{d\sigma_0} \equiv 0, \qquad \theta_{\lambda_0} \equiv u_0.$$

To complete the proof of the theorem it remains to show that $(\lambda, \theta_\lambda)$ is the unique linearly stable positive solution of (9.3) for every $\lambda \in (d\sigma_0, \lambda_0]$ and that θ_λ is the minimal positive solution of (9.3).

Consider the arc of curve

$$\mathfrak{C}_s^+ := \{(\lambda, \theta_\lambda) : d\sigma_0 \leq \lambda \leq \lambda_0\}.$$

As the solutions of (9.3) can be viewed as fixed points of a compact operator, $\mathfrak{C}_s^+$ is a compact arc of continuous curve in $\mathbb{R} \times \mathcal{C}_0(\bar{\Omega})$. Thus, combining the local uniqueness given by the main theorem of M. G. Crandall and P. H. Rabinowitz [59] with the uniqueness results established by Propositions 9.6 and 9.7, it becomes apparent that there exists $\delta > 0$ such that the unique linearly stable positive solutions of (9.3) in $\mathfrak{C}_s^+ \cap \bar{B}_\delta(d\sigma_0, 0)$ are those on $\mathfrak{C}_s^+$; $B_\delta(d\sigma_0, 0)$ stands for the ball of radius δ centered at $(d\sigma_0, 0)$.

To prove the uniqueness of the linearly stable positive solution we argue by contradiction assuming that there is a positive solution, $(\tilde{\lambda}, \tilde{u})$, of (9.3) such that

$$\tilde{\lambda} \in (d\sigma_0, \lambda_0], \qquad \lambda_1[\mathcal{L}(\tilde{\lambda}, \tilde{u}), \Omega] \geq 0, \qquad \tilde{u} \neq \theta_{\tilde{\lambda}}.$$

Necessarily,

$$\tilde{u} \notin \mathfrak{C}_s^+ \cap \bar{B}_\delta(d\sigma_0, 0). \tag{9.31}$$

Thanks again to Propositions 9.6 and 9.7, there exist $\epsilon > 0$ and a $\mathcal{C}^2$-map

$$
\begin{array}{ccc}
(\tilde{\lambda} - \epsilon, \tilde{\lambda}] & \longmapsto & \mathcal{C}_0(\bar{\Omega}) \\
\lambda & \mapsto & \xi_\lambda
\end{array}
$$

such that (λ, ξ_λ) is a positive solution of (9.3) with

$$\lambda_1[\mathcal{L}(\lambda, \xi_\lambda), \Omega] > 0$$

for all $\lambda \in (\tilde{\lambda} - \epsilon, \tilde{\lambda})$. Moreover, $\xi_{\tilde{\lambda}} = \tilde{u}$. By global continuation of the curve (λ, ψ_λ) towards the left of $\tilde{\lambda}$, it is clear that some of the following alternatives occur:

(A1) $\xi_\lambda > 0$ and $\lambda_1[\mathcal{L}(\lambda, \xi_\lambda), \Omega] > 0$ for all $\lambda < \tilde{\lambda}$.

(A2) There is a $\lambda_b < \tilde{\lambda}$ such that

$$\xi_\lambda > 0 \quad \text{and} \quad \lambda_1[\mathcal{L}(\lambda, \xi_\lambda), \Omega] > 0 \quad \text{for all} \quad \lambda \in (\lambda_b, \tilde{\lambda}],$$

but $\xi_{\lambda_b} = 0$.

(A3) There is $\lambda_t < \tilde{\lambda}$ such that

$$\xi_\lambda > 0, \quad \lambda_1[\mathcal{L}(\lambda, \xi_\lambda), \Omega] > 0 \quad \text{for all} \quad \lambda \in (\lambda_t, \tilde{\lambda}],$$

$\psi_{\lambda_t} > 0$ and $\lambda_1[\mathcal{L}(\lambda_t, \psi_{\lambda_t}), \Omega] = 0$.

Arguing as above, it becomes apparent that (A2) holds. As $d\sigma_0$ is the unique value of λ for which bifurcation to positive solutions from $u = 0$ occurs,

$$\lambda_b = d\sigma_0, \qquad \lim_{\lambda \downarrow d\sigma_0} \|\xi_\lambda\|_{C(\bar\Omega)} = 0. \qquad (9.32)$$

By (9.31),

$$\xi_{\tilde\lambda} = \tilde u \notin \mathfrak{C}_s^+ \cap \bar B_\delta(d\sigma_0, 0).$$

Consequently, since the unique linearly stable positive solutions of (9.3) in $\mathfrak{C}_s^+ \cap \bar B_\delta(d\sigma_0, 0)$ are those of $\mathfrak{C}_s^+$, the curve (λ, ξ_λ), $d\sigma_0 \leq \lambda \leq \tilde\lambda$, remains outside $\mathfrak{C}_s^+ \cap \bar B_\delta(d\sigma_0, 0)$, which contradicts (9.32) and concludes the proof of the uniqueness.

Finally, by Proposition 9.14, (9.3) possesses a minimal positive solution for each $\lambda \in (d\sigma_0, \tilde\lambda]$. Moreover, it is linearly stable. Therefore, thanks to the uniqueness of the linearly stable positive solution, θ_λ must be the minimal positive solution of (9.3) for all $\lambda \in (d\sigma_0, \lambda_0]$. This ends the proof. $\square$

Theorem 9.15 *Suppose* $\mathfrak{D} > 0$ *and* (9.3) *admits a positive solution,* (λ_0, u_0), *such that*

$$\lambda_0 > d\sigma_0, \qquad \lambda_1[\mathfrak{L}(\lambda_0, u_0), \Omega] = 0. \qquad (9.33)$$

Then, (9.3) *does not admit a positive solution if* $\lambda > \lambda_0$.

Proof: On the contrary, suppose that (9.3) possesses a positive solution, $(\tilde\lambda, \tilde u)$, with $\tilde\lambda > \lambda_0$. By Proposition 9.14, (9.3) has a minimal positive solution at $\lambda = \tilde\lambda$, $\theta_{\tilde\lambda}$. Moreover,

$$\lambda_1[\mathfrak{L}(\tilde\lambda, \theta_{\tilde\lambda}), \Omega] \geq 0.$$

Thus, thanks to Theorem 9.12, (9.3) has a unique linearly stable positive solution for $\lambda = \lambda_0$, θ_0. Moreover, $\lambda_1[\mathfrak{L}(\lambda_0, \theta_0), \Omega] > 0$. Therefore, thanks to (9.33), $u_0 \neq \theta_0$, and hence, (9.3) possesses two linearly stable positive solutions at $\lambda = \lambda_0$, which is impossible. This ends the proof. $\square$

9.5 Curve of stable positive solutions

Subsequently, we will denote by J_λ the set of values of the parameter λ for which (9.3) possesses a linearly stable positive solution. The following result provides us with the structure of J_λ and the global behavior of the associated curve of linearly stable positive solutions. By Theorem 9.9, $J_\lambda \subset (d\sigma_0, \infty)$. Also, $J_\lambda = \emptyset$ if $\mathfrak{D} \leq 0$. Moreover, thanks to Theorem 9.12, J_λ must be an interval. Consequently, the next result holds.

Theorem 9.16 *Suppose a satisfies* (HA) *and* $\mathfrak{D} > 0$. *Then, some of the following options occur:*

(a) *There exists* $\lambda_* \in (d\sigma_0, \lambda_\omega(a)]$ *such that* $J_\lambda = (d\sigma_0, \lambda_*)$.

(b) *There exists* $\lambda_* \in (d\sigma_0, \lambda_\omega(a))$ *such that* $J_\lambda = (d\sigma_0, \lambda_*]$.

Moreover, in both cases the set of linearly stable positive solutions consists of a $\mathcal{C}^2$-*curve*
$$\begin{aligned} J_\lambda &\longmapsto \mathbb{R} \times \mathcal{C}_0(\bar\Omega) \\ \lambda &\mapsto (\lambda, \theta_\lambda) \end{aligned}$$
such that $\lambda \mapsto \theta_\lambda(x)$ *is increasing for all* $x \in \Omega$. *Moreover,*

$$\lim_{\lambda\downarrow d\sigma_0} \|\theta_\lambda\|_{\mathcal{C}(\bar\Omega)} = 0 \quad \text{and} \quad \lambda_1[\mathfrak{L}(\lambda, \theta_\lambda), \Omega] > 0 \quad \text{for all} \quad \lambda \in \text{int } J_\lambda.$$

Furthermore:

- *In case* (a)
$$\lim_{\lambda\uparrow\lambda_*} \|\theta_\lambda\|_{\mathcal{C}(\bar\Omega)} = \infty \tag{9.34}$$

  and (9.3) *cannot admit a positive solution for* $\lambda \geq \lambda^*$. *Therefore,* $\mathfrak{C}^+$ *consists of an increasing smooth curve filled in by linearly asymptotically stable positive solutions blowing up to infinity as* $\lambda \uparrow \lambda_*$.

- *In case* (b),
$$\lambda_1[\mathfrak{L}(\lambda_*, \theta_{\lambda_*}), \Omega] = 0$$

  and hence, owing to Proposition 9.7, $(\lambda_*, \theta_{\lambda_*})$ *is a quadratic subcritical turning point of* $\mathfrak{C}^+$. *In particular, there exists* $\epsilon > 0$ *such that for every* $\lambda \in (\lambda_* - \epsilon, \lambda_*)$ *the problem* (9.3) *possesses at least two non-degenerate positive solutions: one of them linearly asymptotically stable and another linearly unstable. Moreover, according to Theorem 9.15,* (9.3) *cannot admit a positive solution if* $\lambda > \lambda_*$.

Remark 9.17 Thanks to Theorem 7.1 of H. Amann and J. López-Gómez [13], the case (b) occurs if any set of positive solutions of (9.3), $\mathcal{S} \subset \mathbb{R} \times \mathcal{C}_0(\bar\Omega)$, with $\mathcal{P}_\lambda(\mathcal{S})$ bounded, is bounded in $\mathbb{R} \times \mathcal{C}_0(\bar\Omega)$, where we are denoting by $\mathcal{P}_\lambda$ the λ-projection operator. According to Theorem 4.3 of H. Amann and J. López-Gómez [13], these a priori bounds are available under the next hypothesis.

Hypothesis (HB)

Either $N \in \{1, 2\}$ or $N \geq 3$ and

$$a_+(x) = \alpha_+(x) \left[\text{dist}\,(x, \partial\Omega_+)\right]^\gamma \quad \textit{for all } x \in \Omega_+ \textit{ near } \partial\Omega_+,$$

for some continuous function α_+ positive and bounded away from zero in a neighborhood of $\partial\Omega_+$ and some positive constant $\gamma > 0$ such that

$$p < \min \left\{ \frac{N + 1 + \gamma}{N - 1}, \frac{N + 2}{N - 2} \right\}.$$

Actually, in the presence of a priori bounds, (9.3) possesses at least two positive solutions for each

$$\lambda \in (d\sigma_0, \lambda_*) = J_\lambda \setminus \{\lambda_*\},$$

as illustrated by Figures 9.2 and 9.3. This follows easily by using the fixed point index in cones, as in Section 10.7. Indeed, $(\lambda, 0)$ has local index zero, while $(\lambda, \theta_\lambda)$ has local index one by the Schauder formula because

$$\lambda_1[\mathfrak{L}(\lambda, \theta_\lambda), \Omega] > 0.$$

As the global index is zero, because (9.3) cannot admit a positive solution for $\lambda > \lambda_*$ and we are imposing the existence of a priori bounds, (9.3) should admit a further positive solution, with index -1, if it is non-degenerate. The interested reader can review Theorem 7.4 of H. Amann and J. López-Gómez [13] for further technical details.

Remark 9.18 According to Remark 9.17, in all the numerical experiments carried out in Section 9.1, (9.3) cannot admit any positive solution for $\lambda > \lambda_t = \lambda_*$, where λ_t is the first (subcritical) turning point of $\mathfrak{C}^+$.

Remark 9.19 According to J. López-Gómez [148], in the special case when $p = 1$, it is easily seen that the option (b) occurs provided $\mathfrak{D} > 0$ is sufficiently small, independently of the spatial dimension, N, and the decay rates of a_+ along $\partial\Omega_+$. Moreover,

$$\lim_{\mathfrak{D} \downarrow 0} \lambda_* = d\sigma_0.$$

Proof of Theorem 9.16: Thanks to Proposition 9.4 and Theorems 9.9 and 9.12 there exists $\lambda_* \leq \lambda_\omega(a)$ such that either $J_\lambda = (d\sigma_0, \lambda_*)$, or $J_\lambda = (d\sigma_0, \lambda_*]$.

In case $J_\lambda = (d\sigma_0, \lambda_*]$, the alternative (b) occurs and the result is an easy consequence of Theorem 9.12.

Suppose $J_\lambda = (d\sigma_0, \lambda_*)$. Then, (a) occurs and, according to Theorem 9.12, in order to complete the proof it suffices to show that condition (9.34) holds. On the contrary, assume that there is a constant $M > 0$ and a sequence $\lambda_n \in J_\lambda$, $n \geq 1$, such that

$$\lim_{n \to \infty} \lambda_n = \lambda_*$$

and

$$\|\theta_{\lambda_n}\|_{\mathcal{C}(\bar{\Omega})} \le M, \qquad n \ge 1. \tag{9.35}$$

By Theorem 9.11, the map $\lambda \mapsto \|\theta_\lambda\|_{\mathcal{C}(\bar{\Omega})}$ is increasing. Thus, it follows from (9.35) that

$$\|\theta_\lambda\|_{\mathcal{C}(\bar{\Omega})} \le M \quad \text{for all} \quad \lambda \in J_\lambda.$$

Therefore, the point-wise limit

$$u_*(x) := \lim_{\lambda \uparrow \lambda_*} \theta_\lambda(x), \qquad x \in \Omega,$$

is well defined. A standard bootstrapping argument combined with the fact that the solutions of (9.3) are fixed points of a compact operator shows that (λ_*, u_*) provides us with a positive solution of (9.3), which is impossible, because we are assuming $\lambda_* \notin J_\lambda$. This ends the proof. $\square$

9.6   Dynamics in the presence of a stable steady state

This section studies the dynamics of (9.1) for $\lambda \in (d\sigma_0, \lambda_\omega(a))$. For the validity of the remaining results of this chapter it suffices to impose $p > 0$. The stronger condition $p \ge 1$ was only needed in order to apply the local bifurcation theorem of M. G. Crandall and P. H. Rabinowitz [59]. Actually, adopting the methodology of Chapters 6 and 7 of J. López-Gómez [163], all the results of the chapter are valid for $p > 0$, even the regularities of the curves of positive solutions, (λ, u), because the strong positivity of u actually entails their analyticities outside $(d\sigma_0, 0)$.

9.6.1   Global existence versus blow-up in finite time

The next result shows how in the case when $\lambda \in (d\sigma_0, \lambda_\omega(a))$ the dynamics of (9.1) depend on the relative size of a_+ with respect to λ.

Theorem 9.20 *Suppose $\lambda \in (d\sigma_0, \lambda_\omega(a))$ and u_0 is a positive strict subsolution of (9.4). Then:*

(a) *There exists $\varepsilon > 0$ such that (9.3) has a positive solution if $(a_+)_M \le \varepsilon$. Moreover,*

$$\lim_{t \uparrow \infty} \|u(\cdot, t; \lambda, a, u_0) - \theta_{[\lambda, a]}\|_{\mathcal{C}(\bar{\Omega})} = 0, \tag{9.36}$$

where $\theta_{[\lambda, a]}$ stands for the minimal positive solution of (9.3).

(b) *Assume that there exists a smooth subdomain $D \subset \Omega_+$, with $\bar{D} \subset \Omega_+$, such that*

$$A \equiv \min_{\bar{D}} a_+ > \left(\frac{p}{\alpha}\right)^p \left(\frac{\Sigma}{p+1}\right)^{p+1} \tag{9.37}$$

where

$$\Sigma := \lambda_1[-d\Delta, D] - \lambda, \qquad \alpha := -d \int_{\partial D} \frac{\partial \varphi}{\partial n} \theta_{[\lambda, a_-]} \, d\sigma, \qquad (9.38)$$

$\theta_{[\lambda, a_-]}$ *stands for the unique positive solution of* (9.4), $\varphi \gg 0$ *is the principal eigenfunction of* $\lambda_1[-\Delta, D]$ *normalized so that* $\int_D \varphi = 1$, *and* n *stands for the outward unit normal vector field to* D. *Then,* $u(\cdot, t; \lambda, a, u_0)$ *blows up in* $L^\infty(\Omega)$ *at some time* $T_b = T_b(u_0) > 0$. *In other words,*

$$T_{\max}(u_0) = T_b(u_0) < +\infty.$$

Proof: Fix $\lambda \in (d\sigma_0, \lambda_w(a))$ and let $u_0 > 0$ be a strict subsolution of (9.4). Such subsolution exists because $\lambda > d\sigma_0$. Indeed, it suffices to take $\varepsilon \psi$ for sufficiently small $\varepsilon > 0$, where ψ stands for a positive eigenfunction associated to σ_0 in Ω. On the other hand, by Theorem 4.1, (9.4) admits a (unique) positive solution, $\theta_{[\lambda, a_-]}$, if, and only if, $d\sigma_0 < \lambda < \lambda_w$. By Lemma 1.8,

$$u_0 \ll \theta_{[\lambda, a_-]}. \qquad (9.39)$$

Naturally, as u_0 also is a subsolution of (9.3), $t \mapsto u(\cdot, t; \lambda, a, u_0)$ is increasing. Moreover, for every $t > 0$,

$$u(\cdot, t; \lambda, a, u_0) \geq u(\cdot, t; \lambda, a_-, u_0) \qquad \text{in } \Omega.$$

Hence, thanks to Theorem 5.2(b), we find that

$$\lim_{t \uparrow \infty} u(\cdot, t; \lambda, a, u_0) \geq \lim_{t \uparrow \infty} u(\cdot, t; \lambda, a_-, u_0) = \theta_{[\lambda, a_-]} \qquad (9.40)$$

provided $T_{\max}(u_0) = +\infty$!

Now, we will prove Part (a). The first assertion is a direct consequence of Theorem 9.5. So, suppose (9.3) admits a positive solution and let $\theta_{[\lambda, a]}$ denote its minimal positive solution. By Theorem 9.12, $\theta_{[\lambda, a]}$ is the unique linearly stable non-negative solution of (9.3). Moreover, by Theorem 1.7,

$$\theta_{[\lambda, a_-]} \ll \theta_{[\lambda, a]},$$

because $\theta_{[\lambda, a]}$ provides a positive strict supersolution of the associated sublinear problem (9.4). Therefore, thanks to (9.39), we have that

$$u_0 \ll \theta_{[\lambda, a]}.$$

Thus, (9.36) is a direct consequence from the fact that $\theta_{[\lambda, a]}$ is the minimal positive solution of (9.3), which concludes the proof of Part (a).

Subsequently, we will prove Part (b). Since

$$\lambda < \lambda_w = \lambda_1[-d\Delta, \Omega \setminus \bar{\Omega}_-] \leq \lambda_1[-d\Delta, \Omega_+] < \lambda_1[-d\Delta, D],$$

it is apparent that

$$\Sigma = \lambda_1[-d\Delta, D] - \lambda > 0.$$

Moreover,

$$\alpha = -\int_{\partial D} \frac{\partial\varphi}{\partial n} \theta_{[\lambda,a_-]} \, d\sigma > 0,$$

because $\frac{\partial\varphi}{\partial n} < 0$ on ∂D. Consequently, the constant on the right-hand side of (9.37) is well defined and positive.

Suppose (9.37). To show that $u(\cdot, t; \lambda, a, u_0)$ blows up in a finite time we will argue by contradiction. So, suppose $T_{\max}(u_0) = +\infty$ and set

$$I(t) := \int_D u(x, t; \lambda, a, u_0)\varphi(x) \, dx \in (0, \infty) \quad \text{for all} \quad t > 0.$$

As $I(t)$ is non-decreasing because u_0 is a subsolution of (9.3), the limit

$$L := \lim_{t\uparrow\infty} I(t) \in (0, \infty] \tag{9.41}$$

is well defined. On the other hand, setting

$$u := u(x, t; \lambda, a, u_0),$$

multiplying by φ the nonlinear parabolic equation of (9.1), integrating in D and applying the formula of integration by parts yields

$$I'(t) = d\int_D \varphi\Delta u \, dx + \lambda\int_D \varphi u \, dx + \int_D a u^{p+1}\varphi \, dx$$
$$\geq -d\int_{\partial D} u\frac{\partial\varphi}{\partial n} \, d\sigma + (\lambda - \lambda_1[-d\Delta, D])\, I(t) + A\int_D u^{p+1}\varphi \, dx,$$

where

$$' := \frac{d}{dt}, \qquad A \equiv \min_{\bar{D}} a = \min_{\bar{D}} a_+,$$

because $\bar{D} \subset \Omega_+$. On the other hand,

$$\int_D u\varphi \, dx = \int_D \varphi^{1-\frac{1}{p+1}}\varphi^{\frac{1}{p+1}} u \, dx$$
$$\leq \left(\int_D \varphi\right)^{1-\frac{1}{p+1}} \left(\int_D \varphi u^{p+1} \, dx\right)^{\frac{1}{p+1}}$$
$$= \left(\int_D u^{p+1}\varphi \, dx\right)^{\frac{1}{p+1}}$$

since $\int_D \varphi = 1$. Hence,

$$\int_D u^{p+1}\varphi \, dx \geq \left(\int_D u\varphi \, dx\right)^{p+1}$$

and consequently,

$$I'(t) \geq -d \int_{\partial D} u \frac{\partial \varphi}{\partial n} \, d\sigma - \Sigma I(t) + A I^{p+1}(t) \tag{9.42}$$

for all $t > 0$. Suppose $L < \infty$. Then,

$$\lim_{t \uparrow \infty} I'(t) = 0$$

and letting $t \uparrow \infty$ in (9.42), it follows from (9.40) that

$$0 \geq \alpha - \Sigma L + A L^{p+1}, \tag{9.43}$$

where α is the constant defined in (9.38). Setting

$$f(x) := A x^{p+1} - \Sigma x + \alpha, \qquad x > 0,$$

(9.43) can be equivalently written as

$$f(L) \leq 0. \tag{9.44}$$

The function f satisfies

$$f(0) = \alpha > 0, \quad \lim_{x \uparrow \infty} f(x) = \infty \quad \text{and} \quad f'(x) = (p+1) A x^p - \Sigma \quad \forall \, x > 0.$$

Thus,

$$f'(x) = 0 \quad \text{if and only if} \quad x = x_L := \left(\frac{\Sigma}{(p+1)A} \right)^{\frac{1}{p}}$$

and, due to (9.44), necessarily

$$f(x_L) = A \left(\frac{\Sigma}{(p+1)A} \right)^{\frac{p+1}{p}} - \Sigma \left(\frac{\Sigma}{(p+1)A} \right)^{\frac{1}{p}} + \alpha \leq 0.$$

Equivalently,

$$A \leq \left(\frac{p}{\alpha} \right)^p \left(\frac{\Sigma}{p+1} \right)^{p+1}$$

which contradicts (9.37). Therefore, $L = \infty$.

Subsequently, we will use the estimate

$$I'(t) \geq -\Sigma I(t) + A I^{p+1}(t), \qquad \forall \, t > 0, \tag{9.45}$$

which is a direct consequence from (9.42). Since $\lim_{t \uparrow \infty} I(t) = \infty$, there exists $t_0 > 0$ such that

$$I(t_0) > \left(\frac{\Sigma}{A} \right)^{\frac{1}{p}}. \tag{9.46}$$

On the other hand, the change of variable

$$I(t) = e^{-\Sigma(t-t_0)} J(t), \qquad t \geq t_0 > 0,$$

transforms (9.45), (9.46) into

$$\begin{cases} J'(t) \geq A e^{-p\Sigma(t-t_0)} J^{p+1}(t), & t \geq t_0 > 0, \\ J(t_0) = I(t_0) > \left(\frac{\Sigma}{A}\right)^{\frac{1}{p}}. \end{cases} \tag{9.47}$$

Thus, integrating the differential inequality of (9.47) gives

$$\frac{-1}{p} \left[J^{-p}(t) - J^{-p}(t_0)\right] \geq \frac{-A}{p\Sigma} \left[e^{-p\Sigma(t-t_0)} - 1\right] \qquad \forall\, t > t_0$$

and hence,

$$J^{-p}(t) \leq J^{-p}(t_0) - \frac{A}{\Sigma} \left[1 - e^{-p\Sigma(t-t_0)}\right] \qquad \forall\, t > t_0, \tag{9.48}$$

since $p > 0$. On the other hand, since

$$J(t) = e^{\Sigma(t-t_0)} I(t) \quad \text{for all} \quad t > t_0 \quad \text{and} \quad \lim_{t \to \infty} I(t) = \infty,$$

it is apparent that $\lim_{t \to \infty} J(t) = \infty$ and therefore, letting $t \to \infty$ in (9.48) yields

$$0 \leq J^{-p}(t_0) - \frac{A}{\Sigma}.$$

Equivalently,

$$I(t_0) = J(t_0) \leq \left(\frac{\Sigma}{A}\right)^{\frac{1}{p}},$$

which contradicts (9.46). This contradiction comes from the assumption that u is globally defined in time. Consequently, the solution blows up in a finite time, $T_b(u_0) = T_{\max}(u_0)$.   □

Note that the mapping

$$\lambda \mapsto \Sigma \equiv \lambda_1[-d\Delta, D] - \lambda$$

is decreasing, while

$$\lambda \mapsto \alpha \equiv -d \int_{\partial D} \frac{\partial \varphi}{\partial n} \theta_{[\lambda,a_-]}\, d\sigma$$

is increasing, because $\lambda \mapsto \theta_{[\lambda,a_-]}|_D$ is point-wise increasing. Therefore,

$$\lambda \mapsto \left(\frac{p}{\alpha}\right)^p \left(\frac{\Sigma}{p+1}\right)^{p+1}$$

is decreasing. This is a rather natural feature establishing that the bigger is λ the smaller can be taken $\min_{\bar{D}} a_+$ for $T_{\max}(u_0) < +\infty$.

9.6.2 Complete blow-up in Ω_+

In this section we assume that $u(\cdot, t; \lambda, a, u_0)$ blows up in a finite time, $T_b = T_{\max}(u_0) < +\infty$, for example, in the context of Theorem 9.20(b), to give some sufficient conditions so that it cannot admit a *weak continuation* for $t > T_b$, i.e., so that *life after death* cannot occur. To simplify the exposition as much as possible, through the next discussion we will assume that the nodal behavior of $a(x)$ obeys the general patterns of Figure 9.1 with

$$\sigma_1 := \lambda_1[-\Delta, \Omega_0] < \sigma_2 := \lambda_1[-\Delta, \Omega_+],$$

as in the numerical example of Section 5.3. Then,

$$\lambda_\omega = \lambda_1[-d\Delta, \Omega \setminus \bar{\Omega}_-] = \min\{\lambda_1[-d\Delta, \Omega_0], \lambda_1[-d\Delta, \Omega_+]\} = d\sigma_1.$$

We also introduce the approximating functions

$$f_k(x, u) := \begin{cases} \lambda u + a(x)\min\{u^{p+1}, k\} & \text{if } x \in \Omega_+, \\ \lambda u + a(x)u^{p+1} & \text{if } x \in \Omega_- \cup \Omega_0, \end{cases} \quad k \in \mathbb{N},$$

as well as the associated approximating problems

$$\begin{cases} u_t - d\Delta u = f_k(x, u), & x \in \Omega, \ t > 0, \\ u(x, t) = 0, & x \in \partial\Omega, \ t > 0, \\ u(x, 0) = u_0(x), & x \in \Omega, \end{cases} \quad k \in \mathbb{N}. \quad (9.49)$$

For each $k \in \mathbb{N}$, let

$$u_k := u_k(x, t; \lambda, a, u_0)$$

denote the solution of (9.49). As f_k is bounded in Ω_+, u_k is globally defined in time. Moreover, since $f_k \leq f_{k+1}$, it satisfies $u_k \leq u_{k+1}$. Thus, the limit

$$\tilde{u}(x, t) := \lim_{k \to \infty} u_k(x, t; \lambda, a, u_0) \in (0, \infty] \quad (9.50)$$

is well defined for all $x \in \Omega$ and $t > 0$. Moreover, by construction,

$$\tilde{u}(x, t) = u(x, t; \lambda, a, u_0) \quad \text{for all} \quad x \in \bar{\Omega} \quad \text{and} \quad t < T_b = T_{\max}(u_0),$$

though the problem of finding $\tilde{u}(x, t)$ for $t \geq T_b$ may be a hard task. Nevertheless, $\tilde{u}$ provides us with a weak extension of u for times $t \geq T_b$ in a rather natural way.

Also, we denote by $B(u_0)$ the set of points $x \in \Omega$ for which there is a sequence $(x_k, t_k) \in \Omega \times (0, T_b)$, $k \geq 1$, such that

$$\lim_{k \to \infty} (x_k, t_k) = (x, T_b) \quad \text{and} \quad \lim_{k \to \infty} u(x_k, t_k; \lambda, a, u_0) = \infty,$$

often referred to as the *blow-up set* of $u(\cdot, t; \lambda, a, u_0)$. Finally, we will set

$$\mathbb{E}_{u_0}(t) := I(u(\cdot, t; \lambda, a, u_0)),$$

where

$$I(w) := \int_{\Omega} \left(\frac{d}{2}|\nabla w|^2 - \frac{\lambda}{2}w^2 - \frac{a(x)}{p+2}|w|^{p+2} \right) dx, \qquad w \in H_0^1(\Omega),$$

is the associated *energy functional*. According to P. H. Rabinowitz [215, Pr. B 10], it is well known that, for every $w \geq 0$,

$$I'(w)\xi = \int_{\Omega} \left[d\nabla w \cdot \nabla \xi - (\lambda w + a(x)w^{p+1})\xi(x) \right] dx \quad \text{for all } \xi \in H_0^1(\Omega).$$

Thus,

$$\frac{d\,\mathbb{E}_{u_0}}{dt}(t) = I'(u)u_t = \int_{\Omega} \left[d\nabla u \cdot \nabla u_t - (\lambda u + a(x)u^{p+1})u_t \right] dx$$

$$= \int_{\Omega} \left[-d\Delta u - \lambda u - a(x)u^{p+1} \right] u_t \, dx = - \int_{\Omega} u_t^2 \, dx \leq 0$$

and hence, $\mathbb{E}_{u_0}$ is decreasing in $[0, T_b)$.

The following result shows that, under the general assumptions of Theorem 9.20(b),

$$\tilde{u}(x,t) = \infty \quad \text{for all} \quad x \in \Omega_+ \quad \text{and} \quad t > T_b$$

if $B(u_0) \subset \Omega_+$ and p is sufficiently close to zero. In these circumstances, it is said that $u(x,t;\lambda,a,u_0)$ *completely blows up* in Ω_+ at time T_b. Consequently, it does not admit any weak extension in Ω_+ after time T_b.

Theorem 9.21 *Suppose $\lambda \in (d\sigma_0, d\sigma_1)$, $u_0 > 0$ is a strict subsolution of (9.4) and (9.37) holds in some smooth domain D with $\bar{D} \subset \Omega_+$. Then, $T_{\max}(u_0) < +\infty$ and*

$$\tilde{u} = u \leq L^{\min}_{[\lambda,a_-,\Omega\setminus\bar{\Omega}_+]} \qquad in \ \Omega \setminus \bar{\Omega}_+ .$$

In particular, $B(u_0) \subset \bar{\Omega}_+$. Moreover, if either

$$p + 1 < \begin{cases} +\infty, & if \ n = 1, \\ (3n+8)/(3n-4), & if \ n > 1, \end{cases} \tag{9.51}$$

or $a_+(x)$ satisfies Hypothesis (HB) in Remark 9.17, then

$$\lim_{t \uparrow T_{\max}(u_0)} \mathbb{E}_{u_0}(t) = -\infty. \tag{9.52}$$

If, in addition, $B(u_0) \subset \Omega_+$, then

$$\tilde{u}(x,t) = +\infty \quad for \ all \ x \in \Omega_+, \ t > T_{\max}(u_0), \tag{9.53}$$

i.e., u blows up completely in Ω_+.

Proof: $T_b := T_{\max}(u_0) < +\infty$ follows from Theorem 9.20(b). Since $\lambda < d\sigma_1$, by Theorem 4.7 the large solution $L^{\min}_{[\lambda,a_-,\Omega\setminus\bar\Omega_+]}$ is well defined. Moreover, since u_0 is a subsolution of (9.3), the restriction $u|_{\Omega\setminus\bar\Omega_+}$ provides us with a subsolution of

$$-d\Delta u = \lambda u + a_- u^{p+1}$$

in $\Omega \setminus \bar\Omega_+$ for all $t \in (0, T_b)$. Hence,

$$u(\cdot, t; \lambda, a, u_0) \leq L^{\min}_{[\lambda,a_-,\Omega\setminus\bar\Omega_+]} \quad \text{in} \quad \Omega \setminus \bar\Omega_+$$

for all $t < T_b$. Consequently, by the definition of $\tilde u$,

$$\tilde u(x, t) = \lim_{k\to\infty} u_k(x, t) \leq L^{\min}_{[\lambda,a_-,\Omega\setminus\bar\Omega_+]} \quad \text{for all } t > 0.$$

In particular, $B(u_0) \subset \bar\Omega_+$. It should be noted that this entails $u(x, t; \lambda, a, u_0) = \tilde u(x, t)$ to be defined for all $t > 0$ if $x \in \Omega \setminus \bar\Omega_+$.

Now, suppose (9.51) or Hypothesis (HB). Then, (9.52) is a consequence from the results of J. López-Gómez and P. Quittner [177, Section 5]. Therefore, in case $B(u_0) \subset \Omega_+$, (9.53) holds by applying [177, Th. 1.1] in

$$\Omega_\delta := \{x \in \Omega \ : \ \text{dist}(x, \partial\Omega_+) < \delta\}$$

for a sufficiently small $\delta > 0$. This ends the proof. $\qquad\square$

It should be noted that the point-wise limit

$$\mathfrak{M}_{[\lambda,a,\Omega]}(x) := \begin{cases} +\infty & \text{if } x \in \Omega_+, \\ \lim_{t\uparrow\infty} \tilde u(x, t; u_0) & \text{if } x \in \Omega \setminus \Omega_+, \end{cases}$$

is well defined, and it provides us with the minimal positive solution of

$$\begin{cases} -d\Delta w = \lambda w + a_- w^{p+1} & \text{in } \Omega \setminus \bar\Omega_+, \\ w = \lim_{t\uparrow\infty} \tilde u & \text{on } \partial\Omega_+, \\ w = 0 & \text{on } \partial\Omega. \end{cases}$$

Indeed, as the initial data u_0 which are subsolutions of (9.3) give rise to non-decreasing solutions of (9.1) and (9.49) because u_0 also is a subsolution of (9.49) for $k > \|u_0\|_{\mathcal{C}(\bar\Omega)}$, the extended solution $\tilde u$ also is non-decreasing in t and therefore, $\lim_{t\uparrow\infty} \tilde u$ is well defined in Ω.

In the radially symmetric case, Theorem 9.21 can be sharpened to obtain the next result.

Theorem 9.22 *Suppose* $0 < R_1 < R_2 < R$,

$$\Omega = B_R := \{x \in \mathbb{R}^n \ : \ |x| < R\},$$
$$\Omega_+ = B_{R_1}, \quad \Omega_- = B_{R_2} \setminus \bar B_{R_1}, \quad \Omega_0 = B_R \setminus \bar B_{R_2},$$

$$d\sigma_0 := \lambda_1[-d\Delta, B_R] < \lambda < d\sigma_1 := \lambda_1[-d\Delta, \Omega_0] < d\sigma_2 := \lambda_1[-d\Delta, \Omega_+],$$

the weight function $a \in C^1[0, R]$ is radially symmetric, $a(x) = a(|x|)$, and it satisfies (9.37) for some smooth subdomain D with $\bar{D} \subset \Omega_+$, and $u_0(x) = u_0(|x|)$, $u_0 \in C^1[0, R]$, is a positive strict subsolution of (9.4) in B_R. Then, the solution $u(x, t) := u(x, t; \lambda, a, u_0)$ of the problem

$$
\begin{cases}
u_t - d\Delta u = \lambda u + a(|x|)u^{p+1}, & x \in B_R, \ t > 0, \\
u(x, t) = 0, & x \in \partial B_R, \ t > 0, \\
u(x, 0) = u_0(|x|), & x \in B_R,
\end{cases}
$$

blows up in a finite time $T_b := T_b(u_0) < \infty$ in $L^\infty(\Omega)$. Suppose, in addition, the following:

(H1) $0 < p < (N + 2)/(N - 2) - 1 = 4/(N - 2)$ if $N \geq 3$.

(H2) There exists $\rho \in (R_1, R_2)$ such that $a'(r) \leq 0$ for each $r \in [0, \rho]$, and

$$
\int_0^\rho a(r)r^{n-1}\,dr > 0 \quad \text{and} \quad \lambda_1[-d\Delta, B_\rho] < \lambda < d\sigma_1.
$$

(H3) $u_0'(r) \leq 0$ for all $r \in [0, R]$.

Then, there exists $T_c \geq T_b$ such that

$$
\tilde{u}(x, t) = +\infty \quad \text{for all } (x, t) \in \bar{\Omega}_+ \times (T_c, \infty), \tag{9.54}
$$

while

$$
\tilde{u}(\cdot, t) \leq L^{\min}_{[\lambda, a_-, \Omega \setminus \bar{\Omega}_+]} \quad \text{in } \Omega \setminus \bar{\Omega}_+ \tag{9.55}
$$

is a classical solution for each $t > 0$. If, in addition,

$$
\lim_{t \uparrow T_b} u(R_1, t) = +\infty,
$$

then,

$$
\lim_{t \uparrow \infty} \tilde{u}(\cdot, t) = L^{\min}_{[\lambda, a_-, \Omega \setminus \bar{\Omega}_+]} \quad \text{in } \Omega \setminus \bar{\Omega}_+,
$$

uniformly in compact sets of $\Omega \setminus \bar{\Omega}_+$ and therefore,

$$
\lim_{t \uparrow \infty} \tilde{u}(\cdot, t) = \mathfrak{M}^{\min}_{[\lambda, a_-, \Omega]} := \begin{cases} +\infty & \text{in } \bar{\Omega}_+, \\ L^{\min}_{[\lambda, a_-, \Omega \setminus \bar{\Omega}_+]} & \text{in } \Omega \setminus \bar{\Omega}_+. \end{cases}
$$

Consequently, in such case, the asymptotic behavior of $u(x, t; \lambda, a, u_0)$ is governed by the metasolution $\mathfrak{M}^{\min}_{[\lambda, a_-, \Omega]}$.

To construct examples satisfying the requirements of this theorem, we can proceed as follows. First, fix $p > 0$ satisfying $p < 4/(N - 2)$ if $N \geq 3$ and choose $R > 0$. Then, pick $R_2 < R$ sufficiently close to R so that

$$
\lambda_1[-\Delta, B_{R_2}] < \lambda_1[-\Delta, B_R \setminus \bar{B}_{R_2}] = \lambda_1[-\Delta, \Omega_0]. \tag{9.56}
$$

This is possible because we already know that

$$\lim_{R_2 \uparrow R} \lambda_1[-\Delta, B_{R_2}] = \sigma[-\Delta, B_R] \quad \text{and} \quad \lim_{R_2 \uparrow R} \lambda_1[-\Delta, B_R \setminus \bar{B}_{R_2}] = \infty.$$

After reaching (9.56), pick $R_1 \in (0, R_2)$ sufficiently small so that

$$\sigma_1 := \lambda_1[-\Delta, \Omega_0] < \sigma_2 := \lambda_1[-\Delta, B_{R_1}] = \sigma[-\Delta, \Omega_+],$$

which is possible because

$$\lim_{R_1 \downarrow 0} \lambda_1[-\Delta, B_{R_1}] = \infty.$$

Then, pick a λ satisfying

$$d\sigma_0 = \lambda_1[-d\Delta, B_{R_2}] < \lambda < d\sigma_1 = \lambda_1[-d\Delta, B_R \setminus \bar{B}_{R_2}]$$

and $\rho \in (R_1, R_2)$ sufficiently close to R_2 so that

$$\lambda_1[-d\Delta, B_\rho] < \lambda < \lambda_1[-d\Delta, B_R \setminus \bar{B}_{R_2}].$$

Finally, one should choose appropriate $u_0(r)$ and $a(r)$, satisfying the remaining requirements of the statement.

Proof of Theorem 9.22: By Theorem 9.21, u blows up in $L^\infty(\Omega)$ in a finite time T_b and (9.55) holds.

Subsequently, we assume (H1)–(H3). In particular,

$$a(0) > 0 > a(\rho), \qquad \int_0^\rho a(r) r^{n-1}\, dr > 0,$$

because $0 \in \Omega_+$, where $a > 0$, and any x with $|x| = \rho$ lies in Ω_-, where $a < 0$. Pick $t_0 \in (0, T_b)$ and consider the auxiliary problem

$$\begin{cases} v_t - d\Delta v = \lambda v + a(|x|) v^{p+1}, & x \in B_\rho,\ t > t_0, \\ v(x, t) = 0, & x \in \partial B_\rho,\ t > t_0, \\ v(x, t_0) = \delta\varphi, & x \in B_\rho, \end{cases} \tag{9.57}$$

where $\varphi > 0$ stands for a principal eigenfunction of $\lambda_1[-\Delta, B_\rho]$ and $\delta > 0$ is sufficiently small so that $\delta\varphi$ can be a strict subsolution of

$$-d\Delta v = \lambda v + a_- v^{p+1} \quad \text{in} \quad B_\rho.$$

It may be accomplished because $\lambda > \lambda_1[-d\Delta, B_\rho]$. Shortening δ, if necessary, one can also get

$$\delta\varphi < u(\cdot, t_0; \lambda, a, u_0) \quad \text{in} \quad \bar{B}_\delta. \tag{9.58}$$

Let $v(x, t; \delta\varphi)$ denote the solution of (9.57) and consider $\tilde{v}(x, t; \delta\varphi)$, its associated extended function through the approximating process (9.49), (9.50). By (9.58), the parabolic maximum principle implies that

$$\tilde{v}(x, t; \delta\varphi) \le \tilde{u}(x, t + t_0; u_0) \quad \text{for all} \quad x \in B_\rho,\ t > 0. \tag{9.59}$$

Moreover, thanks to Theorem 9.20, $v(x, t; \delta\varphi)$ blows up in $L^\infty(B_\rho)$ in a finite time $\hat{T}_b \geq T_b - t_0$. Thus, owing to [177, Th. 1.3], we find that

$$\tilde{v}(x, t; \delta v) = +\infty \qquad \text{for all } (x, t) \in \bar{B}_{R_1} \times (\hat{T}_b, \infty).$$

Therefore, setting

$$T_c := \hat{T}_b + t_0 \geq T_b,$$

we find from (9.59) that

$$\tilde{u}(x, t; u_0) = +\infty \qquad \text{for all } (x, t) \in \bar{B}_{R_1} \times (T_c, \infty),$$

which concludes the proof of (9.54).

Now, set $u := u(r, t)$ for each $r \in [0, R]$ and $t \in [0, T_b)$, and suppose

$$\lim_{k \to \infty} u(R_1, t_k) = \infty \tag{9.60}$$

for some sequence $t_k \uparrow T_b$ as $k \to \infty$. Then, the auxiliary function

$$w(r, t) := r^{N-1} u_r(r, t), \qquad (r, t) \in [0, R] \times [0, T_b),$$

satisfies

$$w(0, t) = 0, \quad w(R, t) < 0, \quad w(r, 0) \leq 0,$$

and, differentiating with respect to r the differential equation, multiplying by r^{N-1} and rearranging terms yields

$$w_t = d\left(w_{rr} - \frac{N-1}{r} w_r\right) + \lambda w + a(p+1)u^p w + a_r r^{N-1} u^{p+1}$$

$$\leq d\left(w_{rr} - \frac{N-1}{r} w_r\right) + (\lambda + a(p+1)u^p) w \qquad \text{in } (0, \rho) \times (0, T_b),$$

because $a_r \leq 0$ in $[0, \rho]$. Thus, $w \leq 0$ and hence, $u_r \leq 0$ in $(0, \rho) \times (0, T_b)$. Therefore, it follows from (9.60) that

$$\lim_{t \uparrow T_b} u(r, t) = +\infty \quad \text{uniformly in } [0, R_1],$$

since $t \mapsto u(r, t)$ is increasing, because u_0 is a subsolution of (9.4). The remaining assertions of the theorem follow straight from the fact that

$$L(x) := \lim_{t \uparrow \infty} \tilde{u}(x, t; u_0), \qquad x \in \Omega \setminus \bar{\Omega}_+,$$

provides us with a solution of

$$\begin{cases} -d\Delta u = \lambda u + a_- u^{p+1} & \text{in } \Omega \setminus \bar{\Omega}_+, \\ u = \infty & \text{on } \partial\Omega_+, \\ u = 0 & \text{on } \partial\Omega. \end{cases}$$

This ends the proof. □

In the contrary case when there exists a constant $C > 0$ such that

$$u(R_1, t; \lambda, a, u_0) \leq C \qquad \text{for all } t > 0, \tag{9.61}$$

the limit

$$g(R_1) := \lim_{t \uparrow \infty} u(R_1, t; \lambda, a, u_0)$$

is well defined and hence,

$$L := \lim_{t \uparrow \infty} u(\cdot, t; \lambda, a, u_0) \quad \text{in} \quad \Omega \setminus \bar{\Omega}_+$$

provides us with a positive strong solution of

$$\begin{cases} -d\Delta w = \lambda w + a_- w^{p+1} & \text{in} \quad \Omega \setminus \bar{\Omega}_+, \\ w = g & \text{on} \quad \partial \Omega_+, \\ w = 0 & \text{on} \quad \partial \Omega. \end{cases}$$

Therefore, it is of great interest to ascertain whether (9.60) or (9.61) holds. In Figure 9.7 we have represented the two possible limiting profiles of $\tilde{u}(x, t)$ as $t \uparrow \infty$, according to either case (9.60), left, and (9.61), right. It should be noted that $\tilde{u} = \infty$ in Ω_+ for sufficiently large $t > 0$.

As a consequence of Theorem 4.9 we already know that

$$\lim_{\lambda \uparrow d\sigma_1} \mathfrak{M}^{\min}_{[\lambda, a_-, \Omega \setminus \bar{\Omega}_+]} = \mathfrak{M}^{\min}_{[d\sigma_1, a_-, \Omega_-]}.$$

Consequently, the profile of $L^{\min}_{[\lambda, a_-, \Omega \setminus \bar{\Omega}_+]}$ for λ close to $d\sigma_1$ looks like Figure 9.7. On the other hand, according to the proof of Theorem 9.22, $u_r \leq 0$ in

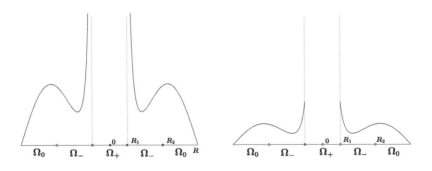

FIGURE 9.7: The asymptotic profiles of $\tilde{u}$ in cases (9.60) and (9.61).

$(0, \rho)$ for all $t < T_b$, which might be incompatible with the graphs shown in Figure 9.7. Therefore, it seems that the matasolution cannot be reached in a finite time. Of course, such a possibility cannot be excluded when λ is separated from $d\sigma_1$. By comparing the gradients from both sides, the interior of Ω_+ and the interior of $\Omega \setminus \bar{\Omega}_+$, it should be possible to prove that the situation illustrated by the right picture of Figure 9.7 cannot occur.

9.7   Dynamics for $\lambda \in [d\sigma_1, d\sigma_2)$ and a_+ small

Throughout the rest of this chapter, to simplify the notations as much as possible, we will assume that the nodal behavior of $a(x)$ obeys the general patterns of Figure 9.1 with

$$\sigma_1 := \lambda_1[-\Delta, \Omega_0] < \sigma_2 := \lambda_1[-\Delta, \Omega_+].$$

Then, the next result holds.

Theorem 9.23 *Suppose $\lambda < d\sigma_2$ and the singular problem*

$$\begin{cases} -d\Delta u = \lambda u + a_- u^{p+1} & \text{in } \Omega \setminus \bar{\Omega}_0, \\ u = +\infty, & \text{on } \partial(\Omega \setminus \bar{\Omega}_0), \end{cases} \tag{9.62}$$

possesses a unique positive solution, L. Then, there exists $\varepsilon > 0$ such that

$$\begin{cases} -d\Delta u = \lambda u + a u^{p+1} & \text{in } \Omega \setminus \bar{\Omega}_0, \\ u = +\infty, & \text{on } \partial(\Omega \setminus \bar{\Omega}_0), \end{cases} \tag{9.63}$$

possesses a minimal positive solution if $(a_+)_M \leq \varepsilon$.

Proof: Some sufficient conditions for the uniqueness of the large solution of (9.62) had been already given in Part II. According to the Hardy inequality, it follows from M. Bertsch and R. Rostamian [27] that the principal eigenvalue of the linearization of (9.62) at L is well defined and satisfies

$$\lambda_1[-d\Delta - (p+1)a_- L^p - \lambda, \Omega \setminus \bar{\Omega}_0] > 0, \tag{9.64}$$

because

$$\lambda_1[-d\Delta - a_- L^p - \lambda, \Omega \setminus \bar{\Omega}_0] = 0.$$

Actually, the existence of these eigenvalues for general singular potentials, as well as some of their most pivotal properties, will be rigourously established in Section 10.8, so we refrain from giving any further detail here.

The *non-degeneracy condition* (9.64) allows us to apply the implicit function theorem to construct a solution of (9.63) close to the unique solution of

(9.62). Once we establish the existence of a solution, the minimal one can be constructed as in Chapter 4. This ends the proof. □

The following result ascertains the dynamics of (9.1) when u_0 is a subsolution of (9.4) and $(a_+)_M \leq \varepsilon$.

Theorem 9.24 *Suppose* $\lambda \in [d\sigma_1, d\sigma_2)$, u_0 *is a positive strict subsolution of* (9.4), *and* (9.63) *possesses a minimal positive solution,* $L^{\min}_{[\lambda,a,\Omega\backslash\bar{\Omega}_0]}$. *Then, the solution of* (9.1), $u(x,t;\lambda,a,u_0)$, *is globally defined in time, i.e.,* $T_{\max}(u_0) = \infty$. *Moreover,*

$$\lim_{t\uparrow\infty} u(x,t;\lambda,a,u_0) = L^{\min}_{[\lambda,a,\Omega\backslash\bar{\Omega}_0]}(x) \quad \text{for all} \quad x \in \Omega \backslash \bar{\Omega}_0, \tag{9.65}$$

while

$$\lim_{t\uparrow\infty} u(x,t;\lambda,a,u_0) = \infty \quad \text{for all} \quad x \in \bar{\Omega}_0 \backslash \partial\Omega. \tag{9.66}$$

Proof: The subsolution u_0 exists because $\lambda > d\sigma_0$. Moreover, u_0 is a subsolution of (9.3) because $a_- \leq a$. Thus, $t \mapsto u(\cdot,t;\lambda,a,u_0)$ is point-wise increasing, i.e., $u(\cdot,t;\lambda,a,u_0)$ is a subsolution of (9.3) for all $t \in [0,T_{\max})$. Actually, as u_0 is a positive strict subsolution of (9.4) and $L^{\min}_{[\lambda,a,\Omega\backslash\bar{\Omega}_0]}$ provides us with a positive strict supersolution of (9.63), it follows from Lemma 1.8 that

$$u_0 < \theta_{[\lambda,a_-,\Omega]} < L_{[\lambda,a_-,\Omega\backslash\bar{\Omega}_0]} < L^{\min}_{[\lambda,a,\Omega\backslash\bar{\Omega}_0]} \quad \text{in} \quad \Omega \backslash \bar{\Omega}_0.$$

Thus, by the the maximum principle,

$$u(x,t;\lambda,a,u_0) \leq L^{\min}_{[\lambda,a,\Omega\backslash\bar{\Omega}_0]}(x) \quad \text{for all} \quad (x,t) \in (\Omega \backslash \bar{\Omega}_0) \times (0,T_{\max}). \tag{9.67}$$

Subsequently, for sufficiently small $\delta > 0$ with $\partial\Omega^\delta_+ \subset \Omega_-$ we consider the open set

$$\Omega^\delta_+ := \{\, x \in \Omega \,:\, \text{dist}\,(x,\Omega_+) < \delta \,\}.$$

Thanks to (9.67), there is a constant $M > 0$ such that

$$u(x,t;\lambda,a,u_0) \leq M \quad \text{for all} \quad (x,t) \in \partial\Omega^\delta_+ \times (0,T_{\max}).$$

Let U denote the unique solution of

$$\begin{cases} \frac{\partial u}{\partial t} - d\Delta u = \lambda u + a_- u^{p+1} \leq \lambda u & \text{in} \quad (\Omega \backslash \Omega^\delta_+) \times (0,\infty) \\ u = 0 & \text{on} \quad \partial\Omega \times (0,\infty) \\ u = M & \text{on} \quad \partial\Omega^\delta_+ \times (0,\infty) \\ u(\cdot,0) = u_0 & \text{in} \quad \Omega \backslash \Omega^\delta_+. \end{cases}$$

Note that U is globally defined in time. Moreover, by the parabolic maximum principle,

$$u(x,t;\lambda,a,u_0) \leq U(x,t) \quad \text{for all} \quad (x,t) \in (\Omega \backslash \Omega^\delta_+) \times (0,T_{\max}).$$

Therefore, $u(\cdot, t; \lambda, a, u_0)$ is bounded above in $\bar{\Omega}$ for all $t \in (0, T_{\max})$ and consequently, $T_{\max} = +\infty$.

On the other hand, for every $t > 0$,

$$u(\cdot, t; \lambda, a_-, u_0) \leq u(\cdot, t; \lambda, a, u_0) \qquad \text{in } \Omega,$$

and hence, (9.66) follows from Theorem 5.2(c). In particular,

$$\lim_{t \uparrow \infty} u(x, t; \lambda, a, u_0) = \infty \quad \text{for all } (x, t) \in \partial(\Omega \setminus \bar{\Omega}_0) \times (0, \infty).$$

Therefore, $u(\cdot, t; \lambda, a, u_0)$ must approach a positive solution of (9.63) bounded above by $L^{\min}_{[\lambda, a, \Omega \setminus \bar{\Omega}_0]}$. By the minimality of $L^{\min}_{[\lambda, a, \Omega \setminus \bar{\Omega}_0]}$, (9.65) holds and the proof is complete. $\quad \square$

9.8   Dynamics for $\lambda \in [d\sigma_1, d\sigma_2)$ and a_+ large

The next counterpart of Theorem 9.20(b) establishes that $T_b = T_{\max} < +\infty$ if $u_0 > 0$ is a strict subsolution of (9.4) and $(a_+)_M$ is sufficiently large.

Theorem 9.25 *Suppose* $\lambda \in [d\sigma_1, d\sigma_2)$, $u_0 > 0$ *is a strict subsolution of* (9.4) *and there exists a smooth domain* D *such that* $\bar{D} \subset \Omega_+$ *and*

$$A := \min_{\bar{D}} a_+ > \left(\frac{p}{\alpha} \right)^p \left(\frac{\Sigma}{p+1} \right)^{p+1} \tag{9.68}$$

where

$$\Sigma := \lambda_1[-d\Delta, D] - \lambda, \qquad \alpha := -d \int_{\partial D} \frac{\partial \varphi}{\partial n} L^{\min}_{[\lambda, a_-, \Omega \setminus \bar{\Omega}_0]} \, d\sigma,$$

likewise in Theorem 9.20. Then, $T_{\max}(u_0) < +\infty$, *i.e.,* $u(\cdot, t; \lambda, a, u_0)$ *blows up in a finite time* $T_b := T_{\max}$ *in* $L^\infty(\Omega_+)$.

Proof: Fix $\lambda \in [d\sigma_1, d\sigma_2)$ and let $u_0 > 0$ be a strict subsolution of (9.4); it exists because $\lambda > d\sigma_0$. Then, u_0 provides us with a subsolution of (9.3) and hence the mapping $t \mapsto u(\cdot, t; \lambda, a, u_0)$ is increasing. Moreover, for any $t > 0$,

$$u(\cdot, t; \lambda, a, u_0) \geq u(\cdot, t; \lambda, a_-, u_0) \qquad \text{in } \Omega.$$

Thus, by Theorem 5.2(c),

$$\lim_{t \uparrow \infty} u(\cdot, t; \lambda, a, u_0) \geq \mathfrak{M}^{\min}_{[\lambda, a_-, \Omega \setminus \bar{\Omega}_0]}.$$

In particular,

$$\lim_{t \uparrow \infty} u(\cdot, t; \lambda, a, u_0) \geq L^{\min}_{[\lambda, a_-, \Omega \setminus \bar{\Omega}_0]} \qquad \text{in } D \subset \Omega_+. \tag{9.69}$$

Note that, since

$$\lambda < d\sigma_2 = \lambda_1[-d\Delta, \Omega_+] < \lambda_1[-d\Delta, D],$$

$\Sigma > 0$. Moreover, $\alpha > 0$, because $\frac{\partial\varphi}{\partial n} \ll 0$ on ∂D. So, the constant on the right hand side of (9.68) is positive. The rest of the proof follows by adapting the argument given in the proof of Theorem 9.20(b). Now, instead of (9.40), we should use (9.69). The technical details are omitted here. □

As for every $\lambda \in [d\sigma_0, d\sigma_1)$ and $\mu, \varrho \in [d\sigma_1, d\sigma_2)$ with $\mu < \varrho$, we have that

$$\theta_{[\lambda,a_-,\Omega]} < L^{\min}_{[\mu,a_-,\Omega\setminus\bar{\Omega}_0]} < L^{\min}_{[\varrho,a_-,\Omega\setminus\bar{\Omega}_0]} \quad \text{in} \quad D \subset \Omega_+,$$

it becomes apparent that (9.37) implies (9.68).

As a consequence from Theorem 9.25, the next counterpart of Theorem 9.21 holds.

Theorem 9.26 *Suppose* $\lambda \in [d\sigma_1, d\sigma_2)$, $u_0 > 0$ *is a strict subsolution of* (9.4), *and* (9.68) *holds for some smooth domain* D *with* $\bar{D} \subset \Omega_+$. *Then,* $u := u(x, t; \lambda, a, u_0)$ *blows up in* $L^\infty(\Omega_+)$ *in a finite time* $T_b = T_{\max} < +\infty$, *and the following assertions are true:*

(a) $\tilde{u} = u \le L^{\min}_{[\lambda,a_-,\Omega_-]}$ *in* Ω_- *is a classical solution for each* $t > 0$, *where* $L^{\min}_{[\lambda,a_-,\Omega_-]}$ *stands for the minimal positive solution of*

$$\begin{cases} -d\Delta w = \lambda w + a_- w^{p+1} & \text{in} \quad \Omega_-, \\ w = \infty & \text{on} \quad \partial\Omega_-. \end{cases} \tag{9.70}$$

(b) *For sufficiently small* $\delta > 0$,

$$u(\cdot, t; \lambda, a_-, u_0) \le \tilde{u}(\cdot, t; u_0) \le U_\delta(\cdot, t) \quad \text{in} \quad \bar{\Omega}_0, \tag{9.71}$$

where U_δ *stands for the unique solution of*

$$\begin{cases} \frac{\partial u}{\partial t} - d\Delta u = \lambda u + a_- u^{p+1} & \text{in} \quad \Omega_\delta \times (0, \infty) \\ u = 0 & \text{on} \quad \partial\Omega \times (0, \infty) \\ u = L^{\min}_{[\lambda,a_-,\Omega_-]} & \text{on} \quad (\partial\Omega_\delta \cap \Omega_-) \times (0, \infty) \\ u(\cdot, 0) = u_0 & \text{in} \quad \Omega_\delta \end{cases} \tag{9.72}$$

with

$$\Omega_\delta := \{ x \in \Omega : \ \text{dist}(x, \Omega_0) < \delta \}.$$

Thus, $\tilde{u}(\cdot, t; u_0)|_{\bar{\Omega}_0}$ *is a classical solution for each* $t > 0$ *and*

$$\lim_{t\uparrow\infty} \tilde{u}(\cdot, t; u_0) = \infty \quad \text{uniformly in compact subsets of } \bar{\Omega}_0 \setminus \partial\Omega. \tag{9.73}$$

In particular, $B(u_0) \subset \bar{\Omega}_+$. *If one further assumes that either* (9.51) *or* $a_+(x)$ *satisfies Hypothesis* (HB) *in Remark 9.17, then* (9.52) *holds. If, in addition,* $B(u_0) \subset \Omega_+$, *then* (9.53) *also holds and hence* $u(x, t; \lambda, a, u_0)$ *blows up completely in* Ω_+.

Proof: The fact that u blows up in $L^\infty(\Omega_+)$ in a finite time T_b is guaranteed by Theorem 9.25. Part (a) follows from the fact that $\tilde{u}(\cdot, t; u_0)|_{\Omega_-}$ provides us with a subsolution of the singular problem (9.70) for all $t > 0$. Now, choose $\delta > 0$ sufficiently small so that $\partial\Omega_\delta \setminus \partial\Omega \subset \Omega_-$. Then, thanks to Part (a), $\tilde{u}$ provides us with a subsolution of (9.72) and therefore the upper estimate of (9.71) holds. The lower estimate follows from the fact that $u_{[\lambda, a_-, \Omega]}$ is a subsolution of (9.1). The property (9.73) follows from (9.71) and Theorem 5.2. The remaining assertions of the theorem are easy consequences from the previous features by adapting the proof of Theorem 9.21.   □

Adapting Theorem 9.22 to the present situation, one can obtain some further sufficient conditions ensuring that

$$\tilde{u}(\cdot, t; u_0) = \infty \quad \text{in} \quad \bar{\Omega}_+ \quad \text{for all} \quad t > T_b,$$

while

$$\lim_{t \uparrow \infty} \tilde{u}(\cdot, t; u_0) = \begin{cases} +\infty & \text{in } \bar{\Omega}_0 \setminus \partial\Omega, \\ L^{\min}_{[\lambda, a_-, \Omega_-]} & \text{in } \Omega_-, \end{cases}$$

though no additional details will be given here.

9.9   Dynamics for $\lambda \geq d\sigma_2$

For this range of values of λ no restriction on the size of $(a_+)_M$ is needed to get the blow-up in Ω_+. Consequently, the following result holds.

Theorem 9.27 *Suppose $\lambda \geq d\sigma_2$ and $u_0 > 0$ is a strict subsolution of (9.4). Then, $u(x, t; \lambda, a, u_0)$ blows up in $L^\infty(\Omega_+)$ in a finite time T_b and the following assertions are true:*

(a) $\tilde{u}(\cdot, t; u_0) \leq L^{\min}_{[\lambda, a_-, \Omega_-]}$ *in Ω_- is a classical solution for all $t > 0$.*

(b) *For sufficiently small $\delta > 0$,*

$$u(\cdot, t; \lambda, -a, u_0) \leq \tilde{u}(\cdot, t; \lambda, a, u_0) \leq U_\delta(\cdot, t) \quad \text{in } \bar{\Omega}_0,$$

where $U_\delta(x, t)$ stands for the unique solution of (9.72). Thus, $\tilde{u}(\cdot, t; u_0)|_{\bar{\Omega}_0}$ is a classical solution for each $t > 0$ and (9.73) holds.

In particular, $B(u_0) \subset \bar{\Omega}_+$. If one further assumes that either (9.51) or $a_+(x)$ satisfies Hypothesis (HB) in Remark 9.17, then (9.52) holds. If, in addition, $B(u_0) \subset \Omega_+$, then (9.53) holds, i.e., u blows up completely in Ω_+.

9.10 Comments on Chapter 9

This chapter has been elaborated from R. Gómez-Reñasco and J. López-Gómez [106], [107], [108], J. López-Gómez [148], [157] and J. López-Gómez and P. Quittner [177].

Proposition 9.2 goes back to H. Berestycki, I. Capuzzo-Dolcetta and L. Nirenberg [23], [24], as well as its proof through the Picone indentity, [209]. Theorem 9.5 goes back to J. López-Gómez [148]. The numerical experiments of Section 9.1 were carried out by R. Gómez-Reñasco and J. López-Gómez [106]. Originally, they were intended for the *Proceedings of the International Conference on Operator Theory* held in October 1998 in Winnipeg (Manitoba, Canada), but [106] was rejected and never re-submitted. So, the results of the numerical experiments with the example (9.14) have been published in this book for the first time.

The conjecture on the number of solutions of (9.3) in the one-dimensional setting goes back to R. Gómez-Reñasco and J. López-Gómez [106], [107], [108] and it has been solved very recently in an extremely elegant paper by G. Feltrin and F. Zanolin, [84]. Proposition 9.6 is folklore and Proposition 9.7 was inspired by H. Amann [11, Pr. 20.8].

The uniqueness of the stable positive solution established in Sections 9.3 and 9.4, as well as Theorem 9.16, goes back to R. Gómez-Reñasco and J. López-Gómez [107], [108], and it was part of the Ph.D. thesis of R. Gómez-Reñasco [105] under the supervision of the author. The nice global continuation argument used in the proof of Theorem 9.12 has been invoked in a number of different contexts by other authors to obtain similar stability and multiplicity results. Among others, it was adapted by J. López-Gómez, M. Molina-Meyer and A. Tellini [173] to establish the existence and the uniqueness of the linearly stable positive solution in a general class of non-homogeneous superlinear indefinite boundary value problems, by J. García-Melián et al. [102] to establish some stability and multiplicity results in an elliptic problem with nonlinear absorption and a nonlinear incoming flux on the boundary, and by J. García-Melián [99] to establish the stability of the minimal large solution of (9.63) at $\lambda = 0$ as well as the existence of a further large solution.

Proposition 9.14 goes back, at least, to H. Amann [11, Th. 20.3], where it was shown that (9.3) possesses a minimal positive solution if it has some, and H. Amann [11, Pr. 20.4], where the linear stability of the minimal positive solution was established. Our proof of Proposition 9.14 goes back to J. López-Gómez, M. Molina-Meyer and A. Tellini [175].

The analysis of the dynamics of (9.1) in Sections 9.6–9.9 goes back to J. López-Gómez [157], [159] and J. López-Gómez and P. Quittner [177]. The concept of complete blow-up was coined by P. Baras and L. Cohen [20]. The results of these sections show the crucial role played by the metasolutions to describe the asymptotic behavior of the positive solutions not only in the context of sublinear problems such as those studied in Parts I and II, but also for superlinear indefinite problems such as those analyzed in this chapter.

Theorem 9.23 was the first available result establishing the existence of large solutions in a class of semilinear elliptic equations of superlinear indefinite type. It goes back to J. López-Gómez [157]. Seven years later it was re-discovered by J. García-Melián [99]. Similarly, the multiplicity result of [99] also is a direct consequence from the analysis carried out in this chapter. In his master's thesis submitted to the University of Udine, A. Tellini [227] introduced the one-dimensional prototype model

$$\begin{cases} -u'' = \lambda u + a_b(x)u^{p+1}, & x \in (0,1), \\ u(0) = u(1) = \infty, \end{cases} \tag{9.74}$$

for the choice

$$a_b(x) := \begin{cases} -c & \text{if } x \in [0,\alpha) \cup (1-\alpha,1], \\ b & \text{if } x \in [\alpha,1-\alpha], \end{cases} \tag{9.75}$$

with $\alpha \in (0,0.5)$, $b > 0$ and $c > 0$, and found an extremely interesting multiplicity result. Namely, using b as the main continuation parameter and $\lambda \leq 0$ as the secondary one, A. Tellini [227] established that there is a value of the parameter b, b^*, for which (9.74) has an arbitrarily large number of positive solutions provided $-\lambda > 0$ is sufficiently large. This multiplicity result was substantially sharpened by J. López-Gómez, A. Tellini and F. Zanolin in [184]; by using some sophisticated phase portrait techniques all the possible global bifurcation diagrams in b of (9.74) were ascertained according to the different ranges of values of the secondary parameter λ. As a consequence of the analysis carried out in [184], for any integer $n \geq 0$ the problem (9.74) possesses solutions with n strict critical points in the interval $(\alpha, 1-\alpha)$ if $-\lambda > 0$ is sufficiently large. Moreover, these solutions are asymmetric if $n = 2m \geq 2$, in spite of the symmetry of the problem. The solution with 0 strict critical points, referred to as the *trivial solution*, plays a pivotal role in understanding the complexity of the global bifurcation diagrams of (9.74). Actually, it plays a similar role as an *organizing center* in the context of singularity theory (see, e.g., M. Golubitsky and D. G. Shaeffer [104]). It can be constructed as follows. Let $\ell(x)$ denote the unique solution of

$$\begin{cases} -u'' = \lambda u - cu^p & \text{in } (0,\alpha), \\ u(0) = +\infty, \quad u'(\alpha) = 0, \end{cases}$$

and set $m_0 := \ell(\alpha)$. Then, the constant m_0 solves $-u'' = \lambda u + bu^{p+1}$ if, and only if,

$$b = b^* := -\lambda/m_0^p.$$

For this value of b, the trivial solution u_t is defined through

$$u_t(x) = \begin{cases} \ell(x), & x \in [0,\alpha), \\ m_0, & x \in [\alpha,1-\alpha], \\ \ell(1-x), & x \in (1-\alpha,1]. \end{cases}$$

Note that u_t is symmetric around $x = 0.5$. Subsequently, given a solution u of Problem (9.74) with $n \geq 0$ strict critical points in $(\alpha, 1-\alpha)$, it is said that u is of *type* (n, a) if u is asymmetric around $x = 0.5$, while it is said that it is of type (n, s) if it is symmetric.

Basically, as illustrated by Figure 9.8, for $\lambda \leq 0$ sufficiently close to 0, the global structure of the bifurcation diagram of positive solutions of (9.74) consists of a *primary curve* establishing a homotopy between the (unique) solution of the sublinear problem

$$\begin{cases} -u'' = \lambda u + a_0(x)u^{p+1} & \text{in } (0,1), \\ u(0) = u(1) = +\infty, \end{cases} \tag{9.76}$$

and the metasolution

$$\mathfrak{M}(x) := \begin{cases} L(x), & x \in [0, \alpha), \\ \infty, & x \in [\alpha, 1-\alpha], \\ L(1-x), & x \in (1-\alpha, 1], \end{cases} \tag{9.77}$$

where L stands for the unique solution of the singular problem

$$\begin{cases} -u'' = \lambda u - cu^{p+1} & \text{in } [0, \alpha), \\ u(0) = u(\alpha) = +\infty. \end{cases} \tag{9.78}$$

The plots of Figure 9.8, as well as all subsequent plots of this section, were computed by J. López-Gómez, M. Molina-Meyer and A. Tellini [176] for the special values of the parameters

$$\alpha = 1/3, \qquad p = 1, \qquad c = 1.$$

As in all the remaining bifurcation diagrams, in the left picture we are representing the value of b in abscissas versus the value of $u(\alpha)$ in ordinates. The bifurcation diagram shows a single *primary curve* which emanates from

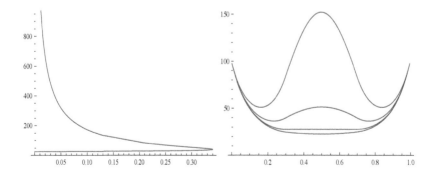

FIGURE 9.8: Global bifurcation diagram for $\lambda = -5$ and plots of some solutions along it.

the unique positive solution of (9.76), whose existence and uniqueness can be established as in Parts I and II, and can be path-followed toward the right to reach the critical value $b = 0.3401$, where it turns backward exhibiting a sub-critical turning point. Once switched to this turning point, the solutions on the upper half-branch can be continued for all $0 < b < 0.3401$. The numerical experiments of [176] reveal that along the upper half-branch the solutions of (9.74) blow up in $[\alpha, 1-\alpha]$ as $b \downarrow 0$, while they approximate L in $[0, \alpha)$. Conse-quently, as $b \downarrow 0$, these solutions approximate the *metasolution* $\mathfrak{M}(x)$ defined in (9.77). Therefore, the global bifurcation diagram establishes a homotopy between the unique classical solution of the sublinear problem (9.76) and the unique metasolution of

$$-u'' = \lambda u + a_b(x)u^{p+1}$$

supported in $(0, \alpha) \cup (1-\alpha, 1)$. In particular, for every $b \in (0, 0.3401)$, the prob-lem (9.74) possesses at least two positive solutions. Moreover, the solutions on the lower half-branch are linearly asymptotically stable, while those on the upper half-branch are unstable with one-dimensional unstable manifolds.

Essentially, as $-\lambda > 0$ increases, a piece of the primary curve rotates counterclockwise around the trivial solution u_t and, almost after every half rotation, a closed loop of positive solutions bifurcates from it. The loop consists of solutions of asymmetric type and it persists for all further values of $-\lambda$. The left plot of Figure 9.9 shows the global bifurcation diagram of the positive solutions of (9.74) for $\lambda = -300$. It consists of two curves: the continuous line, which is the primary branch, and the dashed line, which is the first solution loop bifurcating from the primary curve at $b = 12.8294$ and $b = 526.4099$.

The subcritical turning point of the primary curve occurs at $b = 527.4319$. The central plot of Figure 9.9 shows a magnification of this turning point capturing the subcritical bifurcation point of the closed loop. All solutions on the primary curve between the two squares marked in the left plot are of type

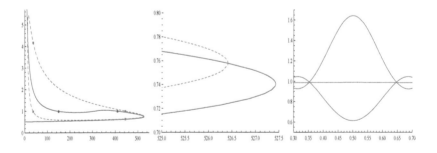

FIGURE 9.9: Global bifurcation diagram for $\lambda = -300$ (left picture), magni-fication of the turning point along the primary curve exhibiting the subcritical bifurcation of the first closed loop (central picture), and plots of u_t and two solutions of type $(3, s)$ (right picture).

$(3, s)$, while the remaining ones are of type $(1, s)$. The changes of type on the primary branch occur at the level of the trivial solution, u_t. The readers can consult [176] for a detailed discussion about the types of the solutions along each solution arc in the global bifurcation diagram. The lower half-branch of the primary curve consists of linearly stable solutions, while the upper one is filled by unstable solutions with one-dimensional unstable manifolds if they are outside the loop and two-dimensional unstable manifolds if they are encircled by the loop. All the solutions on the loop have one-dimensional unstable manifolds. The solutions on the right picture of Figure 9.9 have been plotted in the central interval $(\alpha, 1 - \alpha)$, instead of in $(0, 1)$, because $u(0) = u(1) = +\infty$.

As λ decreases from $\lambda = -300$ up to $\lambda = -750$, we get the global bifurcation diagram plotted at the top left in Figure 9.10. Now, the previous arc of curve with the solutions of type $(3, s)$ along the primary branch rotates counterclockwise around u_t generating two new turning points on it: one supercritical and another subcritical. As in the previous cases, in the first picture of Figure 9.10 we are plotting with a dashed line the first loop bifurcated from the solutions of type $(1, s)$ on the primary curve, and the small squares along it mark the values of b where the type of the solutions changed from $(1, a)$

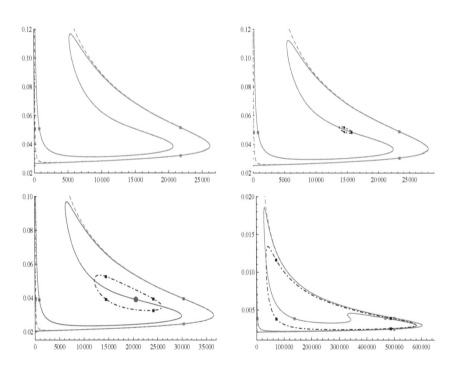

FIGURE 9.10: Global bifurcation diagram for $\lambda = -750$ (upper left), $\lambda = -760.3$ (upper right), $\lambda = -800$ (bottom left) and $\lambda = -1300$ (bottom right).

to $(2, a)$, or vice versa. The maximal subcritical turning point on the primary branch arises at $b = 2.6184588 \times 10^4$ and the two bifurcation points of the first closed loop from the primary curve occur at $b = 0.0011 \times 10^4$ and $b = 2.6184574 \times 10^4$.

At the scale used to print the global bifurcation diagrams of Figure 9.10, there was not enough room to plot the first closed loop of asymmetric solutions emanated from the primary branch. Indeed, in the first plot, the values of $u(\alpha)$ along the loop reached 2.15, which is substantially larger than 0.12. Changing the scale on the vertical axis to plot the entire loop would not have allowed us to appreciate the counterclockwise rotation of the branch of solutions of type $(3, s)$ on the primary curve, because the twist would have been compressed to a $1/18$ fraction of the vertical axis.

As λ decreased from $\lambda = -750$ and reached the value $\lambda = -760.3$, a second closed loop emanates from the primary curve. As λ decreases, the loop grows, approximating u_t. The upper right picture of Figure 9.10 shows the global bifurcation diagram of (9.74) for $\lambda = -760.3$. In Figure 9.10, the second closed loops bifurcated from the primary curves have been represented with a dot-dashed black line.

The second loop of the upper right diagram of Figure 9.10 bifurcates from the primary branch at $b = 1.3656 \times 10^4$ and $b = 1.5716 \times 10^4$. The trivial solution arises at $b^* = 1.5791 \times 10^4$. The types of the solutions along the second loop change at $b = 1.4522 \times 10^4$ and $b = 1.5695 \times 10^4$. As the range of values of b for which (9.74) admits a positive solution is very large, the last value looks very close to b^* in the diagram. Hence, two of the small squares marking the change of type along the second loop are almost superimposed with the dot marking u_t.

As λ decreases from $\lambda = -760.3$, the loop grows until it reaches u_t and encircles it for any smaller value of λ. The bottom left plot of Figure 9.10 shows the global bifurcation diagram of (9.74) for $\lambda = -800$. At this value, the second loop bifurcates from the primary branch at $b = 1.2110 \times 10^4$ and $b = 2.5941 \times 10^4$ and u_t arises at $b^* = 2.0530 \times 10^4$.

Finally, in Figure 9.11 we have plotted the global bifurcation diagram computed for $\lambda = -2000$, where one can clearly differentiate the third solution loop bifurcating from the primary curve. It should be noted that, for every b between the two bifurcation points of the third bifurcated loop from the primary curve, the problem (9.74) possesses at least 12 solutions: 6 symmetric and 6 asymmetric.

An extremely remarkable feature is that, in all the computed bifurcation diagrams, the dimensions of the unstable manifolds of all solutions on the primary curve increase by one at each bifurcation and turning point, until reaching the interior of the last closed bifurcated loop, where they start to decrease according to the same patterns to become one-dimensional again.

As far as the behavior of the n-th loop is concerned, it consists entirely of solutions of type $(2n-1, a)$ as occurs soon after it bifurcates from the primary branch, or it consists of solutions of type $(2n-1, a)$ near the bifurcation points from the primary curve together with two central arcs of solutions of type $(2n, a)$, for each $n = 1, 2, 3, \ldots$. But interested readers are referred to [176] for further technical or computational details.

Actually, according to J. López-Gómez, A. Tellini and F. Zanolin [184], there is a further sequence of values of λ,

$$\cdots < \lambda_{n+1} < \lambda_n < \lambda_{n-1} < \cdots < \lambda_5 < \lambda_4 < -2000$$

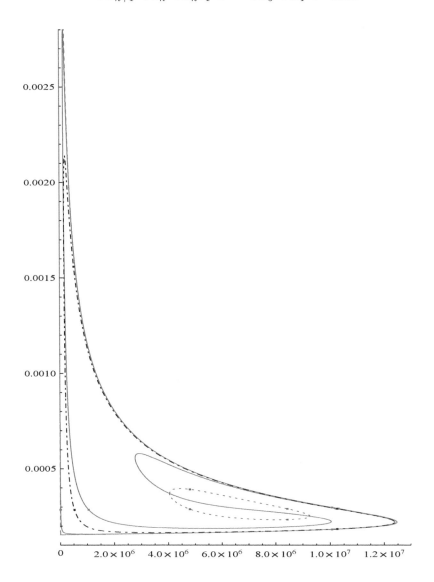

FIGURE 9.11: Global bifurcation diagram for $\lambda = -2000$.

such that whenever λ crosses λ_k, a new closed loop of positive solutions bifurcates from the primary branch at some value of b close to b^*, and all these closed loops persist for all smaller values of λ.

Naturally, to approximate the solutions of (9.74), J. López-Gómez, M. Molina-Meyer and A. Tellini computed the solutions of

$$\begin{cases} -u'' = \lambda u + a_b(x)u^{p+1}, & x \in (0,1), \\ u(0) = u(1) = M, \end{cases} \quad (9.79)$$

for sufficiently large $M > 0$, rather than the solutions of (9.74). Thanks to the results of J. López-Gómez, A. Tellini and F. Zanolin [184] the solutions of (9.79) indeed approximate the solutions of (9.74) as $M \uparrow \infty$.

Rather astonishingly, when a_b loses its symmetry about 0.5, all the previous global bifurcation diagrams spread out into an arbitrarily large number of global components. Essentially, each closed loop generates an isola as illustrated by Figure 9.12. Most precisely, choosing

$$a_b(x) := \begin{cases} -c_0 & \text{if } x \in [0,\alpha), \\ b & \text{if } x \in [\alpha, 1-\alpha], \\ -c_1 & \text{if } x \in (1-\alpha, 1], \end{cases} \quad (9.80)$$

with $c_0 \neq c_1$, instead of (9.75), provokes a bifurcation diagram like the one plotted on the left of Figure 9.9 to perturb into some of the bifurcation diagrams plotted in Figure 9.12 (A), if $c_0 > c_1$, or in Figure 9.12 (B), if $c_0 < c_1$, where the isolas provoked by the symmetry breaking have been plotted with a dashed line, while the perturbed primary branches have been plotted by a continuous line.

Actually, as a consequence of the work of J. López-Gómez and A. Tellini [183], there is a sequence of values of λ,

$$\cdots < \tilde{\lambda}_{n+1} < \tilde{\lambda}_n < \tilde{\lambda}_{n-1} < \cdots < \tilde{\lambda}_2 < \tilde{\lambda}_1 < 0$$

such that for every $\lambda < \tilde{\lambda}_n$ the set of positive solutions of problem (9.74) with

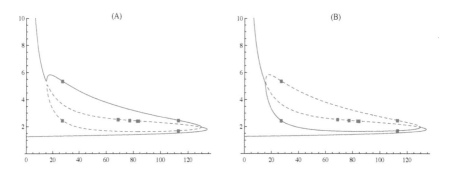

FIGURE 9.12: A genuine symmetry breaking of a bifurcation diagram.

the choice (9.80) possesses at least $n + 1$ components: one unbounded, the primary one, and the remaining n bounded. Naturally, $\tilde{\lambda}_n \to \lambda_n$ if $c_0 \to c_1 = c$.

Essentially, when the symmetry of the weight function $a_b(x)$ is broken, by whatever mechanism, each of the closed loops generates an isola, like the ones shown in Figure 9.12, changing the structure of the primary branch accordingly. The plots of Figure 9.12 go back to J. López-Gómez, M. Molina-Meyer and A. Tellini [174] where the interested reader is sent for any further detail.

In general multi-dimensional problems, we should not expect to get a global bifurcation diagram with the closed loops of asymmetric solutions bifurcating from the primary curve, but rather a finite number of compact components filled in by asymmetric solutions plus an additional (unbounded) primary curve, as in Figure 9.12 unless the problem is radially symmetric. More precisely, we conjecture that in the absence of radial symmetry for any given superlinear indefinite problem of the type

$$\begin{cases} -\Delta u = \lambda u + a(x)u^{p+1}, & x \in \Omega, \\ u = M, & x \in \partial\Omega, \end{cases} \tag{9.81}$$

with sufficiently large $0 < M \leq \infty$, there exists λ_n, $n \geq 1$, with $\lambda_{n+1} < \lambda_n < 0$ such that for all $\lambda \in (\lambda_{n+1}, \lambda_n)$ the solution set of positive solutions of (9.81) consists of at least $n + 1$ components; among them, one unbounded and n bounded.

9.10.1 Abiotic stress hypothesis

In population dynamics, when $a(x)$ changes sign in Ω and $\lambda < 0$ the parabolic problem

$$\begin{cases} \frac{\partial u}{\partial t} - d\Delta u = \lambda u + a(x)|u|^p u & \text{in} \quad \Omega \times (0, \infty), \\ u = M & \text{on} \quad \partial\Omega \times (0, \infty), \\ u(\cdot, 0) = u_0 > 0 & \text{in} \quad \Omega, \end{cases} \tag{9.82}$$

models the evolution of a single species in a harsh inhabiting region, Ω, surrounded by territories where the population density equals $M > 0$. In these models, $u(x, t)$ stands for the density of the species, $\lambda < 0$ measures the neat death rate of the species in Ω, and $u_0 > 0$ is the distribution of the initial population. In nature, λ is negative when pesticides are used in high concentrations or a certain patch of the natural environment is polluted by introducing chemicals, waste products, or poisonous substances. Suppose

$$a = a_b = a_- + ba_+.$$

Then, the parameter $b \geq 0$ measures the intensity of the intra-specific facilitative effects of the species u in Ω_+, the interior of the support of a_+. In these generalized logistic prototypes, the individuals of the species u compete for the resources in Ω_-, the interior of the support of a_-, but are facilitated by the others in Ω_+.

Although there are extensive empirical studies on interspecific competition (see, e.g., T. W. Shoener [222] and J. H. Connell [58]) and positive interactions are well documented among organisms from different kingdoms as they can make significant contributions to each other's needs without sharing the same resources (see, e.g., G. E. Hutchinson [119], J. L. Wulff [237], M. B. Saffo [219]), finding positive interactions between similar organisms seems to be a huge task in empirical studies, since they do not arise alone but in combination with competition. However, according to the abiotic stress hypothesis of M. D. Bertness and R. M. Callaway [26], the importance of positive interactions in plant communities increases with abiotic stress or consumer pressure. Several empirical studies support the validity of the abiotic stress hypothesis and, actually, a substantial number of positive interactions in plant communities have been isolated in harsh environmental conditions (see, e.g., R. M. Callaway and L. R. Walker [33] and F. I. Pugnaire [211]). Consequently, (9.81) might be a rather reasonable mathematical model for analyzing the combined effects of facilitation and competition under abiotic stress.

According to the discussion in the previous section, under facilitative effects in competitive media, the harsher the environmental conditions measured by the size of $\lambda < 0$, the richer the dynamics of the species measured by the number of steady-state solutions of (9.81) in complete agreement with the abiotic stress hypothesis of M. D. Bertness and R. M. Callaway [26].

Chapter 10

Spatially heterogeneous competitions

This chapter analyzes the dynamics of the parabolic problem

$$\begin{cases} \partial_t u - d_1\Delta u = \lambda u - \mathfrak{a}(x)\mathfrak{f}_1(x, u)u - \mathfrak{b}(x)uv & \text{in} \quad \Omega \times (0, \infty), \\ \partial_t v - d_2\Delta v = \mu v - \mathfrak{d}(x)\mathfrak{f}_2(x, v)v - \mathfrak{c}(x)uv \\ u = v = 0 & \text{on} \quad \partial\Omega \times (0, \infty), \\ u(\cdot, 0) = u_0 \geq 0 \ \text{ and } \ v(\cdot, 0) = v_0 \geq 0 & \text{in} \quad \Omega, \end{cases}$$

$$(10.1)$$

where Ω is a bounded domain of $\mathbb{R}^N$, $N \geq 1$, with smooth boundary $\partial\Omega$ of class $\mathcal{C}^{2+\nu}$ for some $\nu \in (0, 1]$, $\lambda > 0$, $\mu > 0$, $d_1 > 0$, $d_2 > 0$, and $\mathfrak{a}$, $\mathfrak{b}$, $\mathfrak{c}$, $\mathfrak{d} \in C^\nu(\bar{\Omega})$ are non-negative functions satisfying the following assumptions:

Hypothesis (HC)

The function $a \equiv -\mathfrak{a}$ satisfies Hypothesis (Ha), $\mathfrak{b}(x) > 0$ and $\mathfrak{d}(x) > 0$ for all $x \in \bar{\Omega}$, and $\mathfrak{c}(x) > 0$ for all $x \in \Omega_0$.

As for the nonlinearities $\mathfrak{f}_1$ and $\mathfrak{f}_2$ we assume the following:

Hypothesis (HF)

For each $j = 1, 2$, $\mathfrak{f}_j \in \mathcal{C}^{\nu,1+\nu}(\bar{\Omega} \times [0, \infty))$, $\mathfrak{f}_1$ satisfies Hypothesis (KO) and $\mathfrak{f}_2$ satisfies (Hf) and (Hg).

Under these assumptions, (10.1) provides us with a model for the evolution of two competing species with densities u and v in the habitat Ω. In the absence of v, the species u grows according to the (non-classical) generalized logistic parabolic problem

$$\begin{cases} \frac{\partial u}{\partial t} - d_1 \Delta u = \lambda u - \mathfrak{a}(x)\mathfrak{f}_1(x,u)u & \text{in } \Omega \times (0,\infty), \\ u = 0 & \text{on } \partial\Omega \times (0,\infty), \\ u(\cdot,0) = u_0 \geq 0 & \text{in } \Omega, \end{cases} \qquad (10.2)$$

whose dynamics were described in Chapter 5. Similarly, in the absence of u, the species v grows according to the (classical) generalized logistic problem

$$\begin{cases} \frac{\partial v}{\partial t} - d_2 \Delta v = \mu v - \mathfrak{d}(x)\mathfrak{f}_2(x,v)v & \text{in } \Omega \times (0,\infty), \\ v = 0 & \text{on } \partial\Omega \times (0,\infty), \\ v(\cdot,0) = v_0 \geq 0 & \text{in } \Omega. \end{cases} \qquad (10.3)$$

By the abstract theory of Chapter 2, the dynamics of (10.3) are described by its maximal non-negative steady state, because $\mathfrak{d}(x) > 0$ for all $x \in \bar{\Omega}$.

Thanks to the abstract theory of D. Daners and P. Koch [68], for any $u_0, v_0 \in \mathcal{C}_0(\bar{\Omega})$ with $u_0 \geq 0$ and $v_0 \geq 0$, (10.1) has a unique global strong solution, $(u(x,t;u_0,v_0),v(x,t;u_0,v_0))$, such that

$$u,v \in \mathcal{C}(\bar{\Omega} \times [0,\infty)) \cap \mathcal{C}^{2+\nu,1+\frac{\nu}{2}}(\bar{\Omega} \times (0,\infty)). \qquad (10.4)$$

Indeed, by the parabolic maximum principle, for every $t > 0$,

$$0 \leq u(\cdot,t;u_0,v_0) \leq T_1(t)u_0 \quad \text{and} \quad 0 \leq v(\cdot,t;u_0,v_0) \leq T_2(t)v_0,$$

where

$$T_1(t) = e^{t(d_1\Delta+\lambda)}, \qquad T_2(t) = e^{t(d_2\Delta+\mu)}, \qquad t \geq 0,$$

are the evolution operators associated with $d_j\Delta + \lambda$, $j \in \{1,2\}$, under homogeneous Dirichlet boundary conditions. In particular, the solutions of (10.1) are globally defined in time, $t > 0$.

This chapter establishes that most of the limiting profiles of the positive solutions of (10.1) as $t \uparrow \infty$ are given by a certain *generalized class of non-negative equilibria*. The classical equilibria of (10.1) are the non-negative solutions of

$$\begin{cases} -d_1 \Delta u = \lambda u - \mathfrak{a}(x)\mathfrak{f}_1(x,u)u - \mathfrak{b}(x)uv & \\ -d_2 \Delta v = \mu v - \mathfrak{d}(x)\mathfrak{f}_2(x,v)v - \mathfrak{c}(x)uv & \text{in } \Omega, \\ u = v = 0 & \text{on } \partial\Omega. \end{cases} \qquad (10.5)$$

The generalized class of equilibria consists of the non-negative *metasolutions* of (10.5) supported in one or several components of $\Omega_0 := \text{int } \mathfrak{a}^{-1}(0)$. The next definition describes the most appropriate concept of metasolution for (10.1).

Definition 10.1 *Given* $1 \leq q \leq q_0$, *a subset* $\{i_1, ..., i_q\} \subset \{1, ..., q_0\}$, *and*

$$D := \Omega \setminus \bigcup_{k=1}^{q} \bar{\Omega}_{0,i_k},$$

a couple

$$(\mathfrak{M}_u, \mathfrak{M}_v) : \bar{\Omega} \to ([0, \infty], [0, \infty))$$

is said to be a positive metasolution of (10.5) supported in D if there exists a classical large solution (L_u, L_v) *of the singular elliptic problem*

$$\begin{cases} -d_1 \Delta u = \lambda u - \mathfrak{a}(x)\mathfrak{f}_1(x, u)u - \mathfrak{b}(x)uv & \\ -d_2 \Delta v = \mu v - \mathfrak{d}(x)\mathfrak{f}_2(x, v)v - \mathfrak{c}(x)uv & in \quad D, \\ u = v = 0 & on \quad \partial D \cap \partial \Omega, \\ u = \infty \quad and \quad 0 \leq v < \infty & on \quad \partial D \cap \Omega, \end{cases} \quad (10.6)$$

such that

$$(\mathfrak{M}_u, \mathfrak{M}_v) = (L_u, L_v) \quad in \quad D$$

and

$$\mathfrak{M}_u = \infty \quad in \quad (\Omega \setminus \bar{D}) \cup (\partial D \cap \Omega).$$

It should be noted that, adopting the notation introduced in Chapter 1, we are denoting

$$\Omega_{0,j} := \bigcup_{i=1}^{m_j} \Omega_{0,j}^i, \quad 1 \leq j \leq q_0.$$

By a large solution of (10.6), we mean a solution of the system in D, (u, v), such that

$$u \in C^{2+\nu}(D) \cap C(D \cup (\partial D \cap \partial \Omega)), \qquad v \in C^{2+\nu}(D) \cap C(\bar{D}),$$

and

$$\lim_{d(x)\downarrow 0} u(x) = \infty, \qquad d(x) := \text{dist}(x, \partial D \cap \Omega).$$

Besides $(0, 0)$, there are two types of classical non-negative solutions of (10.5): those with one component vanishing and the other positive, $(u, 0)$ or $(0, v)$, where u and v solve

$$\begin{cases} -d_1 \Delta u = \lambda u - \mathfrak{a}(x)\mathfrak{f}_1(x, u)u & in \quad \Omega, \\ u = 0 & on \quad \partial\Omega, \end{cases} \quad (10.7)$$

and

$$\begin{cases} -d_2 \Delta v = \mu v - \mathfrak{d}(x)\mathfrak{f}_2(x, v)v & in \quad \Omega, \\ v = 0 & on \quad \partial\Omega, \end{cases} \quad (10.8)$$

respectively, usually referred to as the *semi-trivial positive solutions*, and those with both components positive, called *coexistence states*, though those may be linearly unstable.

Similarly, there are two types of non-negative metasolutions: those of the form $(u, 0)$, where u is a metasolution of (10.7), as discussed by Definition 5.1, referred to as the *semi-trivial positive metasolutions*, and those of the form (u, v) with $v > 0$, which will be called *meta-coexistence states*.

Problem (10.5) might also have large solutions of the form (u, v) with u and v large, of course, but those solutions are of no interest for describing the dynamics of (10.1), because they cannot be limiting profiles of a positive solution of (10.1). Indeed, if (u, v) is a classical solution of (10.1), then v is a subsolution of (10.3) and hence, by the parabolic maximum principle,

$$v(\cdot, t; u_0, v_0) \leq v_\mu(\cdot, t; v_0) \qquad \text{for all } t \geq 0,$$

where v_μ stands for the unique solution of (10.3). Thus, by Theorem 2.2, there exists a constant $C > 0$ such that

$$v(\cdot, t; u_0, v_0) \leq C \qquad \text{for all } t \geq 0.$$

Consequently, we have decided to exclude all these solutions from the statement of Definition 10.1, though they are mathematically admissible, because they do not play a role in describing the dynamics of (10.1).

From the point of view of population dynamics, λ and μ represent the intrinsic birth rates of the species u and v, d_1 and d_2 are their respective diffusivity rates in the habitat Ω, which is assumed to be surrounded by inhospitable regions, as a consequence of the homogeneous Dirichlet boundary conditions, and u_0 and v_0 are the initial population densities of the species.

As $a = -\mathfrak{a}$ satisfies Hypothesis (Ha), in the absence of v, in the non-spatial model obtained by switching off to zero the diffusivity d_1 of u, the species u grows according to the Malthus low in Ω_0, while it has logistic growth in each $x \in \Omega_-$ where $\mathfrak{a}(x) > 0$. According to Theorem 5.2, as an effect attributable to the diffusivity $d_1 > 0$, in the spatial model the species u can exhibit a logistic growth in some of the smaller components of $\mathfrak{a}^{-1}(0)$ provided λ is sufficiently small. Nevertheless, the components of $\Omega_0 = \text{int } \mathfrak{a}^{-1}(0)$ are protection zones where u is free from intra-specific competition. Contrarily, as we are assuming that $\mathfrak{d}(x) > 0$ for all $x \in \bar{\Omega}$, according to Theorem 2.2 the species v exhibits a genuine logistic type of behavior, as it must afford the intra-specific competition effects throughout the entire habitat Ω.

The coefficient functions $\mathfrak{b}(x)$ and $\mathfrak{c}(x)$ measure the level of the aggressions of the species v on u and the response of the species v, respectively. As we are assuming that $|\mathfrak{b}^{-1}(0)| = 0$, u receives aggressions from v almost everywhere in Ω. Moreover, since $\mathfrak{c}(x) > 0$ for all $x \in \Omega_0 = \text{int } \mathfrak{a}^{-1}(0)$, the species v is aggressed by u in all the spatial locations where the non-spatial model predicts an exponential growth of u as time passes. Naturally, the main problem addressed in this chapter is ascertaining the output of the competition according to the possible values of the set of parameters arising in (10.1).

Next we are going to introduce some of the notations used throughout this chapter. The set of components of Ω_0, or protection patches of the species u, will be denoted by $\mathcal{R}_\mathfrak{a}$. According to Hypothesis (Ha),

$$\mathcal{R}_\mathfrak{a} := \left\{ \Omega_{0,j}^i \; : \; 1 \leq i \leq m_j, \quad 1 \leq j \leq q_0 \right\}, \tag{10.9}$$

where the components of Ω_0 are labeled so that (1.3) holds, i.e.,

$$\begin{aligned} &\sigma_j := \lambda_1[-\Delta, \Omega_{0,j}^i], && 1 \leq i \leq m_j, \; 1 \leq j \leq q_0, \\ &\sigma_j < \sigma_{j+1}, && 1 \leq j \leq q_0 - 1. \end{aligned} \tag{10.10}$$

It should be remembered that, for every $1 \leq j \leq q_0$,

$$\Omega_{0,j} := \bigcup_{i=1}^{m_j} \Omega_{0,j}^i, \qquad \lambda_1[-\Delta, \Omega_{0,j}] := \sigma_j = \lambda_1[-\Delta, \Omega_{0,j}^i], \; 1 \leq i \leq m_j.$$

Throughout this chapter, for any given $d > 0$ and $V \in L^\infty(\Omega)$, the components of $\mathcal{R}_\mathfrak{a}$ will be re-labeled, if necessary, so that

$$\mathcal{R}_\mathfrak{a} = \left\{ \Omega_{0,j}^i(d,V) \; : \; 1 \leq i \leq m_j(d,V), \quad 1 \leq j \leq q_0(d,V) \right\}$$

with

$$\begin{aligned} &\sigma_j[d,V] := \lambda_1[-d\Delta + V, \Omega_{0,j}^i(d,V)], && 1 \leq i \leq \tilde{m}_j, \; 1 \leq j \leq \tilde{q}_0, \\ &\sigma_j[d,V] < \sigma_{j+1}[d,V], && 1 \leq j \leq \tilde{q}_0 - 1, \end{aligned} \tag{10.11}$$

where

$$\tilde{m}_j := m_j(d,V), \quad \tilde{q}_0 := q_0(d,V), \quad 1 \leq j \leq q_0(d,V).$$

So, according to (1.3),

$$m_j = m_j(1,0), \quad q_0 = q_0(1,0), \quad 1 \leq j \leq q_0.$$

By (10.11), we are imposing

$$\lambda_1[-d\Delta + V, \Omega_{0,j}^i(d,V)] < \lambda_1[-d\Delta + V, \Omega_{0,j+1}^i(d,V)] \tag{10.12}$$

for all $1 \leq j \leq \tilde{q}_0 - 1$ and $1 \leq i \leq \tilde{m}_j$, and

$$\lambda_1[-d\Delta + V, \Omega_{0,j}^i(d,V)] = \lambda_1[-d\Delta + V, \Omega_{0,j}^{i+1}(d,V)] \tag{10.13}$$

for all $1 \leq i \leq \tilde{m}_j - 1$ and $1 \leq j \leq \tilde{q}_0$. As the total number of components of Ω_0 cannot vary with the pair (d,V), it is apparent that

$$\operatorname{card} \mathcal{R}_\mathfrak{a} = \sum_{j=1}^{q_0} m_j = \sum_{j=1}^{q_0(d,V)} m_j(d,V) \quad \text{for all} \quad d > 0 \quad \text{and} \quad V \in L^\infty(\Omega).$$

As in Chapter 1, when we use these notations we shall denote

$$\Omega_{0,j}(d,V) := \bigcup_{i=1}^{m_j(d,V)} \Omega_{0,j}^i(d,V), \qquad 1 \leq j \leq q_0(d,V), \tag{10.14}$$

$$\lambda_1[-d\Delta + V, \Omega_{0,j}(d, V)] := \lambda_1[-d\Delta + V, \Omega_{0,j}^i(d, V)] \qquad (10.15)$$

for all $1 \le j \le q_0(d, V)$ and $1 \le i \le m_j(d, V)$, and

$$\Omega_j(d, V) := \Omega \setminus \left(\bar{\Omega}_{0,1}(d, V) \cup \cdots \cup \bar{\Omega}_{0,j}(d, V) \right), \quad 1 \le j \le q_0(d, V). \quad (10.16)$$

Note that, for every $d > 0$ and $V \in L^\infty(\Omega)$,

$$\Omega_{q_0(d,V)} = \Omega_-.$$

Thanks to (10.11) it is apparent that

$$\sigma_0[d, V] < \sigma_1[d, V] < \sigma_2[d, V]$$
$$< \cdots < \sigma_{q_0(d,V)}[d, V] < \sigma_{q_0(d,V)+1}[d, V] := \infty, \qquad (10.17)$$

where, as in Chapter 1, we have denoted

$$\sigma_0[d, V] := \lambda_1[-d\Delta + V, \Omega].$$

So, $\sigma_0 = \sigma_0[1, 0]$.

10.1   Preliminaries

10.1.1   Dynamics of the semi-trivial positive solutions

The dynamics of u in the absence of v, $v_0 = 0$, are regulated by the parabolic problem (10.2) and described by the next result, which is a direct consequence of Theorem 5.2. Indeed, $v_0 = 0$ implies $v(\cdot, t; u_0, v_0) = 0$ for all $t \ge 0$ and hence $u(\cdot, t; u_0, v_0)$ solves (10.2).

Theorem 10.2 *Suppose $u_0 \in \mathcal{C}_0(\bar{\Omega})$ with $u_0 > 0$ and let $u(x, t) := u_\lambda(x, t; u_0)$ denote the unique solution of (10.2). Then, the following assertions are true:*

(a) *If $\lambda \le d_1\sigma_0$, then*

$$\lim_{t \uparrow \infty} u(\cdot, t) = 0 \qquad in \ \mathcal{C}(\bar{\Omega}).$$

(b) *If $d_1\sigma_0 < \lambda < d_1\sigma_1$, then*

$$\lim_{t \uparrow \infty} u(\cdot, t) = \theta_{\lambda,u} \qquad in \ \mathcal{C}(\bar{\Omega}),$$

where $\theta_{\lambda,u}$ stands for the unique positive solution of (10.7).

(c) *If $d_1\sigma_j \le \lambda < d_1\sigma_{j+1}$ for some $1 \le j \le q_0$, then*

$$\mathfrak{M}_{[\lambda,\Omega_j]}^{\min} \le \liminf_{t \uparrow \infty} u(\cdot, t) \le \limsup_{t \uparrow \infty} u(\cdot, t) \le \mathfrak{M}_{[\lambda,\Omega_j]}^{\max} \qquad in \ \bar{\Omega}, \quad (10.18)$$

where $\mathfrak{M}^{\min}_{[\lambda,\Omega_j]}$ *(resp.* $\mathfrak{M}^{\max}_{[\lambda,\Omega_j]}$) *stands for the minimal (resp. maximal) metasolution of*

$$-d_1 \Delta u = \lambda u - \mathfrak{a}(x)\mathfrak{f}_1(x,u)u$$

supported in Ω_j. *If, in addition,* u_0 *is a subsolution of* (10.7) *in* Ω, *then*

$$\lim_{t \uparrow \infty} u(\cdot, t) = \mathfrak{M}^{\min}_{[\lambda,\Omega_j]} \qquad in \quad \bar{\Omega}. \tag{10.19}$$

It should not be forgotten that

$$\sigma_{q_0+1} := \infty, \qquad \Omega_{q_0} := \Omega_-.$$

Similarly, according to Theorem 2.2, the next result provides us with the dynamics of the species v in the absence of u, i.e., when $u_0 = 0$, as in this case $u(\cdot, t; u_0, v_0) = 0$ for all $t \geq 0$ and $v(\cdot, t; u_0, v_0)$ solves (10.3).

Theorem 10.3 *Suppose* $v_0 \in \mathcal{C}_0(\bar{\Omega})$ *with* $v_0 > 0$ *and let* $v(x, t) := v_\mu(x, t; v_0)$ *denote the unique solution of* (10.3). *Then,*

$$\lim_{t \uparrow \infty} v(\cdot, t) = \begin{cases} 0, & \text{if } \mu \leq d_2\sigma_0, \\ \\ \theta_{\mu,v}, & \text{if } \mu > d_2\sigma_0, \end{cases} \qquad in \ \mathcal{C}(\bar{D}),$$

where $\theta_{\mu,v}$ *stands for the unique positive solution of* (10.8).

The existence and the uniqueness of $\theta_{\lambda,u}$ and of $\theta_{\mu,v}$ in these results are guaranteed by Theorems 2.2 and 4.1, respectively. If necessary, we might adopt the following convention

$$\theta_{\lambda,u} \equiv 0 \ \text{ if } \lambda \leq d_1\sigma_0 \qquad \text{and} \qquad \theta_{\mu,v} \equiv 0 \ \text{ if } \mu \leq d_2\sigma_0. \tag{10.20}$$

10.1.2 Dynamics when $\lambda \leq d_1\sigma_0$ or $\mu \leq d_2\sigma_0$

The next result provides us with the dynamics of (10.1) when $\lambda \leq d_1\sigma_0$, or $\mu \leq d_2\sigma_0$. So, throughout the rest of this chapter we will assume

$$\lambda > d_1\sigma_0, \qquad \mu > d_2\sigma_0. \tag{10.21}$$

Theorem 10.4 *Suppose* $u_0, v_0 \in \mathcal{C}_0(\bar{\Omega})$ *with* $u_0 > 0$ *and* $v_0 > 0$ *and let* $(u(x, t), v(x, t))$ *denote the unique (classical) solution of* (10.1). *Then, the following assertions are true:*

(a) *If* $\lambda \leq d_1\sigma_0$ *and* $\mu \leq d_2\sigma_0$, *then*

$$\lim_{t \uparrow \infty} u(\cdot, t) = 0 \quad \text{and} \quad \lim_{t \uparrow \infty} v(\cdot, t) = 0 \quad in \ \mathcal{C}(\bar{\Omega}).$$

(b) *if $\lambda \leq d_1\sigma_0$ and $\mu > d_2\sigma_0$, then,*

$$\lim_{t\uparrow\infty} u(\cdot,t) = 0 \quad and \quad \lim_{t\uparrow\infty} v(\cdot,t) = \theta_{\mu,v} \quad in \ \mathcal{C}(\bar{\Omega}),$$

where $\theta_{\mu,v}$ is the unique (classical) positive solution of (10.8).

(c) *if $\mu \leq d_2\sigma_0$ and $d_1\sigma_0 < \lambda < d_1\sigma_1$, then*

$$\lim_{t\uparrow\infty} u(\cdot,t) = \theta_{\lambda,u} \quad and \quad \lim_{t\uparrow\infty} v(\cdot,t) = 0 \quad in \ \mathcal{C}(\bar{\Omega}),$$

where $\theta_{\lambda,u}$ is the unique (classical) positive solution of (10.7).

(d) *if $\mu \leq d_2\sigma_0$ and $d_1\sigma_j \leq \lambda < d_1\sigma_{j+1}$ for some $1 \leq j \leq q_0$, then*

$$\mathfrak{M}_{[\lambda,\Omega_j]}^{\min} \leq \liminf_{t\uparrow\infty} u(\cdot,t) \leq \limsup_{t\uparrow\infty} u(\cdot,t) \leq \mathfrak{M}_{[\lambda,\Omega_j]}^{\max} \quad in \quad \bar{\Omega}$$

and

$$\lim_{t\uparrow\infty} v(\cdot,t) = 0 \quad in \ \mathcal{C}(\bar{\Omega}).$$

Proof: Since (10.1) is of Kolmogorov type, $u_0 > 0$ and $v_0 > 0$ imply $u(x,t) > 0$ and $v(x,t) > 0$ for all $x \in \Omega$ and $t > 0$. Thus, by the general assumptions on the coefficients, $u(x,t)$ and $v(x,t)$ are positive subsolutions of the parabolic problems (10.2) and (10.3), respectively, whose solutions are going to be denoted by $U(x,t)$ and $V(x,t)$. Hence, by the parabolic maximum principle,

$$u(x,t) \leq U(x,t) \quad and \quad v(x,t) \leq V(x,t) \tag{10.22}$$

for all $x \in \Omega$ and $t > 0$. According to Theorem 10.2, we already know that

$$\lim_{t\uparrow\infty} U(\cdot,t) = 0 \quad in \ \mathcal{C}(\bar{\Omega})$$

if $\lambda \leq d_1\sigma_0$. So, by (10.22), we find that, for every $\lambda \leq d_1\sigma_0$,

$$\lim_{t\uparrow\infty} u(\cdot,t) = 0 \quad in \ \mathcal{C}(\bar{\Omega}). \tag{10.23}$$

Similarly, according to Theorem 10.3, for every $\mu \leq d_2\sigma_0$,

$$\lim_{t\uparrow\infty} v(\cdot,t) = 0 \quad in \ \mathcal{C}(\bar{\Omega}), \tag{10.24}$$

which completes the proof of Part (a).

Suppose $\lambda \leq d_1\sigma_0$ and $\mu > d_2\sigma_0$. We already know that (10.23) holds. Moreover, by Theorem 10.3,

$$\lim_{t\uparrow\infty} V(\cdot,t) = \theta_{\mu,v} \quad in \ \mathcal{C}(\bar{\Omega}).$$

Thus, by (10.22), we find that

$$\limsup_{t \uparrow \infty} v(\cdot, t) \le \theta_{\mu, v}. \tag{10.25}$$

On the other hand, due to (10.23), for any given $\varepsilon > 0$, there exists $T = t(\varepsilon) > 0$ such that $0 \le \mathfrak{c}(x)u(x,t) \le \varepsilon$ for all $x \in \Omega$ and $t \ge T$. Thus,

$$\frac{\partial v}{\partial t} - d_2 \Delta v = \mu v - \mathfrak{d}(x)\mathfrak{f}_2(x, v)v - \mathfrak{c}(x)uv \ge (\mu - \varepsilon)v - \mathfrak{d}(x)\mathfrak{f}_2(x, v)v$$

for all $x \in \Omega$ and $t \ge T$. Hence, $v(x, t)$ provides us with a supersolution of the parabolic problem

$$\begin{cases} \frac{\partial w}{\partial t} - d_2 \Delta w = (\mu - \varepsilon)w - \mathfrak{d}(x)\mathfrak{f}_2(x, w)w & \text{in } \Omega \times (T, \infty), \\ w = 0 & \text{on } \partial\Omega \times (T, \infty), \\ w(\cdot, T) = v(\cdot, T) > 0 & \text{in } \Omega, \end{cases} \tag{10.26}$$

whose unique solution will be denoted by $w(x, t)$. Suppose $\varepsilon > 0$ has been chosen sufficiently small so that $\mu - \varepsilon > d_2 \sigma_0$. Then, by Theorem 10.3,

$$\lim_{t \uparrow \infty} w(\cdot, t) = \theta_{\mu-\varepsilon, v} \quad \text{in } \mathcal{C}(\bar{\Omega}).$$

Moreover, by the parabolic maximum principle, it is apparent that

$$v(x, t) \ge w(x, t) \qquad \text{for all } x \in \Omega, \ t \ge T.$$

Consequently,

$$\liminf_{t \uparrow \infty} v(\cdot, t) \ge \lim_{t \uparrow \infty} w(\cdot, t) = \theta_{\mu-\varepsilon, v}.$$

Therefore, for sufficiently small $\varepsilon > 0$, we find that

$$\theta_{\mu-\varepsilon, v} \le \liminf_{t \uparrow \infty} v(\cdot, t) \le \limsup_{t \uparrow \infty} v(\cdot, t) = \theta_{\mu, v}.$$

According to Theorem 2.3, $\theta_{\mu-\varepsilon, v}$ approximates $\theta_{\mu, v}$ in $\mathcal{C}(\bar{\Omega})$ as $\varepsilon \downarrow 0$ and these estimates complete the proof of Part (b).

The proof of Part (c) follows similar patterns as the proof of Part (b), except that in this case one should invoke to Theorem 4.1, rather than Theorem 2.3, to establish the continuity of the map $\lambda \mapsto \theta_{\lambda, u}$. So, the technical details of the proof are omitted here.

Lastly, suppose $\mu \le d_2\sigma_0$ and $d_1\sigma_j \le \lambda < d_1\sigma_{j+1}$ for some $1 \le j \le q_0$. Then, (10.24) holds. Moreover, by (10.22) it follows from Theorem 10.2 that

$$\limsup_{t \uparrow \infty} u(\cdot, t) \le \limsup_{t \uparrow \infty} U(\cdot, t) \le \mathfrak{M}_{[\lambda, \Omega_j]}^{\max}.$$

On the other hand, by (10.24), for every $\varepsilon > 0$ there exists $T = t(\varepsilon) > 0$ such that $0 \le \mathfrak{b}(x)v(x, t) \le \varepsilon$ for all $x \in \Omega$ and $t \ge T$. Hence,

$$\frac{\partial u}{\partial t} - d_1 \Delta u = \lambda u - \mathfrak{a}(x)\mathfrak{f}_1(x, u)u - \mathfrak{b}(x)uv \ge (\lambda - \varepsilon)u - \mathfrak{a}(x)\mathfrak{f}_1(x, u)u$$

for all $x \in \Omega$ and $t \geq T$. Consequently, $u(x, t)$ is a supersolution of the problem

$$\begin{cases} \frac{\partial z}{\partial t} - d_1 \Delta z = (\lambda - \varepsilon)z - \mathfrak{a}(x)\mathfrak{f}_1(x, z)z & \text{in } \Omega \times (T, \infty), \\ z = 0 & \text{on } \partial\Omega \times (T, \infty), \\ z(\cdot, T) = u(\cdot, T) > 0 & \text{in } \Omega. \end{cases} \quad (10.27)$$

Let $z(x, t)$ denote the unique solution of (10.27) and suppose, in addition, that $d_1\sigma_j < \lambda < d_1\sigma_{j+1}$. Then, for sufficiently small $\varepsilon > 0$,

$$d_1\sigma_j < \lambda - \varepsilon < d_1\sigma_{j+1}.$$

Thus, by the parabolic maximum principle, we find from Theorem 10.2 that

$$\liminf_{t\uparrow\infty} u(\cdot, t) \geq \liminf_{t\uparrow\infty} z(\cdot, t) \geq \mathfrak{M}^{\min}_{[\lambda-\varepsilon,\Omega_j]}, \quad (10.28)$$

which entails

$$\liminf_{t\uparrow\infty} u(\cdot, t) = \infty \quad \text{in } \Omega \setminus \Omega_j$$

and

$$\liminf_{t\uparrow\infty} u(\cdot, t) \geq \liminf_{t\uparrow\infty} z(\cdot, t) \geq L^{\min}_{[\lambda-\varepsilon,\Omega_j]} \quad \text{in } \Omega_j.$$

As, according to Theorem 4.7, we already know that

$$L^{\min}_{[\lambda-\varepsilon,\Omega_j]} := \lim_{M\uparrow\infty} \Theta_{[\lambda-\varepsilon,\Omega_j,M]}, \qquad L^{\min}_{[\lambda-\varepsilon,\Omega_j]} \geq \Theta_{[\lambda-\varepsilon,\Omega_j,M]},$$

it becomes apparent that, for sufficiently small $\varepsilon > 0$ and every $M > 0$,

$$\liminf_{t\uparrow\infty} u(\cdot, t) \geq \Theta_{[\lambda-\varepsilon,\Omega_j,M]} \quad \text{in } \Omega_j.$$

Therefore, letting $\varepsilon \downarrow 0$ and then $M \uparrow \infty$ yields

$$\liminf_{t\uparrow\infty} u(\cdot, t) \geq L^{\min}_{[\lambda,\Omega_j]} \quad \text{in } \Omega_j,$$

which ends the proof of Part (d) in this case.

In the limiting case when $\lambda = d_1\sigma_j$, one has that

$$d_1\sigma_{j-1} < \lambda - \varepsilon < d_1\sigma_j$$

for small $\varepsilon > 0$ and hence, instead of (10.28), we find that

$$\liminf_{t\uparrow\infty} u(\cdot, t) \geq \liminf_{t\uparrow\infty} z(\cdot, t) \geq \mathfrak{M}^{\min}_{[d_1\sigma_j-\varepsilon,\Omega_{j-1}]},$$

which implies

$$\liminf_{t\uparrow\infty} u(\cdot, t) = \infty \quad \text{in } \Omega \setminus \Omega_{j-1}$$

and

$$\liminf_{t\uparrow\infty} u(\cdot, t) \geq L^{\min}_{[d_1\sigma_j-\varepsilon,\Omega_{j-1}]} \quad \text{in } \Omega_{j-1}. \quad (10.29)$$

As, according to Theorem 4.9,

$$\lim_{\varepsilon\downarrow 0} L^{\min}_{[d_1\sigma_j-\varepsilon,\Omega_{j-1}]} = \begin{cases} \infty & \text{in } \bar{\Omega}_{0,j} \setminus \partial\Omega, \\ L^{\min}_{[d_1\sigma_j,\Omega_j]} & \text{in } \Omega_j = \Omega_{j-1} \setminus \bar{\Omega}_{0,j}, \end{cases}$$

the proof is complete. □

10.2 Dynamics of the model when $\mathfrak{c} = 0$

Throughout this section we impose $\mathfrak{c} = 0$. Then, the evolution of the species v is unaltered by u, which simplifies substantially the mathematical analysis of (10.1) and illuminates the analysis of the general case when $\mathfrak{c} > 0$. When $\mathfrak{c} = 0$, the species u and v get uncoupled and hence studying the system reduces to the problem of analyzing a semilinear parabolic equation with a varying potential. The following result characterizes the existence of coexistence and meta-coexistence states.

Theorem 10.5 *Suppose* $\mathfrak{c} = 0$. *Then, the following assertions are true:*

(a) *The elliptic problem (10.5) possesses a coexistence state if, and only if,*

$$\mu > d_2\sigma_0 = \sigma_0[d_2, 0], \qquad \sigma_0[d_1, \mathfrak{b}\theta_{\mu,v}] < \lambda < \sigma_1[d_1, \mathfrak{b}\theta_{\mu,v}], \qquad (10.30)$$

where $\theta_{\mu,v}$ *is the unique positive solution of (10.8). Moreover, it is unique and linearly asymptotically stable if it exists.*

(b) *For every* $1 \leq j \leq q_0(d_1, \mathfrak{b}\theta_{\mu,v})$, *(10.5) possesses a meta-coexistence state of the form* $(\mathfrak{M}, \theta_{\mu,v})$ *supported in* $\Omega_j(d_1, \mathfrak{b}\theta_{\mu,v})$ *if, and only if,*

$$\mu > d_2\sigma_0, \qquad \lambda < \sigma_{j+1}[d_1, \mathfrak{b}\theta_{\mu,v}]. \qquad (10.31)$$

Proof: It should be noted that

$$\sigma_{q_0(d_1, \mathfrak{b}\theta_{\mu,v})+1}[d_1, \mathfrak{b}\theta_{\mu,v}] = \infty,$$

by definition. Thus, the second estimate of (10.31) is not imposing any real restriction on the size of λ if $j = q_0(d_1, \mathfrak{b}\theta_{\mu,v})$.

According to Theorem 2.2, (10.5) admits a solution (u, v) with $v > 0$ if and only if $\mu > d_2\sigma_0$, which has been already denoted by $\theta_{\mu,v}$. Thus, (10.5) possesses a coexistence state if, and only if, the nonlinear elliptic problem

$$\begin{cases} (-d_1\Delta + \mathfrak{b}\theta_{\mu,v}) u = \lambda u - \mathfrak{a}(x)\mathfrak{f}_1(x, u)u & \text{in } \Omega, \\ u = 0 & \text{on } \partial\Omega, \end{cases} \qquad (10.32)$$

possesses a positive solution. According to Theorem 4.1 and Remark 5.3, (10.32) admits a positive solution if, and only if, the second estimate of (10.30) holds, which ends the proof of the first assertion of Part (a).

Subsequently, we will denote by $\theta_{\lambda,u}[\mathfrak{b}\theta_{\mu,v}]$ the unique positive solution of (10.32) under condition (10.30). If it exists, the coexistence state of (10.5) is given by the solution couple

$$(u_0, v_0) := (\theta_{\lambda,u}[\mathfrak{b}\theta_{\mu,v}], \theta_{\mu,v}).$$

By construction, it is unique. Moreover, the associated spectral problem of the linearized system of (10.5) at the coexistence state (u_0, v_0) can be expressed in the form

$$\begin{cases} \mathcal{L}_1 u = -\mathfrak{b} u_0 v + \tau u, & \text{in } \Omega, \\ \mathcal{L}_2 v = \tau v, & \\ u = v = 0 & \text{on } \partial\Omega, \end{cases} \tag{10.33}$$

where

$$\mathcal{L}_1 := -d_1 \Delta + \mathfrak{a} \frac{\partial f_1}{\partial u}(x, u_0) u_0 + \mathfrak{a} f_1(x, u_0) + \mathfrak{b} v_0 - \lambda,$$

$$\mathcal{L}_2 := -d_2 \Delta + \mathfrak{d} \frac{\partial f_2}{\partial v}(x, v_0) v_0 + \mathfrak{d} f_2(x, v_0) - \mu.$$

According to (HC) and (HF), we have that

$$\mathfrak{a} \frac{\partial f_1}{\partial u}(x, u_0) u_0 \gneq 0 \quad \text{and} \quad \mathfrak{d} \frac{\partial f_2}{\partial v}(x, v_0) v_0 \gneq 0 \qquad \text{in } \Omega.$$

Thus, by the monotonicity properties of the principal eigenvalues,

$$\begin{aligned} \lambda_1[\mathcal{L}_1, \Omega] &> \lambda_1[-d_1 \Delta + \mathfrak{a} f_1(x, u_0) + \mathfrak{b} v_0 - \lambda, \Omega], \\ \lambda_1[\mathcal{L}_2, \Omega] &> \lambda_1[-d_2 \Delta + \mathfrak{d} f_2(x, v_0) - \mu, \Omega]. \end{aligned} \tag{10.34}$$

As $v_0 = \theta_{\mu, v}$ is the unique positive solution of (10.8), Lemma 1.6 implies that

$$\mu = \lambda_1[-d_2 \Delta + \mathfrak{d} f_2(x, v_0), \Omega]$$

and hence it follows from (10.34) that $\lambda_1[\mathcal{L}_2, \Omega] > 0$. Similarly, since u_0 satisfies

$$\begin{cases} (-d_1 \Delta + \mathfrak{a} f_1(x, u_0) + \mathfrak{b} v_0) u_0 = \lambda u_0 & \text{in } \Omega, \\ u_0 = 0 & \text{on } \partial\Omega, \end{cases}$$

necessarily

$$\lambda = \lambda_1[-d_1 \Delta + \mathfrak{a} f_1(x, u_0) + \mathfrak{b} v_0, \Omega],$$

which implies $\lambda_1[\mathcal{L}_1, \Omega] > 0$, by (10.34). Consequently, owing to Theorem 1.1, $\mathcal{L}_1$ and $\mathcal{L}_2$ satisfy the strong maximum principle.

Let $(u, v) \neq (0, 0)$ be a solution of (10.33) for some $\tau \in \mathbb{C}$. Suppose $v \neq 0$. Then, $\tau \in \mathbb{R}$ and, actually, since the principal eigenvalue is dominant, as discussed by Theorem 7.9 of [163], we find that

$$\tau \geq \lambda_1[\mathcal{L}_2, \Omega] > 0.$$

Similarly, if $v = 0$, then $\mathcal{L}_1 u = \tau u$ and hence

$$\tau \geq \lambda_1[\mathcal{L}_1, \Omega] > 0.$$

Therefore, all the eigenvalues of (10.33) are real and positive and, consequently, by the linearized stability principle of Lyapunov (e.g., Lunardi [186]), the coexistence state (u_0, v_0) is linearly exponentially asymptotically stable. This ends the proof of Part (a).

Part (b) is an immediate consequence of Theorem 4.7 and Remark 5.3 applied to the problem

$$\begin{cases} (-d_1\Delta + \mathfrak{b}\theta_{\mu,v})\,u = \lambda u + \mathfrak{a}(x)\mathfrak{f}_1(x, u)u & \text{in } \Omega_j, \\ u = 0 & \text{on } \partial\Omega_j \cap \partial\Omega, \\ u = \infty & \text{on } \partial\Omega_j \setminus \partial\Omega, \end{cases} \quad (10.35)$$

where

$$\Omega_j = \Omega_j(d_1, \theta_{\mu,v}), \qquad 1 \leq j \leq q_0(d_1, \theta_{\mu,v}).$$

The proof is complete $\quad\square$

Subsequently, we will look at the first quadrant of the (λ, μ)-plane, centered at $(d_1\sigma_0, d_2\sigma_0)$, to ascertain the shapes of the regions described by (10.30) and (10.31). According to Theorem 10.4, outside this quadrant the dynamics of (10.1) are governed by the semitrivial solutions and metasolutions. Set

$$\Sigma_0(\mu) := \sigma_0[d_1, \mathfrak{b}\theta_{\mu,v}] = \lambda_1[-d_1\Delta + \mathfrak{b}\theta_{\mu,v}, \Omega], \qquad \mu > d_2\sigma_0. \quad (10.36)$$

By Theorem 2.3, $\mu \mapsto \theta_{\mu,v}$ is increasing. Thus, by the monotonicity of the principal eigenvalue with respect to the potential, $\mu \mapsto \Sigma_0(\mu)$ also is increasing. Consequently, the limit

$$\lim_{\mu\uparrow\infty} \Sigma_0(\mu) \in (0, \infty] \quad (10.37)$$

is well defined. The next result establishes that it equals infinity.

Lemma 10.6 *The following holds*

$$\lim_{\mu\downarrow d_2\sigma_0} \Sigma_0(\mu) = d_1\sigma_0, \qquad \lim_{\mu\uparrow\infty} \Sigma_0(\mu) = \infty. \quad (10.38)$$

Proof: The first limit holds because $\theta_{\mu,v}$ bifurcates from $v = 0$ at $\mu = d_2\sigma_0$, by the continuity of the principal eigenvalue with respect to the potential. The proof of the second limit is more delicate and it relies on the fact that

$$\lim_{\mu\uparrow\infty} \theta_{\mu,v} = \infty \quad \text{uniformly on compact subsets of } \Omega \quad (10.39)$$

and on Theorem 9.5 of [163]. Let denote by φ the principal eigenfunction of $-\Delta$ in Ω under Dirichlet boundary conditions, normalized so that

$$\max_\Omega \varphi = 1.$$

The property (10.39) follows from the fact that, for every $\mu > d_2\sigma_0$, there is a positive constant $\alpha(\mu)$ such that $\alpha(\mu)\varphi \leq \theta_{\mu,v}$ in Ω and

$$\lim_{\mu\uparrow\infty} \alpha(\mu) = \infty. \quad (10.40)$$

Indeed, $\alpha\varphi$ provides us with a subsolution of (10.8) if, and only if,

$$\mathfrak{d}(x)\mathfrak{f}_2(x, \alpha\varphi(x)) \leq \mu - d_2\sigma_0 \quad \text{for all } x \in \bar{\Omega},$$

which can be accomplished by taking any $\alpha = \alpha(\mu) > 0$ such that

$$\mathfrak{f}_2(x, \alpha) \leq \frac{\mu - d_2\sigma_0}{\max_{\bar{\Omega}} \mathfrak{d}} \quad \text{for all } x \in \bar{\Omega},$$

because $\varphi \leq 1$. As the right hand side of this inequality grows to infinity as $\mu \uparrow \infty$, it becomes apparent that $\alpha(\mu)$ can be chosen so that (10.40) holds, as required. By Theorem 1.7, we have that $\alpha(\mu)\varphi \leq \theta_{\mu,v}$ for all $\mu > d_2\sigma_0$. Therefore, (10.39) holds, since $\min_K \varphi > 0$ in any compact subset $K \subset \Omega$.

Next, for sufficiently small $\varepsilon > 0$, we will consider the open subset of Ω, Ω_ε, defined by

$$\Omega_\varepsilon := \{x \in \Omega \; : \; \operatorname{dist}(x, \partial\Omega) > \varepsilon\}.$$

As $\partial\Omega$ is smooth, we have that

$$\lim_{\varepsilon\downarrow 0} |\Omega \setminus \bar{\Omega}_\varepsilon| = 0.$$

Consequently, as in the beginning of Chapter 1,

$$\lim_{\varepsilon\downarrow 0} \lambda_1[-\Delta, \Omega \setminus \bar{\Omega}_\varepsilon] = \infty. \tag{10.41}$$

On the other hand, by the monotonicity of the principal eigenvalue with respect to the potential, we find that

$$\Sigma_0(\mu) = \lambda_1[-d_2\Delta + \mathfrak{b}\theta_{\mu,v}, \Omega] \geq \lambda_1[-d_2\Delta + \mathfrak{b}\theta_{\mu,v}\chi_{\Omega\setminus\bar{\Omega}_\varepsilon}, \Omega],$$

for every $\mu > d_2\sigma_0$ and sufficiently small $\varepsilon > 0$, where $\chi_{\Omega\setminus\bar{\Omega}_\varepsilon}$ stands for the characteristic function of $\Omega \setminus \bar{\Omega}_\varepsilon$. Thus, by (10.39), we find from [163, Th. 9.5] that

$$\lim_{\mu\uparrow\infty} \Sigma_0(\mu) \geq \lim_{\mu\uparrow\infty} \lambda_1[-d_2\Delta + \mathfrak{b}\theta_{\mu,v}\chi_{\Omega\setminus\bar{\Omega}_\varepsilon}, \Omega] = d_2\lambda_1[-\Delta, \Omega \setminus \bar{\Omega}_\varepsilon].$$

As this estimate holds for all $0 < \varepsilon < \varepsilon_0$, (10.41) implies (10.38). □

But the behavior of the curves

$$\Sigma_j(\mu) := \sigma_j[d_1, \mathfrak{b}\theta_{\mu,v}] = \lambda_1[-d_1\Delta + \mathfrak{b}\theta_{\mu,v}, \Omega^i_{0,j}(d_1, \mathfrak{b}\theta_{\mu,v})], \quad \mu > d_2\sigma_0,$$

for each $1 \leq j \leq q_0(d_1, \mathfrak{b}\theta_{\mu,v})$, is substantially more involved than the behavior of $\Sigma_0(\mu)$, because even the number of these curves, $q_0(d_1, \mathfrak{b}\theta_{\mu,v})$, might vary with the parameter μ. However, by the monotonicity of the principal eigenvalue with respect to the domain,

$$\Sigma_0(\mu) < \Sigma_j(\mu), \quad \mu > d_2\sigma_0, \qquad 1 \leq j \leq q_0(d_1, \mathfrak{b}\theta_{\mu,v}), \tag{10.42}$$

because $\Omega_{0,j}^i \subsetneqq \Omega$. Moreover, by construction,

$$d_1\sigma_j = \sigma_j[d_1, 0] = \lambda_1[-d_1\Delta, \Omega_{0,j}^i], \quad 1 \le j \le q_0, \quad 1 \le i \le m_j, \quad (10.43)$$

$$\sigma_1 = \sigma_1[d_1, 0] < \cdots < \sigma_{q_0} = \sigma_{q_0}[d_1, 0]. \quad (10.44)$$

Thus, since $\theta_{\mu,v}$ bifurcates from $v = 0$ at $\mu = d_2\sigma_0$, by the continuity of the principal eigenvalue with respect to the potential, it is apparent that

$$\lim_{\mu\downarrow d_2\sigma_0} \lambda_1[-d_1\Delta + \mathfrak{b}\theta_{\mu,v}, \Omega_{0,j}^i] = \lambda_1[-d_1\Delta, \Omega_{0,j}^i] = \sigma_j[d_1, 0] = d_1\sigma_j \quad (10.45)$$

for all $1 \le j \le q_0$ and $1 \le i \le m_j$. Therefore, owing to (10.44) and (10.45), there exists $\varepsilon > 0$ such that

$$\lambda_1[-d_1\Delta + \mathfrak{b}\theta_{\mu,v}, \Omega_{0,j}^{i_1}] < \lambda_1[-d_1\Delta + \mathfrak{b}\theta_{\mu,v}, \Omega_{0,j+1}^{i_2}]$$

for all $\mu \in (d_2\sigma_0, d_2\sigma_0 + \varepsilon]$ and

$$1 \le i_1 \le m_j, \quad 1 \le i_2 \le m_{j+1}, \quad 1 \le j \le q_0 - 1.$$

Hence, any of these eigenvalues may be separated from each other. Consequently, the next result holds.

Proposition 10.7 *Suppose $m_j = 1$ for all $1 \le j \le q_0$, i.e., q_0 equals the total number of components of Ω_0. Then, there exists $\mu_0 > d_2\sigma_0$ such that, for every $\mu \in (d_2\sigma_0, \mu_0)$ and $1 \le j \le q_0$,*

$$m_j(d_1, \mathfrak{b}\theta_{\mu,v}) = 1 \quad and \quad \Omega_{0,j}^1(d_1, \mathfrak{b}\theta_{\mu,v}) = \Omega_{0,j}^1(d_1, 0) = \Omega_{0,j}^1.$$

Moreover,

$$\Sigma_j(\mu) = \sigma_j[d_1, \mathfrak{b}\theta_{\mu,v}] < \Sigma_{j+1}(\mu) = \sigma_{j+1}[d_1, \mathfrak{b}\theta_{\mu,v}] \quad for \ all \ 1 \le j \le q_0 - 1$$

and

$$\lim_{\mu\downarrow d_2\sigma_0} \Sigma_j(\mu) = d_1\sigma_j \quad for \ all \ 1 \le j \le q_0.$$

However, even in the simplest situation covered by Proposition 10.7, the curves $\Sigma_j(\mu)$ and $\Sigma_{j+1}(\mu)$ might meet as the parameter μ separates away from $d_2\sigma_0$. Indeed, the next result holds.

Proposition 10.8 *Suppose $q_0 > 1$, $m_j = 1$ for all $1 \le j \le q_0$,*

$$\mathfrak{f}_2(x, v) = v \quad for \ all \quad (x, v) \in \bar{\Omega} \times [0, \infty), \quad (10.46)$$

and there are $j_1, j_2 \in \{1, ..., q_0\}$ with $j_1 < j_2$ such that

$$\min_{\bar{\Omega}_{0,j_2}^1} \frac{\mathfrak{b}}{\mathfrak{d}} < \min_{\bar{\Omega}_{0,j_1}^1} \frac{\mathfrak{b}}{\mathfrak{d}}. \quad (10.47)$$

Then, there exists $\mu_0 > d_2\sigma_0$ such that, for every $\mu \in (d_2\sigma_0, \mu_0)$,

$$\Sigma_{j_1}(\mu) = \lambda_1[-d_1\Delta + \mathfrak{b}\theta_{\mu,v}, \Omega^1_{0,j_1}] < \Sigma_{j_2}(\mu) = \lambda_1[-d_1\Delta + \mathfrak{b}\theta_{\mu,v}, \Omega^1_{0,j_2}], \quad (10.48)$$

whereas

$$\Sigma_{j_1}(\mu_0) = \Sigma_{j_2}(\mu_0). \tag{10.49}$$

Moreover, there exists $\mu_1 > \mu_0$ such that

$$\lambda_1[-d_1\Delta + \mathfrak{b}\theta_{\mu,v}, \Omega^1_{0,j_1}] > \lambda_1[-d_1\Delta + \mathfrak{b}\theta_{\mu,v}, \Omega^1_{0,j_2}] \quad \text{for all } \mu > \mu_1. \tag{10.50}$$

Suppose $j_1 < j_2$. Then, by definition,

$$\lambda_1[-d_1\Delta + \mathfrak{b}\theta_{\mu,v}, \Omega^1_{0,j_1}(d_1, \mathfrak{b}\theta_{\mu,v})] < \lambda_1[-d_1\Delta + \mathfrak{b}\theta_{\mu,v}, \Omega^1_{0,j_2}(d_1, \mathfrak{b}\theta_{\mu,v})]$$

for all $\mu > d_2\sigma_0$ (see (10.12) if necessary). If we assume, in addition, that $q_0 = 2$ and $m_1 = m_2 = 1$, then $j_1 = 1$, $j_2 = 2$, and, for every $\mu < \mu_0$,

$$\Omega^1_{0,1}(d_1, \mathfrak{b}\theta_{\mu,v}) = \Omega^1_{0,1}(d_1, 0) = \Omega^1_{0,1},$$
$$\Omega^1_{0,2}(d_1, \mathfrak{b}\theta_{\mu,v}) = \Omega^1_{0,2}(d_1, 0) = \Omega^1_{0,2},$$

while, for every $\mu > \mu_1$, we have the *inversion of the protection patches*

$$\Omega^1_{0,1}(d_1, \mathfrak{b}\theta_{\mu,v}) = \Omega^1_{0,2},$$
$$\Omega^1_{0,2}(d_1, \mathfrak{b}\theta_{\mu,v}) = \Omega^1_{0,1}.$$

Which one is the most beneficial for the species u depends on the size of birth rate of the competitor v.

Proof of Proposition 10.8: By Proposition 10.7, (10.48) holds for sufficiently close $\mu > d_2\sigma_0$. Thus, for these μ's we have that

$$\Omega^1_{0,j_1} = \Omega^1_{0,j_1}(d_1, \mathfrak{b}\theta_{\mu,v}), \quad \Omega^1_{0,j_2} = \Omega^1_{0,j_2}(d_1, \mathfrak{b}\theta_{\mu,v}).$$

Subsequently, for every $j \in \{j_1, j_2\}$, we consider the eigenvalue curves

$$S_j(\mu) := \lambda_1[-d_1\Delta + \mathfrak{b}\theta_{\mu,v}, \Omega^1_{0,j}], \quad \mu > d_2\sigma_0. \tag{10.51}$$

The proof of the proposition will proceed by contradiction. Suppose

$$S_{j_1}(\mu) \le S_{j_2}(\mu) \quad \text{for all } \mu > d_2\sigma_0. \tag{10.52}$$

By (10.46), it follows from M. Delgado, J. López-Gómez and A. Suárez [72, Th. 3.4] that

$$\lim_{\mu\uparrow\infty} \frac{\theta_{\mu,v}}{\mu} = \mathfrak{d}^{-1} \quad \text{uniformly in compact subsets of } \Omega.$$

Thus, according to J. E. Furter and J. López-Gómez [92, Le. 3.1], it becomes apparent that

$$\lim_{\mu\uparrow\infty}\frac{S_j(\mu)}{\mu} = \lim_{\mu\uparrow\infty}\lambda_1[-\frac{d_1}{\mu}\Delta + \mathfrak{b}\frac{\theta_{\mu,v}}{\mu}, \Omega_{0,j}^1] = \min_{\bar{\Omega}_{0,j}^1}\frac{\mathfrak{b}}{\mathfrak{d}} \qquad (10.53)$$

for each $j \in \{j_1, j_2\}$. Thus, dividing (10.52) by μ and letting $\mu \uparrow \infty$, we find from (10.53) that

$$\min_{\bar{\Omega}_{0,j_1}^1}\frac{\mathfrak{b}}{\mathfrak{d}} \leq \min_{\bar{\Omega}_{0,j_2}^1}\frac{\mathfrak{b}}{\mathfrak{d}},$$

which contradicts (10.47) and ends the proof. □

When $m_j \geq 2$ for some $1 \leq j \leq q_0 = q_0(d_1, 0)$, the behavior of the curves

$$\Sigma_j(\mu) := \sigma_j[d_1, \mathfrak{b}\theta_{\mu,v}] = \lambda_1[-d_1\Delta + \mathfrak{b}\theta_{\mu,v}, \Omega_{0,j}^i(d_1, \mathfrak{b}\theta_{\mu,v})], \quad \mu > d_2\sigma_0,$$

where $1 \leq j \leq q_0(d_1, \mathfrak{b}\theta_{\mu,v})$, can be more intricate. Indeed, according to (10.46), the map $\mathfrak{F}$ constructed in the proof of Theorem 2.3 is analytic and hence $\mu \mapsto \theta_{\mu,v}$ also is analytic. Thus, by the simplicity of the principal eigenvalues, adapting the proof of [163, Th. 9.1], it follows from Theorem 2.6 on page 377 and Remark 2.9 on page 379 of T. Kato [122] that each of the functions

$$S_j^i(\mu) := \lambda_1[-d_1\Delta + \mathfrak{b}\theta_{\mu,v}, \Omega_{0,j}^i], \quad 1 \leq i \leq m_j, \qquad \mu > d_2\sigma_0,$$

for $1 \leq j \leq q_0$, is analytic. As generically the zeroes of analytic functions should be simple, when the parameter μ perturbs from $d_2\sigma_0$, each of the eigenvalues

$$d_1\sigma_j = \sigma_j[d_1, 0] = \lambda_1[-d_1\Delta, \Omega_{0,j}^i], \qquad 1 \leq i \leq m_j,$$

will generically perturb into m_j simple eigenvalues, $S_j^i(\mu)$, $1 \leq i \leq m_j$. Consequently, for each $\mu > d_2\sigma_0$ we find that

$$\text{card } \mathcal{R}_\mathfrak{a} = q_0(d_1, \mathfrak{b}\theta_{\mu,v})$$

equals the total number of components of Ω_0, i.e.,

$$q_0(d_1, \mathfrak{b}\theta_{\mu,v}) = \sum_{j=1}^{q_0(d_1,0)} m_j,$$

or, equivalently,

$$m_j(d_1, \mathfrak{b}\theta_{\mu,v}) = 1 \quad \text{for all } 1 \leq j \leq q_0(d_1, \mathfrak{b}\theta_{\mu,v}).$$

This should occur for all $\mu > d_2\sigma_0$, except at most at a discrete set of values of μ where at least two of these curves meet. The next result provides us with the simplest example exhibiting such phenomenology. Subsequently, given any smooth subdomain $D \subset \Omega$, we will denote by $\varphi_D > 0$ the principal eigenfunction of $\lambda_1[-\Delta; D]$ with

$$\int_D \varphi_D^2 = 1.$$

Also, we will set $\varphi := \varphi_\Omega$.

Proposition 10.9 *Suppose* (10.46) *and for every*

$$j_1, \ j_2 \in \{1, ..., q_0\}, \quad i_1 \in \{1, ..., m_{j_1}\}, \quad i_2 \in \{1, ..., m_{i_2}\},$$

with $(j_1, i_1) \neq (j_2, i_2)$ *the following holds*

$$\int_{\Omega_{0,j_1}^{i_1}} \mathfrak{b} \, \varphi \, \varphi_{\Omega_{0,j_1}^{i_1}}^2 \neq \int_{\Omega_{0,j_2}^{i_2}} \mathfrak{b} \, \varphi \, \varphi_{\Omega_{0,j_2}^{i_2}}^2 . \tag{10.54}$$

Then, there exists $\mu_0 > d_2\sigma_0$ *such that*

$$q_0(d_1, \mathfrak{b}\theta_{\mu,v}) = \sum_{j=1}^{q_0} m_j \quad \text{for all } \mu \in (d_2\sigma_0, \mu_0), \tag{10.55}$$

i.e., $q_0(d_1, \mathfrak{b}\theta_{\mu,v})$ *equals the total number of components of* Ω_0.

Conditions (10.54) can be easily reached by choosing an adequate $\mathfrak{b}(x)$. The proof of Proposition 10.9 relies on the next perturbation result.

Lemma 10.10 *Suppose* (10.46). *Then, as* $\mu \downarrow d_2\sigma_0$, *we have that*

$$\theta_{\mu,v} = \frac{1}{\int_\Omega \partial\varphi^3}(\mu - d_2\sigma_0)\varphi + O((\mu - d_2\sigma_0)^2) \tag{10.56}$$

and

$$\lambda_1[-d_1\Delta + \mathfrak{b}\theta_{\mu,v}, D] = d_1\sigma_0^D + \frac{\int_D \mathfrak{b}\varphi\varphi_D^2}{\int_\Omega \partial\varphi^3}(\mu - d_2\sigma_0) + O((\mu - d_2\sigma_0)^2) \tag{10.57}$$

for all smooth subdomains $D \subset \Omega$.

Proof: The asymptotic expansion (10.56) is folklore (see, e.g., Lemma 4.3 of M. Delgado, J. López-Gómez and A. Suárez [72]). To prove (10.57), we adapt the proof of [72, Le 4.3]). As we have already discussed before stating Proposition 10.9, by a classical result in perturbation theory, the map

$$\mu \mapsto S_D(\mu) := \lambda_1[-d_1\Delta + \mathfrak{b}\theta_{\mu,v}, D]$$

is analytic. Thus, there exists a constant $K \in \mathbb{R}$ such that

$$S_D(\mu) = d_1\sigma_0^D + K(\mu - d_2\sigma_0) + O((\mu - d_2\sigma_0)^2) \qquad (10.58)$$

as $\mu \downarrow d_2\sigma_0$. Let $\Phi(\mu)$ denote the unique principal eigenfunction associated to $S_D(\mu)$ for which

$$\int_D \Phi(\mu)\varphi_D = 1. \qquad (10.59)$$

According to T. Kato [122, Section VII.2], $\mu \mapsto \Phi(\mu)$ is analytic and hence, it possesses a unique expansion of the form

$$\Phi(\mu) = \Phi_0 + (\mu - d_2\sigma_0)\Phi_1 + O((\mu - d_2\sigma_0)^2) \quad \text{as } \mu \downarrow d_2\sigma_0. \qquad (10.60)$$

Using (10.59) yields

$$\Phi_0 = \varphi_D \quad \text{and} \quad \int_D \Phi_1\varphi_D = 0. \qquad (10.61)$$

Now, substituting (10.56), (10.58) and (10.60) in the identity

$$(-d_1\Delta + \mathfrak{b}\theta_{\mu,v})\Phi(\mu) = S_D(\mu)\Phi(\mu)$$

using (10.61) and identifying terms of order one in $\mu - d_2\sigma_0$ we find that

$$(-d_1\Delta - d_1\sigma_0^D)\Phi_1 = K\varphi_D - \frac{\mathfrak{b}}{\int_\Omega \partial\varphi^3}\,\varphi\,\varphi_D.$$

Finally, multiplying this identity by φ_D and integrating by parts in D yields

$$K = \frac{\int_D \mathfrak{b}\varphi\varphi_D^2}{\int_\Omega \partial\varphi^3},$$

which ends the proof of (10.57). $\qquad \square$

Proof of Proposition 10.9: Thanks to Lemma 10.10, for every $1 \le j \le q_0$ and $1 \le i \le m_j$, we have that

$$S_j^i(\mu) := \lambda_1[-d_1\Delta + \mathfrak{b}\theta_{\mu,v}, \Omega_{0,j}^i]$$
$$= d_1\sigma_0^{\Omega_{0,j}^i} + K_{i,j}(\mu - d_2\sigma_0) + O((\mu - d_2\sigma_0)^2), \qquad \mu \downarrow d_2\sigma_0, \qquad (10.62)$$

where

$$K_{i,j} := \frac{\int_{\Omega_{0,j}^i} \mathfrak{b}\varphi\varphi_{\Omega_{0,j}^i}^2}{\int_\Omega \partial\varphi^3}.$$

Owing to (10.54),

$$K_{i_1,j_1} \ne K_{i_2,j_2} \quad \text{if} \quad (i_1, j_1) \ne (i_2, j_2).$$

Therefore, according to (10.62), for every $1 \leq j \leq q_0$ the eigenvalue

$$\sigma_0^{\Omega_{0,j}^1} = \cdots = \sigma_0^{\Omega_{0,j}^{m_j}}$$

perturbs into m_j (different) eigenvalues. Therefore, (10.55) holds. This ends the proof. □

In Figure 10.1 we have represented the curves arising in the statement of Theorem 10.5 in the special case when

$$q_0 = 2, \qquad m_1 = 3, \qquad m_2 = 2.$$

As suggested by Proposition 10.9, for each $\mu > d_2\sigma_0$ we have that

$$q_0(d_1, \mathfrak{b}\theta_{\mu,v}) = 5 \quad \text{and} \quad m_j(d_1, \mathfrak{b}\theta_{\mu,v}) = 1 \quad \text{for all } 1 \leq j \leq 5,$$

except, at most, at a discrete set of values of the parameter μ, for which

$$q_0(d_1, \mathfrak{b}\theta_{\mu,v}) \in \{1, ..., 4\}.$$

Figure 10.1 shows one of these exceptional values with $q_0(d_1, \mathfrak{b}\theta_{\mu,v}) = 4$. It is the μ coordinate of the crossing point between the second and third curves

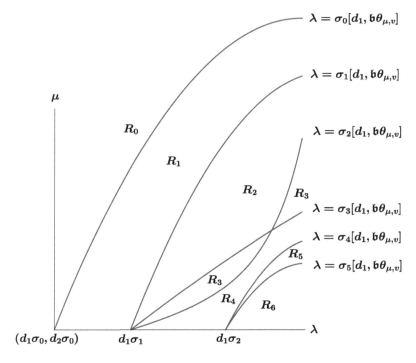

FIGURE 10.1: The μ curves arising in Theorem 10.5.

emanating from $\lambda = d_1\sigma_1$ at $\mu = d_2\sigma_0$. The existence of these three (different) curves is guaranteed by Proposition 10.9. More generally, the exceptional values of μ are the μ coordinates of the crossing points between the eigenvalue curves represented in Figure 10.1, which have been listed on the right of their graphs. These curves divide the first quadrant of the (λ, μ) plane into several regions. For the special configuration shown in Figure 10.1 the first quadrant,

$$Q := \{(\lambda, \mu) \in \mathbb{R}^2 \ : \ \lambda > d_1\sigma_0, \ \mu > d_2\sigma_0\},$$

is divided into the regions R_j, $0 \le j \le 6$, defined by

$$R_0 := \{\,(\lambda, \mu) \in Q \ : \ d_1\sigma_0 < \lambda < \sigma_0[d_1, \mathfrak{b}\theta_{\mu,v}] = \lambda_1[-d_1\Delta + \mathfrak{b}\theta_{\mu,v}, \Omega]\,\},$$
$$R_1 := \{\,(\lambda, \mu) \in Q \ : \ \sigma_0[d_1, \mathfrak{b}\theta_{\mu,v}] < \lambda < \sigma_1[d_1, \mathfrak{b}\theta_{\mu,v}]\},$$
$$R_j := \{\,(\lambda, \mu) \in Q \ : \ \sigma_{j-1}[d_1, \mathfrak{b}\theta_{\mu,v}] \le \lambda < \sigma_j[d_1, \mathfrak{b}\theta_{\mu,v}]\,\}, \qquad 2 \le j \le 5,$$
$$R_6 := \{\,(\lambda, \mu) \in Q \ : \ \lambda \ge \sigma_5[d_1, \mathfrak{b}\theta_{\mu,v}]\}.$$

Thanks to Theorem 10.5(a), (10.5) possesses a coexistence state if, and only if, $(\lambda, \mu) \in R_1$. Moreover, by Theorem 10.5(b), for every $1 \le k \le 5$, it possesses a metacoexistence state of the form $(\mathfrak{M}, \theta_{\mu,v})$ supported in $\Omega_k(d_1, \mathfrak{b}\theta_{\mu,v})$ if, and only if,

$$(\lambda, \mu) \in \bar{R}_0 \cup \bigcup_{j=1}^{k+1} R_j.$$

Thanks to Theorems 4.1, 4.8, 4.9 and Remark 5.3, fixing $\mu > d_2\sigma_0$ and varying λ as the main bifurcation parameter, the u component of the curve of coexistence states grows to infinity in $\Omega_{0,1}[d_1, \mathfrak{b}\theta_{\mu,v}]$ as $\lambda \uparrow \sigma_1[d_1, \mathfrak{b}\theta_{\mu,v}]$. Moreover, for each $1 \le k \le 4$, the minimal metasolution of the form $(\mathfrak{M}, \theta_{\mu,v})$ supported in $\Omega_k[d_1, \mathfrak{b}\theta_{\mu,v}]$ grows towards the minimal metasolution supported in $\Omega_{k+1}[d_1, \mathfrak{b}\theta_{\mu,v}]$ as $\lambda \uparrow \sigma_{k+1}[d_1, \mathfrak{b}\theta_{\mu,v}]$. Owing to Theorem 5.2 and Remark 5.3, the next result holds. It provides us with the dynamics of (10.1) when $\mathfrak{c} = 0$.

Theorem 10.11 *Assume* (HC), (HF), $\mathfrak{c} = 0$, $u_0, v_0 \in C_0(\bar{\Omega})$, $u_0 > 0$ *and* $v_0 > 0$. *Then,*

(a) *If $\mu > d_2\sigma_0$ and $d_1\sigma_0 \le \lambda \le \sigma_0[d_1, \mathfrak{b}\theta_{\mu,v}]$, the semi-trivial state $(0, \theta_{\mu,v})$ is a global attractor for the (classical) positive solutions of* (10.1).

(b) *If $\mu > d_2\sigma_0$ and*

$$\sigma_0[d_1, \mathfrak{b}\theta_{\mu,v}] < \lambda < \sigma_1[d_1, \mathfrak{b}\theta_{\mu,v}],$$

then, the (unique) coexistence state of (10.5) attracts to all (classical) positive solutions of (10.1).

(c) *If $\mu > d_2\sigma_0$ and, for some $k \in \{1, ..., q_0(d_1, \mathfrak{b}\theta_{\mu,v}) - 1\}$,*

$$\sigma_k[d_1, \mathfrak{b}\theta_{\mu,v}] \le \lambda < \sigma_{k+1}[d_1, \mathfrak{b}\theta_{\mu,v}],$$

then,

$$\mathfrak{M}^{\min}_{[\lambda,\Omega_k(d_1,\mathfrak{b}\theta_{\mu,v})]} \leq \liminf_{t\uparrow\infty} u(\cdot,t) \leq \limsup_{t\uparrow\infty} u(\cdot,t) \leq \mathfrak{M}^{\max}_{[\lambda,\Omega_k(d_1,\mathfrak{b}\theta_{\mu,v})]}$$

in $\bar{\Omega}$, *and*

$$\lim_{t\uparrow\infty} v(\cdot,t) = \theta_{\mu,v} \qquad \textit{uniformly in } \bar{\Omega}, \tag{10.63}$$

where (u,v) *stands for the unique solution of* (10.1). *If, in addition,* u_0 *is a subsolution of* (10.32) *in* Ω, *then*

$$\lim_{t\uparrow\infty} u(\cdot,t) = \mathfrak{M}^{\min}_{[\lambda,\Omega_k(d_1,\mathfrak{b}\theta_{\mu,v})]} \qquad \textit{in} \quad \bar{\Omega}.$$

(d) *If* $\mu > d_2\sigma_0$ *and*

$$\lambda \geq \sigma_{q_0(d_1,\mathfrak{b}\theta_{\mu,v})}[d_1,\mathfrak{b}\theta_{\mu,v}],$$

then,

$$\mathfrak{M}^{\min}_{[\lambda,\Omega_-]} \leq \liminf_{t\uparrow\infty} u(\cdot,t) \leq \limsup_{t\uparrow\infty} u(\cdot,t) \leq \mathfrak{M}^{\max}_{[\lambda,\Omega_-]}$$

in $\bar{\Omega}$, *and* (10.63) *holds. Further, if* u_0 *is a subsolution of* (10.32) *in* Ω, *then*

$$\lim_{t\uparrow\infty} u(\cdot,t) = \mathfrak{M}^{\min}_{[\lambda,\Omega_-]} \qquad \textit{in} \quad \bar{\Omega}.$$

Most of the mathematical discussion in this section does not make sense when $q_0 = 1$ and $m_1 = 1$, i.e., when Ω_0 is connected.

10.3   A priori bounds for the coexistence states

In this section we establish the existence of uniform a priori bounds for the coexistence states of (10.5) when (λ,μ) varies in compact subsets of

$$B_{\lambda,\mu} := [(d_1\sigma_0,\infty) \setminus \{d_1\sigma_j, \ 1 \leq j \leq q_0\}] \times (d_2\sigma_0,\infty). \tag{10.64}$$

Throughout the rest of this chapter, it is said that (λ,μ,u,v) is a coexistence state of (10.5) if (u,v) is a coexistence state of (10.5). Moreover, we use the notation

$$\|w\|_\infty := \|w\|_{L^\infty(\Omega)} = \sup_\Omega |w| < +\infty$$

for any function $w \in L^\infty(\Omega)$, and denote

$$H_0^1 := W_0^{1,2}(\Omega).$$

The main result of this section follows.

Theorem 10.12 *Suppose* (HC) *and* (HF) *and let* $(\lambda_n, \mu_n, u_n, v_n)$, $n \geq 1$, *be a sequence of coexistence states of* (10.5) *with*

$$\max_{n \geq 1} \{|\lambda_n|, |\mu_n|\} \leq \Lambda \tag{10.65}$$

for some positive constant Λ. *Then,*

$$v_n \leq \theta_{\Lambda, v} \qquad \text{for all } n \geq 1,$$

where $\theta_{\Lambda, v}$ *stands for the unique positive solution of* (10.8) *for* $\mu = \Lambda$.
  If, in addition,

$$\limsup_{n \to \infty} \|u_n\|_\infty = \infty,$$

there exist $1 \leq j \leq q_0$, $1 \leq i \leq m_j$, *and a subsequence of* $(\lambda_n, \mu_n, u_n, v_n)$, $n \geq 1$, *relabeled by* n, *such that*

$$\lim_{n \to \infty} \lambda_n = d_1 \sigma_j, \qquad \lim_{n \to \infty} \|u_n\|_{L^\infty(\Omega^i_{0,j})} = \infty.$$

Therefore, the coexistence states are uniformly bounded for (λ, μ) *varying in compact subsets of* $B_{\lambda, \mu}$.

The proof of this theorem relies on the next result of technical nature.

Lemma 10.13 *Suppose* $V \in L^\infty(\Omega)$ *and let* $u_n \in H_0^1(\Omega) \cap L^\infty(\Omega)$, $n \geq 1$, *be a sequence of non-negative functions such that*

$$(-d\Delta + V)u_n \leq C\, u_n, \qquad \|u_n\|_\infty \leq C, \qquad n \geq 1, \tag{10.66}$$

for some constants $C > 0$ *and* $d > 0$. *Then there exist a subsequence of* u_n, $n \geq 1$, *labeled again by* n, *and a function* $u \in H_0^1(\Omega) \cap L^\infty(\Omega)$ *such that*

$$\lim_{n \to \infty} u_n = u \quad \text{weakly in } H_0^1(\Omega) \text{ and strongly in } L^p(\Omega) \text{ for all } p > 1.$$

Necessarily, $u \neq 0$ *if there exists* $\omega > 0$ *such that* $\|u_n\|_\infty \geq \omega$ *for sufficiently large* $n \geq 1$.

Proof: Multiplying the differential inequality in (10.66) by u_n and integrating in Ω yields

$$d \int_\Omega |\nabla u_n|^2 + \int_\Omega V u_n^2 \leq C \int_\Omega u_n^2, \qquad n \geq 1.$$

Thus, by the second estimate in (10.66),

$$d \int_\Omega |\nabla u_n|^2 \leq (C - \inf V) \int_\Omega u_n^2 \leq (C - \inf V) C^2 |\Omega|, \qquad n \geq 1.$$

Therefore, the sequence u_n, $n \geq 1$, is bounded in $H_0^1(\Omega) \cap L^\infty(\Omega)$ and hence, there exist $u \in H_0^1(\Omega) \cap L^\infty(\Omega)$ and a subsequence of u_n, $n \geq 1$, labeled again by n, satisfying all the requirements of the statement. Indeed, since H_0^1 is a Hilbert space, it is reflexive. Hence, by Theorem III.16 of H. Brezis [32], any bounded subset of $H_0^1(\Omega)$ must be relatively compact with respect to the weak topology of $H_0^1(\Omega)$. Thus, there exists $u \in H_0^1(\Omega)$ such that, along some subsequence, relabeled by n, we have that $u_n \rightharpoonup u$ in $H_0^1(\Omega)$. Moreover, by a celebrated compactness result of F. Rellich and V. I. Kondrachov [163, Th. 4.5], the imbedding $H_0^1(\Omega) \hookrightarrow L^2(\Omega)$ is compact. So, $u \in L^2(\Omega)$, and u_n can be chosen so that, in addition, $\lim_{n\to\infty} u_n = u$ in $L^2(\Omega)$. On the other hand, owing to Theorem IV.9 of [32], the subsequence can be refined so that $\lim_{n\to\infty} u_n(x) = u(x)$ almost everywhere in Ω. Therefore, letting $n \to \infty$ in $|u_n(x)| \le C$, $n \ge 1$, also yields $\|u\|_\infty \le C$. In particular, $u \in L^\infty(\Omega)$. As for every $p > 2$,

$$\int_\Omega |u_n - u|^p = \int_\Omega |u_n - u|^2 |u_n - u|^{p-2} \le (2C)^{p-2} \int_\Omega |u_n - u|^2,$$

the L^2 convergence of u entails the L^p convergence for all $p > 1$, which ends the proof of the assertion above.

Lastly, suppose $\|u_n\|_\infty \ge \omega$, $n \ge 1$, for some constant $\omega > 0$, and $u = 0$. Then, since

$$\lim_{n\to\infty} \|u_n\|_{L^p(\Omega)} = 0 \qquad \text{for all } p > 1$$

we have that

$$\lim_{n\to\infty} u_n = 0 \quad \text{almost everywhere in } \Omega.$$

This is impossible, because there should exist a subset $D \subset \Omega$ with positive measure such that

$$|u_n(x)| \ge \frac{\omega}{2}$$

for almost all $x \in D$, which ends the proof. $\square$

Proof of Theorem 10.12: Let $(\lambda_n, \mu_n, u_n, v_n)$, $n \ge 1$, be a sequence of coexistence states of (10.5) satisfying (10.65) for some $\Lambda > 0$. Then, it follows from the v equation of (10.5) that

$$-d_2 \Delta v_n \le \mu_n v_n - \partial f_2(\cdot, v_n) v_n \le \Lambda v_n - \partial f_2(\cdot, v_n) v_n, \qquad n \ge 1.$$

Hence, for every $n \ge 1$, the function v_n is a positive subsolution of

$$\begin{cases} -d_2 \Delta v = \Lambda v - \partial f_2(\cdot, v) v & \text{in } \Omega, \\ v = 0 & \text{on } \partial\Omega. \end{cases}$$

Thus, thanks to Theorem 1.7, $v_n \le \theta_{\Lambda, v}$ for all $n \ge 1$. So, also

$$-d_2 \Delta v_n \le \Lambda v_n - \partial f_2(\cdot, v_n) v_n \le \Lambda \theta_{\Lambda, v}$$

for all $n \ge 1$. So, the sequence v_n satisfies (10.66). Consequently, according to Lemma 10.13, there exists $v \in H_0^1(\Omega) \cap L^\infty(\Omega)$ such that, along some subsequence of v_n, relabeled by n,

$$\lim_{n\to\infty} v_n = v \quad \text{weakly in } H_0^1(\Omega) \text{ and strongly in } L^p(\Omega) \text{ for all } p > 1.$$

Subsequently, we will focus our attention on this subsequence. Note that $v_n > 0$ implies $v \geq 0$. As λ_n and μ_n, $n \geq 1$, are bounded, without loss of generality we can also assume that

$$\lim_{n \to \infty} (\lambda_n, \mu_n) = (\lambda, \mu)$$

for some (λ, μ). Suppose

$$\limsup_{n \to \infty} \|u_n\|_\infty = \infty.$$

Then, we can consider a subsequence, labeled again by n, for which

$$\lim_{n \to \infty} \|u_n\|_\infty = \infty.$$

Setting

$$\hat{u}_n := \frac{u_n}{\|u_n\|_\infty}, \qquad n \geq 1,$$

it follows from the u-equation of (10.5) that

$$-d_1 \Delta \hat{u}_n \leq \Lambda \hat{u}_n \qquad \text{for all } n \geq 1.$$

Hence, owing to Lemma 10.13, there exists $\hat{u} \in H_0^1(\Omega) \cap L^\infty(\Omega)$, with $\hat{u} > 0$, such that

$$\lim_{n \to \infty} \hat{u}_n = \hat{u} \quad \text{weakly in } H_0^1(\Omega) \text{ and strongly in } L^p(\Omega) \text{ for all } p > 1.$$

Note that $\hat{u} \neq 0$, because $\|\hat{u}_n\|_\infty = 1$ for all $n \geq 1$. Moreover, since

$$-d_1 \Delta u_n \leq \Lambda u_n - \mathfrak{a}\mathfrak{f}_1(\cdot, u_n)u_n, \qquad n \geq 1,$$

and $\mathfrak{f}_1$ satisfies (KO), it follows from Theorem 1.7 and Proposition 3.3, that for any compact subset $K \subset \Omega_-$ there exists a constant $C_K > 0$ such that

$$\|u_n\|_{L^\infty(K)} \leq C_K \qquad \text{for all } n \geq 1.$$

Consequently, $\hat{u} = 0$ in any compact subset of Ω_-. Thus, setting

$$\Omega_\delta := \Omega_0 \cup \{ x \in \Omega \ : \ \text{dist}(x, \partial\Omega_0 \cap \Omega) < \delta \}$$

for sufficiently small $\delta > 0$, we find that

$$\hat{u} \in \bigcap_{\delta > 0} H_0^1(\Omega_\delta).$$

As $\partial\Omega_0$ satisfies the segment property, it is stable, as discussed at the end of the proof of [144, Th. 4.2] and in [163, Th. 8.4]. Therefore, $\hat{u} \in H_0^1(\Omega_0)$.

Let $\varphi \in C_0^\infty(\Omega_0)$ be a test function supported in Ω_0. Then, multiplying the u_n-equation by φ, dividing by $\|u_n\|_\infty$, and integrating by parts in Ω_0 yields

$$d_1 \int_{\Omega_0} \nabla \hat{u}_n \nabla \varphi = \lambda_n \int_{\Omega_0} \hat{u}_n \varphi - \int_{\Omega_0} \mathfrak{b} \hat{u}_n v_n \varphi, \qquad n \geq 1,$$

because $\mathfrak{a} = 0$ in Ω_0. Hence, letting $n \to \infty$ yields

$$d_1 \int_{\Omega_0} \nabla \hat{u} \nabla \varphi = \lambda \int_{\Omega_0} \hat{u} \varphi - \int_{\Omega_0} \mathfrak{b} \hat{u} v \varphi.$$

Therefore, $\hat{u}$ provides us with a weak solution of

$$\begin{cases} -d_1 \Delta \hat{u} = (\lambda - \mathfrak{b}v)\hat{u} & \text{in } \Omega_0, \\ \hat{u} = 0 & \text{on } \partial\Omega_0. \end{cases} \tag{10.67}$$

As $\mathfrak{b}v \in L^\infty(\Omega)$, by elliptic regularity, $\hat{u} \in C_0^1(\bar{\Omega}_{0,j}^i)$ for all $1 \leq j \leq q_0$ and $1 \leq i \leq m_j$. Moreover, since $\hat{u} \neq 0$ and $\hat{u} = 0$ in Ω_-, there exist $1 \leq j_0 \leq q_0$ and $1 \leq i_0 \leq m_{j_0}$ such that

$$\hat{u} > 0 \quad \text{in} \quad D_0 := \Omega_{0,j_0}^{i_0}.$$

Therefore, by (10.67), $\hat{u}$ is a principal eigenfunction of

$$\lambda = \lambda_1[-d_1\Delta + \mathfrak{b}v, D_0]. \tag{10.68}$$

In particular, $\hat{u}$ lies in the interior of the cone of positive functions of $C_0^1(\bar{D}_0)$. By construction,

$$\lim_{n \to \infty} \|u_n\|_{L^\infty(D_0)} = \infty.$$

Let $\varphi \in C_0^\infty(D_0)$ be a test function in D_0. Multiplying the v_n equation by φ and integrating by parts in D_0 shows that

$$d_2 \int_{D_0} \nabla v_n \nabla \varphi = \mu_n \int_{D_0} v_n \varphi - \int_{D_0} \partial \mathfrak{f}_2(\cdot, v_n) v_n \varphi - \|u_n\|_\infty \int_{D_0} \mathfrak{c} \varphi v_n \hat{u}_n,$$

for all $n \geq 1$. So, dividing each of these identities by $\|u_n\|_\infty$ and letting $n \to \infty$ yields

$$\int_{D_0} \mathfrak{c} \varphi v \hat{u} = 0.$$

Consequently, $v = 0$ in D_0, because $v(x) \geq 0$, $\hat{u}(x) > 0$ and $\mathfrak{c}(x) > 0$ for all $x \in D_0$. Consequently, (10.68) becomes

$$\lambda = \lambda_1[-d_1\Delta, D_0] = d_1 \sigma_{j_0},$$

which ends the proof. $\square$

10.4 Global continua of coexistence states

This section applies the abstract theory of J. López-Gómez [149] to get some necessary and sufficient conditions for coexistence states of (10.5). In order to state the results, we need to introduce some notations. For each $j \in \{1, 2\}$, we will denote by e_j the unique solution of

$$
\begin{cases}
-d_j \Delta e_j = 1 & \text{in } \Omega, \\
e_j = 0 & \text{on } \partial\Omega.
\end{cases}
$$

Thanks to Theorem 1.1, e_1 and e_2 are strongly positive in $C_0^1(\bar\Omega)$. Subsequently, for every $j \in \{1, 2\}$, we will denote by

$$
X_j := C_{e_j}(\bar\Omega)
$$

the ordered Banach space consisting of all functions $u \in C(\bar\Omega)$ for which there is a positive constant $\kappa > 0$ such that

$$
-\kappa e_j \leq u \leq \kappa e_j,
$$

equipped with the Minkowski norm

$$
\|u\|_{e_j} := \inf\{\kappa > 0 \; : \; -\kappa e_j \leq u \leq \kappa e_j\}
$$

and ordered by its cone of positive functions, P_j; the interior of the cone P_j will be denoted by $\text{int}\, P_j$, $j = 1, 2$ (see Section 7.2 of [163], if necessary). As the Minkowski norm is monotone, according to [163, Th. 6.1], the cones P_j are normal (see Section 6.1 of [163] to get familiar with the most basic concepts in ordered Banach spaces).

Using these notations, for any $K \geq 0$, the solutions of (10.5) are the fixed points of the compact operator

$$
\mathcal{K}_{\lambda,\mu} : X_1 \times X_2 \to X_1 \times X_2
$$

defined by

$$
\mathcal{K}_{\lambda,\mu}(u, v) := \begin{pmatrix} (-d_1\Delta + K)^{-1}[(\lambda + K)u - \mathfrak{a}f_1(\cdot, u)u - \mathfrak{b}uv] \\ (-d_2\Delta + K)^{-1}[(\mu + K)v - \mathfrak{d}f_2(\cdot, v)v - \mathfrak{c}uv] \end{pmatrix}. \tag{10.69}
$$

Therefore, the abstract theory of [142] and Chapter 7 of [163] can be applied to establish the existence of a component of the set of coexistence states of (10.5) emanating from each of the semi-trivial positive solutions.

10.4.1 Regarding μ as the main bifurcation parameter

In the next result, λ is fixed and μ is regarded as the main bifurcation parameter. Later, the roles of these parameters will be exchanged. Throughout the rest of this book, a *component* is defined as a closed and connected subset which is maximal for the inclusion; $\mathcal{P}_\mu$ stands for the μ projection of (μ, u, v).

Theorem 10.14 *Suppose* (HC) *and* (HF), *and regard* $\mu > d_2\sigma_0$ *as the main bifurcation parameter. Then, for every*

$$d_1\sigma_0 < \lambda < d_1\sigma_1, \tag{10.70}$$

$\mu = \sigma_0[d_2, \mathfrak{c}\theta_{\lambda,u}]$ *is the unique bifurcation value to coexistence states of* (10.5) *from* $(\theta_{\lambda,u}, 0)$. *Moreover, the component of the set of coexistence states emanating from* $(\theta_{\lambda,u}, 0)$ *at*

$$\mu = \sigma_0[d_2, \mathfrak{c}\theta_{\lambda,u}] = \lambda_1[-d_2\Delta + \mathfrak{c}\theta_{\lambda,u}, \Omega],$$

denoted by

$$\mathfrak{C}^+_{(\mu,\theta_{\lambda,u},0)} \subset \mathbb{R} \times \operatorname{int} P_1 \times \operatorname{int} P_2,$$

meets the other curve of semi-trivial solutions $(\mu, 0, \theta_{\mu,v})$ *at the unique value of* μ, μ_λ, *for which*

$$\lambda = \sigma_0[d_1, \mathfrak{b}\theta_{\mu_\lambda,v}] = \lambda_1[-d_1\Delta + \mathfrak{b}\theta_{\mu_\lambda,v}, \Omega]. \tag{10.71}$$

In particular, (10.5) *possesses a coexistence state if*

$$(\mu - \sigma_0[d_2, \mathfrak{c}\theta_{\lambda,u}])(\lambda - \sigma_0[d_1, \mathfrak{b}\theta_{\mu,v}]) > 0. \tag{10.72}$$

Now, instead of (10.70), *assume that*

$$\lambda \geq d_1\sigma_1.$$

Then, $(\mu, u, v) = (\mu_\lambda, 0, \theta_{\mu_\lambda,v})$, *where* μ_λ *is the unique value of* μ *for which* (10.71) *holds, is the unique bifurcation point to coexistence states of* (10.5) *from the curve* $(\mu, 0, \theta_{\mu,v})$. *Moreover, the component of coexistence states emanating from* $(\mu, 0, \theta_{\mu,v})$ *at* $\mu = \mu_\lambda$, *denoted by*

$$\mathfrak{C}^+_{(\mu,0,\theta_{\mu,v})} \subset \mathbb{R} \times \operatorname{int} P_1 \times \operatorname{int} P_2,$$

is unbounded in $\mathbb{R} \times X_1 \times X_2$. *If, in addition,*

$$\lambda \in (d_1\sigma_1, \infty) \setminus \{d_1\sigma_j : 2 \leq j \leq q_0\}, \tag{10.73}$$

then

$$(\mu_\lambda, \infty) \subset \mathcal{P}_\mu\left(\mathfrak{C}^+_{(\mu,0,\theta_{\mu,v})}\right) \subset (\mu_\lambda^1, \infty)$$

where μ_λ^1 *stands for the unique value of* μ *for which*

$$\lambda = \sigma_1[d_1, \mathfrak{b}\theta_{\mu_\lambda^1,v}] = \lambda_1[-d_1\Delta + \mathfrak{b}\theta_{\mu_\lambda^1,v}, \Omega_0]. \tag{10.74}$$

Actually, there exists $\mu_* \in (\mu_\lambda^1, \mu_\lambda]$ *such that* (10.5) *cannot admit a coexistence state if* $\mu < \mu_*$.

Theorem 10.14 will be inferred from the next proposition.

Proposition 10.15 *Suppose* (10.70) *is satisfied and* (10.5) *possesses a coexistence state. Then,*

$$\lambda > \sigma_0[d_1, \mathfrak{b}\theta_{\{\mu,v;c\theta_{\lambda,u}\}}],\qquad(10.75)$$

where $\theta_{\lambda,u}$ *is the unique positive solution of* (10.7) *and*

$$\Theta := \theta_{\{\mu,v;c\theta_{\lambda,u}\}}$$

stands for the unique positive solution of

$$\begin{cases} (-d_2\Delta + c\theta_{\lambda,u})\Theta = \mu\Theta - \mathfrak{d}f_2(\cdot,\Theta)\Theta & in\ \ \Omega, \\ \Theta = 0 & on\ \ \partial\Omega. \end{cases}\qquad(10.76)$$

In particular, (10.5) *cannot admit a coexistence state for sufficiently large* μ. *Moreover, if it admits a coexistence state with* $\lambda \geq d_1\sigma_1$, *then*

$$\lambda < \sigma_1[d_1, \mathfrak{b}\theta_{\mu,v}].\qquad(10.77)$$

Proof of Proposition 10.15: Suppose (10.70) holds and (10.5) admits a coexistence state, (u,v). Then, u and v are positive subsolutions of (10.7) and (10.8), respectively. Thus, according to Lemma 1.9, we find that $\lambda > d_1\sigma_0$ and $\mu > d_2\sigma_0$. Moreover, by Lemma 1.8,

$$u \leq \theta_{\lambda,u} \quad and \quad v \leq \theta_{\mu,v}.\qquad(10.78)$$

Then, substituting (10.78) into the v equation of (10.5) yields

$$-d_2\Delta v \geq \mu v - \mathfrak{d}f_2(\cdot,v)v - c\theta_{\lambda,u}v.$$

Hence, v provides us with a positive supersolution of (10.76). So, according to Theorem 1.7,

$$v \geq \theta_{\{\mu,v;c\theta_{\lambda,u}\}}.$$

Substituting this estimate into the u equation of (10.5) shows that

$$-d_1\Delta u \leq \lambda u - \mathfrak{a}f_1(\cdot,u)u - \mathfrak{b}\theta_{\{\mu,v;c\theta_{\lambda,u}\}}u.$$

In other words, u is a positive subsolution of

$$\begin{cases} (-d_1\Delta + \mathfrak{b}\theta_{\{\mu,v;c\theta_{\lambda,u}\}})u = \lambda u - \mathfrak{a}f_1(\cdot,u)u & in\ \ \Omega, \\ u = 0 & on\ \ \partial\Omega. \end{cases}$$

Therefore, thanks to Lemma 1.9, (10.75) holds.

By adapting the proof of (10.39), it is easily seen that

$$\lim_{\mu\uparrow\infty} \theta_{\{\mu,v;c\theta_{\lambda,u}\}} = \infty \quad uniformly\ in\ compact\ subsets\ of\ \ \Omega.$$

Thus, owing to Lemma 10.6,

$$\lim_{\mu\uparrow\infty} \sigma_0[d_1, \mathfrak{b}\theta_{\{\mu,v;c\theta_{\lambda,u}\}}] = \infty.\qquad(10.79)$$

According to (10.79), (10.75) cannot be satisfied for large μ. Therefore, as we already know that (10.75) is necessary for the existence of a coexistence state, (10.5) cannot admit a coexistence state for large μ.

Subsequently, we assume that $\lambda \geq d_1\sigma_1$. Then, substituting (10.78) into the u equation of (10.5) yields

$$-d_1\Delta u \geq \lambda u - \mathfrak{a}\mathfrak{f}_1(\cdot, u)u - \mathfrak{b}\theta_{\mu,v}u$$

and hence, u is a positive supersolution of

$$\begin{cases} (-d_1\Delta + \mathfrak{b}\theta_{\mu,v})u = \lambda u - \mathfrak{a}\mathfrak{f}_1(\cdot, u)u & \text{in} \quad \Omega, \\ u = 0 & \text{on} \quad \partial\Omega. \end{cases} \tag{10.80}$$

In case $\lambda \leq \sigma_0[d_1, \mathfrak{b}\theta_{\mu,v}]$, (10.77) holds from the estimate

$$\lambda \leq \sigma_0[d_1, \mathfrak{b}\theta_{\mu,v}] < \sigma_1[d_1, \mathfrak{b}\theta_{\mu,v}].$$

So, suppose $\lambda > \sigma_0[d_1, \mathfrak{b}\theta_{\mu,v}]$. Then, by adapting Theorem 1.7 it is easily seen that (10.80) possesses a positive solution. Therefore,

$$\lambda = \lambda_1[-d_1\Delta + \mathfrak{b}\theta_{\mu,v} + \mathfrak{a}\mathfrak{f}_1(\cdot, u), \Omega] < \lambda_1[-d_1\Delta + \mathfrak{b}\theta_{\mu,v}, \Omega_{0,j}^i]$$

for all $1 \leq j \leq q_0$ and $1 \leq i \leq m_j$, which entails (10.77) and ends the proof of the proposition. $\square$

Proof of Theorem 10.14: Assume (10.70). The fact that

$$(\mu, u, v) = (\sigma_0[d_2, \mathfrak{c}\theta_{\lambda,u}], \theta_{\lambda,u}, 0)$$

is the unique bifurcation point to coexistence states from $(u, v) = (\theta_{\lambda,u}, 0)$ follows easily by linearizing (10.5) at $(\theta_{\lambda,u}, 0)$ and using the local bifurcation theorem of M. G. Crandall and P. H. Rabinowitz [59], as in the proof of Theorem 7.2.2 of [149]. The technical details are omitted here. Actually, the existence of the component $\mathfrak{C}^+_{(\mu,\theta_{\lambda,u},0)}$ follows from [149, Th. 7.2.2] and it satisfies some of the following alternatives:

(A1) $\mathfrak{C}^+_{(\mu,\theta_{\lambda,u},0)}$ is unbounded in $\mathbb{R} \times X_1 \times X_2$.

(A2) There exists $\mu_\omega \in \mathbb{R}$ such that

$$\lambda = \sigma_0[d_1, \mathfrak{b}\theta_{\mu_\omega,v}] \quad \text{and} \quad (\mu_\omega, 0, \theta_{\mu_\omega,v}) \in \bar{\mathfrak{C}}^+_{(\mu,\theta_{\lambda,u},0)}.$$

(A3) There exists a further positive solution $\hat{\theta}_{\lambda,u} \neq \theta_{\lambda,u}$ of (10.7) with

$$(\sigma_0[d_2, \mathfrak{c}\hat{\theta}_{\lambda,u}], \hat{\theta}_{\lambda,u}, 0) \in \bar{\mathfrak{C}}^+_{(\mu,\theta_{\lambda,u},0)}.$$

(A4) $\lambda = d_1\sigma_0$ and $(d_2\sigma_0, 0, 0) \in \bar{\mathfrak{C}}^+_{(\mu,\theta_{\lambda,u},0)}.$

By (10.70), Alternative (A4) cannot occur. Thanks to Theorem 2.2 and Remark 5.3, $\theta_{\lambda,u}$ is the unique positive solution of (10.7). So, Alternative (A3) cannot occur either. Moreover, owing to Proposition 10.15, (10.5) cannot admit a coexistence state for large μ, and any coexistence state, (u, v), of (10.5) satisfies $u \leq \theta_{\lambda,u}$ and $v \leq \theta_{\mu,v}$. Therefore, $\mathfrak{C}^+_{(\mu,\theta_{\lambda,u},0)}$ is bounded in $\mathbb{R} \times X_1 \times X_2$. Consequently, Alternative (A2) occurs. Necessarily $\mu_\omega = \mu_\lambda$, because the map $\mu \mapsto \sigma_0[d_1, \mathfrak{b}\theta_{\mu,v}]$ is increasing. Moreover, since $\mathcal{P}_\mu$ is continuous and $\mathfrak{C}^+_{(\mu,\theta_{\lambda,u},0)}$ is connected, the projection

$$I_\mu := \mathcal{P}_\mu\left(\mathfrak{C}^+_{(\mu,\theta_{\lambda,u},0)}\right)$$

must be an interval. As a by-product, for any (λ, μ) in the region enclosed by the two curves,

$$\lambda = \sigma_0[d_1, \mathfrak{b}\theta_{\mu,v}] \quad \text{and} \quad \mu = \sigma_0[d_2, \mathfrak{c}\theta_{\lambda,u}], \qquad d_1\sigma_0 < \lambda < d_1\sigma_1,$$

described by (10.72), (10.5) has a coexistence state. This ends the proof of the first part of the theorem.

Subsequently, we assume that $\lambda \geq d_1\sigma_1$. As before, the existence of $\mathfrak{C}^+_{(\mu,0,\theta_{\mu,v})}$ can be derived from [149, Th. 7.2.2]. Similarly, $\mathfrak{C}^+_{(\mu,0,\theta_{\mu,v})}$ satisfies some of the following alternatives:

(B1) $\mathfrak{C}^+_{(\mu,0,\theta_{\mu,v})}$ is unbounded in $\mathbb{R} \times X_1 \times X_2$.

(B2) There exists a semi-trivial positive solution of the form $(u, 0)$ such that $(\mu, u, 0) \in \bar{\mathfrak{C}}^+_{(\mu,0,\theta_{\mu,v})}$.

(B3) There exists a positive solution $\hat{\theta}_{\mu,v} \neq \theta_{\mu,v}$ of (10.8) such that $(\mu, 0, \hat{\theta}_{\mu,v}) \in \bar{\mathfrak{C}}^+_{(\mu,0,\theta_{\mu,v})}$.

(B4) $\lambda = d_1\sigma_0$ and $(d_2\sigma_0, 0, 0) \in \bar{\mathfrak{C}}^+_{(\mu,0,\theta_{\mu,v})}$.

As $\lambda \geq d_1\sigma_1 > d_1\sigma_0$, Alternative (B4) cannot occur. By Theorem 4.1 and Remark 5.3, (10.7) admits a positive solution if, and only if, $d_1\sigma_0 < \lambda < d_1\sigma_1$. Thus, since we are assuming that $\lambda \geq d_1\sigma_1$, (10.7) cannot admit a positive solution and hence Alternative (B2) also fails. Furthermore, since (10.8) possesses a unique positive solution, Alternative (B3) cannot occur either. Therefore, Alternative (B1) holds.

By Proposition 10.15,

$$\lambda < \sigma_1[d_1, \mathfrak{b}\theta_{\mu,v}]. \tag{10.81}$$

Thus, $\mu > \mu_\lambda^1$, where μ_λ^1 stands for the unique value of μ for which

$$\lambda = \sigma_1[d_1, \mathfrak{b}\theta_{\mu_\lambda^1,v}].$$

Hence,

$$\mathcal{P}_\mu\left(\mathfrak{C}^+_{(\mu,0,\theta_{\mu,v})}\right) \subset (\mu_\lambda^1, \infty).$$

Suppose, in addition, condition (10.73) holds. Then, by Theorem 10.12, the coexistence states of (10.5) are uniformly bounded in compact subsets of $\mu \geq d_2\sigma_0$. Consequently, since $\mathfrak{C}^+_{(\mu,0,\theta_{\mu,v})}$ is unbounded in $\mathbb{R} \times X_1 \times X_2$,

$$(\mu_\lambda, \infty) \subset \mathcal{P}_\mu\left(\mathfrak{C}^+_{(\mu,0,\theta_{\mu,v})}\right).$$

It remains to establish the existence of μ_*. Let (μ, u, v) be a coexistence state of (10.5) with $\mu < \mu_\lambda$. Then,

$$\lambda = \sigma_0[d_1, \mathfrak{b}\theta_{\mu_\lambda,v}] > \sigma_0[d_1, \mathfrak{b}\theta_{\mu,v}]. \tag{10.82}$$

Moreover, $v \leq \theta_{\mu,v}$ and u is a positive supersolution of (10.80). Thus, by Remark 5.3, we find from Theorem 1.7 that $u \geq \theta_{[\lambda,\mathfrak{b}\theta_{\mu,v}]}$, where $\theta_{[\lambda,\mathfrak{b}\theta_{\mu,v}]}$ is the unique positive solution of (10.80); it exists because, due to (10.81) and (10.82),

$$\sigma_0[d_1, \mathfrak{b}\theta_{\mu,v}] < \lambda < \sigma_1[d_1, \mathfrak{b}\theta_{\mu,v}].$$

By Theorem 10.12, there exists a positive constant, C, such that

$$\|u\|_\infty \leq C.$$

Thus, since $\theta_{[\lambda,\mathfrak{b}\theta_{\mu,v}]} \leq u$, we can infer from the previous estimate that

$$\theta_{[\lambda,\mathfrak{b}\theta_{\mu,v}]} \leq C. \tag{10.83}$$

Suppose (10.5) admits a sequence of coexistence states, (μ_n, u_n, v_n), $n \geq 1$, with

$$\lim_{n\to\infty} \mu_n = \mu_\lambda^1.$$

Then, according to (10.83), we find that

$$\theta_{[\lambda,\mathfrak{b}\theta_{\mu_n,v}]} \leq C, \qquad n \geq 1. \tag{10.84}$$

On the other hand, by the continuous dependence of the principal eigenvalue with respect to the potential and the continuous dependence of $\theta_{\mu,v}$ with respect to μ, it becomes apparent that

$$\lim_{n\to\infty} \sigma_1[d_1, \mathfrak{b}\theta_{\mu_n,v}] = \sigma_1[d_1, \mathfrak{b}\theta_{\mu_\lambda^1,v}] = \lambda.$$

Consequently, according to Theorem 4.1,

$$\lim_{n\to\infty} \theta_{[\lambda,\mathfrak{b}\theta_{\mu_n,v}]} = \infty$$

uniformly on compact subsets of some of the components of Ω_0, which contradicts (10.84) and ends the proof. $\square$

Theorem 10.14 establishes that under condition (10.70) the problem (10.5) behaves much like its classical counterpart with $\Omega_0 = \emptyset$. Naturally, as $\lambda \uparrow d_1\sigma_1$

the u components of the coexistence states might grow to infinity in some or several components of Ω_0, but within the range

$$d_1\sigma_0 < \lambda < d_1\sigma_1,$$

the bifurcation diagrams of coexistence states of (10.5) look like those found in the pioneering plots of J. C. Eilbeck, J. E. Furter and J. López-Gómez [82].

The behavior of the model for $\lambda > d_1\sigma_1$ is completely different, because, according to Theorem 10.14, (10.5) possesses a coexistence state for every $\mu > \mu_\lambda$ if $\lambda > d_1\sigma_1$ but $\lambda \neq d_1\sigma_j$ for all $2 \leq j \leq q_0$. Actually, we imposed $\mathfrak{c}(x) > 0$ for all $x \in \Omega_0$ from the very beginning in order to get this result. Indeed, thanks to Theorem 10.11, in the special case when $\mathfrak{c} = 0$, the problem (10.5) admits a coexistence state if, and only if,

$$\sigma_0[d_1, \mathfrak{b}\theta_{\mu,v}] < \lambda < \sigma_1[d_1, \mathfrak{b}\theta_{\mu,v}],$$

or, equivalently,

$$\mu_\lambda^1 < \mu < \mu_\lambda,$$

in strong contrast with the case when $\mathfrak{c}(x) > 0$ for all $x \in \Omega_0$.

Figure 10.2 represents an admissible bifurcation diagram of coexistence states in the special case $\mathfrak{c} = 0$ with $\lambda \geq d_1\sigma_1$.

According to Theorem 2.2 and Remark 5.3, for this choice (10.5) cannot admit a semi-trivial positive solution of the form $(u, 0)$. Figure 10.2 plots the

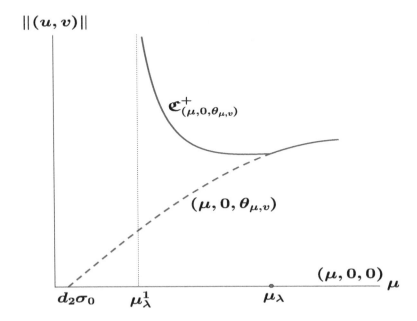

FIGURE 10.2: The component $\mathfrak{C}^+_{(\mu,0,\theta_{\mu,v})}$ in case $\mathfrak{c} = 0$ with $\lambda \geq d_1\sigma_1$.

norm of the solutions

$$\|(u,v)\| := \|u\|_\infty + \|v\|_\infty$$

versus the main bifurcation parameter μ. The horizontal axis consists of the trivial solutions $(\mu, 0, 0)$. The curve of semi-trivial positive solutions $(\mu, 0, \theta_{\mu,v})$ bifurcates from $(\mu, 0, 0)$ at the critical value of the parameter $\mu = d_2\sigma_0$ and it is defined for all $\mu > d_2\sigma_0$. By linearizing (10.5) around it, it is easily seen that it is linearly unstable if $\mu < \mu_\lambda$ and linearly stable if $\mu > \mu_\lambda$. At $\mu = \mu_\lambda$ there is a secondary bifurcation from $(\mu, 0, \theta_{\mu,v})$ to a subcritical curve of coexistence states of (10.5). Namely,

$$\mathfrak{C}^+_{(\mu,0,\theta_{\mu,v})} = \left\{ (\mu, \theta_{[d_1, \mathfrak{b}\theta_{\mu,v}]}, \theta_{\mu,v}) \ : \ \mu_\lambda^1 < \mu < \mu_\lambda \right\}. \tag{10.85}$$

According to Theorem 10.11(b), each of these coexistence states is a global attractor with respect to the positive solutions of (10.1). According to Theorem 4.1 and Remark 5.3, since

$$\lambda = \sigma_1[d_1, \mathfrak{b}\theta_{\mu_\lambda^1, v}],$$

the u component of these coexistence states blows up in $\Omega_{0,1}[d_1, \mathfrak{b}\theta_{\mu,v}]$ as $\mu \downarrow \mu_\lambda^1$. As usual, solid lines in Figure 10.2 are filled in by linearly stable solutions, while dashed lines indicate linearly unstable solutions.

Obviously, the situation sketched in Figure 10.2 for $\mathfrak{c} = 0$ is quite different from the one described by Theorem 10.14 when $\mathfrak{c}(x) > 0$ for all $x \in \Omega_0$ and $\lambda \geq d_1\sigma_1$. Although at first glance the huge differences between both cases might seem paradoxical, since the solution diagrams for $\mathfrak{c} = 0$ and for sufficiently small $\mathfrak{c} > 0$ should be reminiscent, the huge differences are rather natural. Indeed, the next result explains what's going on when $\mathfrak{c} > 0$ perturbs from zero. The notations introduced in Theorem 10.14 are maintained.

Theorem 10.16 *Suppose* (HC) *and* (HF), *set*

$$\mathfrak{c} := \varrho \frac{\mathfrak{c}}{\|\mathfrak{c}\|_\infty}, \qquad \varrho := \|\mathfrak{c}\|_\infty,$$

and regard to ϱ as a parameter. Then, for each $\varepsilon > 0$ there exists $\varrho_\varepsilon > 0$ such that (10.5) *possesses a coexistence state for every $\mu \in (\mu_\lambda^1 + \varepsilon, \infty)$ provided $\varrho \in (0, \varrho_\varepsilon)$. Therefore, if we denote by $\mu_*(\mathfrak{c})$ the minimal value of μ for which* (10.5) *has a coexistence state, whose existence is given by Theorem 10.14, then*

$$\lim_{\varrho \downarrow 0} \mu_*(\mathfrak{c}) = \mu_\lambda^1.$$

Proof: In the special case $\mathfrak{c} = 0$, for each $\mu \in (\mu_\lambda^1, \mu_\lambda)$ the unique coexistence state of (10.5), $(\mu, \theta_{[d_1, \mathfrak{b}\theta_{\mu,v}]}, \theta_{\mu,v})$, is non-degenerate. Therefore, applying the implicit function theorem along the solution curve

$$(\mu, \theta_{[d_1, \mathfrak{b}\theta_{\mu,v}]}, \theta_{\mu,v}), \qquad \mu_\lambda^1 < \mu < \mu_\lambda,$$

with ϱ as the main continuation parameter, yields the result. Outside a neighborhood of $\mu = \mu_\lambda$, one can complete very easily all technical details. In a neighborhood of $\mu = \mu_\lambda$, one should use the technical devices of [148]. □

It turns out that the curve of coexistence states of (10.5) for $\mathfrak{c} = 0$, given by (10.85), perturbs into a compact arc of curve of coexistence states if $\mathfrak{c} \sim 0$. If one further assumes

$$\lambda \in [d_1\sigma_1, \infty) \setminus \{d_1\sigma_j : 1 \leq j \leq q_0\},$$

then, according to Theorem 10.14, the perturbed component $\mathfrak{C}^+_{(\mu,0,\theta_{\mu,v})}$ for $\mathfrak{c} \sim 0$ must turn backward at some turning point, $(\mu_t(\mathfrak{c}), u_t(\mathfrak{c}), v_t(\mathfrak{c}))$, as illustrated in Figure 10.3, where it has been represented a genuine case where

$$\mu_*(\mathfrak{c}) = \mu_t(\mathfrak{c}),$$

though, in general, $\mu_*(\mathfrak{c}) \leq \mu_t(\mathfrak{c})$, and $\mu_*(\mathfrak{c}) < \mu_t(\mathfrak{c})$ might indeed occur. Naturally, the *global continuation* of the component $\mathfrak{C}^+_{(\mu,0,\theta_{\mu,v})}$ for all further $\mu > \mu_t(\mathfrak{c})$ can be accomplished thanks to the existing a priori bounds guaranteed by Theorem 10.12. Arguing as in the proof of Theorem 10.16, it becomes apparent that

$$\lim_{\mathfrak{c}\downarrow 0} \mu_t(\mathfrak{c}) = \mu_\lambda^1$$

and that $u_t(\mathfrak{c})$ must blow up in some component of Ω_0 as $\mathfrak{c} \downarrow 0$. Our perturbation analysis shows that $\mathfrak{C}^+_{(\mu,0,\theta_{\mu,v})}$ looks like Figure 10.3 for sufficiently

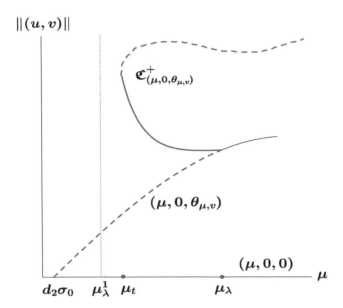

FIGURE 10.3: The component $\mathfrak{C}^+_{(\mu,0,\theta_{\mu,v})}$ for $\mathfrak{c} \sim 0$ and λ satisfying (10.73).

small $\mathfrak{c}$. However, one cannot exclude the existence of a further *global super-critical folding*, or even an *isola*, separated away from $\mathfrak{C}_{(\mu,0,\theta_{\mu,v})}$. Although $\mathfrak{C}^+_{(\mu,0,\theta_{\mu,v})}$ consists of a smooth curve filled in by asymptotically stable coexistence states from the bifurcation point $(\mu_\lambda, 0, \theta_{\mu_\lambda,v})$ to reach the turning point $(\mu_t(\mathfrak{c}), u_t(\mathfrak{c}), v_t(\mathfrak{c}))$, one cannot guarantee anything about the fine structure of $\mathfrak{C}^+_{(\mu,0,\theta_{\mu,v})}$ beyond the turning point. In Section 10.6, we will show that (10.5) possesses at least two coexistence states for each $\mu \in (\mu_*(\mathfrak{c}), \mu_\lambda)$, as illustrated in Figure 10.3. In general, $\mu_*(\mathfrak{c})$ should grow as $\mathfrak{c}$ increases.

Summarizing, as $\mathfrak{c}$ perturbs from zero, the component $\mathfrak{C}^+(\mu, 0, \theta_{\mu,v})$ shown in Figure 10.2 bends backward, looking like Figure 10.3. Consequently, as in Chapter 9, (10.5) or (10.1) exhibits a sort of superlinear indefinite character as $\mathfrak{c}$ perturbs from zero.

10.4.2 Regarding λ as the main bifurcation parameter

In this section, we will carry out the corresponding discussion fixing $\mu > d_2\sigma_0$ and using λ as the main bifurcation parameter, exchanging the roles played by the parameters λ and μ in the previous section. The technical details of the proofs are omitted here, since they can be be easily reconstructed from the proof of Theorem 10.14. As the behavior of the model can be very intricate for large $\mathfrak{c}$, we will focus our attention on the case when $\mathfrak{c}$ is small. How small should $\mathfrak{c}$ be for the validity of our analysis will be discussed later.

By Remark 5.3, we find from Theorem 4.8 that

$$\lim_{\lambda \uparrow d_1\sigma_1} \theta_{\lambda,u} = \mathfrak{M}^{\min}_{[d_1\sigma_1,\Omega_1]} \tag{10.86}$$

(see (10.14)). Subsequently, for sufficiently small $\delta > 0$, we will consider the open δ neighborhoods

$$\Omega_{\delta,1} := \Omega_{0,1} \cup \{x \in \Omega : \operatorname{dist}(x, \Omega_{0,1}) < \delta\} \supsetneq \Omega_{0,1}, \qquad \Omega_1^\delta := \Omega \setminus \bar{\Omega}_{\delta,1} \subsetneq \Omega_1.$$

By construction, the metasolution $\mathfrak{M}^{\min}_{[d_1\sigma_1,\Omega_1]}$ is uniformly bounded above in $\bar{\Omega}_1^\delta$ by some positive constant C_δ. Since

$$\mathfrak{M}^{\min}_{[d_1\sigma_1,\Omega_1]} = \infty \quad \text{in} \quad \bar{\Omega}_{0,1} \setminus \partial\Omega,$$

necessarily

$$\lim_{\delta \downarrow 0} C_\delta = \infty.$$

Thus, for every $\lambda \in (d_1\sigma_0, d_1\sigma_1)$ and sufficiently small $\delta > 0$,

$$\sigma_0[d_2, \mathfrak{c}\theta_{\lambda,u}] < \lambda_1[-d_2\Delta + \mathfrak{c}\theta_{\lambda,u}, \Omega_1^\delta] < \lambda_1[-d_2\Delta + \mathfrak{c}\mathfrak{M}^{\min}_{[d_1\sigma_1,\Omega_1]}, \Omega_1^\delta], \tag{10.87}$$

because $\lambda \mapsto \theta_{\lambda,u}$ is increasing and it approximates the minimal metasolution

$\mathfrak{M}^{min}_{[d_1\sigma_1,\Omega_1]}$ as $\lambda \uparrow d_1\sigma_1$. Thanks to (10.86) and (10.87), the next limit is well defined

$$\lambda_1[-d_2\Delta + c\mathfrak{M}^{min}_{[d_1\sigma_1,\Omega_1]}, \Omega_1] := \lim_{\lambda\uparrow d_1\sigma_1} \sigma_0[d_2, c\theta_{\lambda,u}].$$

Moreover, by (10.87), for sufficiently small $\delta > 0$, we have

$$\lambda_1[-d_2\Delta + c\mathfrak{M}^{min}_{[d_1\sigma_1;\Omega_1]}, \Omega_1] \leq \lambda_1[-d_2\Delta + c\mathfrak{M}^{min}_{[d_1\sigma_1,\Omega_1]}, \Omega_1^\delta]$$
$$\leq \lambda_1[-d_2\Delta, \Omega_1^\delta] + c_M C_\delta, \quad c_M := \max_{\bar\Omega} c.$$

Hence,

$$\lim_{c_M\downarrow 0} \lambda_1[-d_2\Delta + c\mathfrak{M}^{min}_{[d_1\sigma_1,\Omega_1]}, \Omega_1] \leq \lambda_1[-d_2\Delta, \Omega_1^\delta].$$

Therefore, by the continuous dependence of the principal eigenvalue with respect to the domain (see [163, Ch. 8]), letting $\delta \downarrow 0$ yields

$$\lim_{c_M\downarrow 0} \lambda_1[-d_2\Delta + c\mathfrak{M}^{min}_{[d_1\sigma_1,\Omega_1]}, \Omega_1] = \lambda_1[-d_2\Delta, \Omega_1], \tag{10.88}$$

because, since $c\mathfrak{M}^{min}_{[d_1\sigma_1,\Omega_1]} > 0$, we also have that

$$\lambda_1[-d_2\Delta + c\mathfrak{M}^{min}_{[d_1\sigma_1,\Omega_1]}, \Omega_1] \geq \lambda_1[-d_2\Delta, \Omega_1].$$

Throughout the rest of this section we will assume that

$$\lambda_1[-d_2\Delta, \Omega_1] < \mu_{d_1\sigma_1}, \tag{10.89}$$

where, for every $\lambda > d_1\sigma_0$, μ_λ is the unique value of $\mu > d_2\sigma_0$ for which

$$\lambda = \sigma_0[d_1, b\theta_{\mu_\lambda,v}].$$

According to (10.88) and (10.89), there exists $\varrho > 0$ such that

$$\lambda_1[-d_2\Delta + c\mathfrak{M}^{min}_{[d_1\sigma_1,\Omega_1]}, \Omega_1] < \mu_{d_1\sigma_1} \quad \text{if } c_M < \varrho.$$

Subsequently, we will assume $c_M < \varrho$ and denote, to simplify the notation,

$$\mu_0 := \lambda_1[-d_2\Delta + c\mathfrak{M}^{min}_{[d_1\sigma_1,\Omega_1]}, \Omega_1].$$

Then,

$$\mu_0 \equiv \lim_{\lambda\uparrow d_1\sigma_1} \sigma_0[d_2, c\theta_{\lambda,u}] < \mu_{d_1\sigma_1}.$$

Figure 10.4 shows four important curves in the (λ,μ) plane from the point of view of the existence of coexistence states for (10.5). Two of them enclose the regions where, according to Proposition 10.15, (10.5) does not admit a coexistence state,

$$\lambda \leq \sigma_0[d_1, b\theta_{\{\mu,v;c\theta_{\lambda,u}\}}] \quad \text{and} \quad \lambda \geq \sigma_1[d_1, b\theta_{\mu,v}].$$

The other two are the curves $\lambda = \sigma_0[d_1, \mathfrak{b}\theta_{\mu,v}]$, $\mu > d_2\sigma_0$, and $\mu = \sigma_0[d_2, \mathfrak{c}\theta_{\lambda,u}]$, $d_1\sigma_0 < \lambda < d_1\sigma_1$, which provide us with all the bifurcation points to coexistence states from the semi-trivial solutions $(0, \theta_{\mu,v})$ and $(\theta_{\lambda,u}, 0)$, respectively.

Thanks to the assumption (10.89), the graph of the curve $\mu = \sigma_0[d_2, \mathfrak{c}\theta_{\lambda,u}]$ lies below the graph of $\lambda = \sigma_0[d_1, \mathfrak{b}\theta_{\mu,v}]$ for $\mathfrak{c}_M$ sufficiently small, which is the situation illustrated in Figure 10.4.

Subsequently, we consider $d_2\sigma_0 < \mu < \mu_0$ and regard λ as the main bifurcation parameter. Let

$$\mathfrak{C}^+_{(\lambda,0,\theta_{\lambda,u})} \subset \mathbb{R} \times \mathrm{int}\, P_1 \times \mathrm{int}\, P_2$$

denote the component of the set of coexistence states of (10.5) emanating from $(\lambda, 0, \theta_{\mu,v})$ at

$$\lambda = \sigma_0[d_1, \mathfrak{b}\theta_{\mu,v}],$$

whose existence is guaranteed by [149, Th. 7.2.2]. For $\mu \sim d_2\sigma_0$, the local bifurcation analysis of J. C. Eilbeck, J. E. Furter and J. López-Gómez [82] shows that $\mathfrak{C}^+_{(\lambda,0,\theta_{\mu,v})}$ meets the other branch of semi-trivial solutions, $(\lambda, \theta_{\lambda,u}, 0)$, at $\lambda = \lambda_\mu$, the unique value of λ for which

$$\mu = \sigma_0[d_2, \mathfrak{c}\theta_{\lambda_\mu,u}].$$

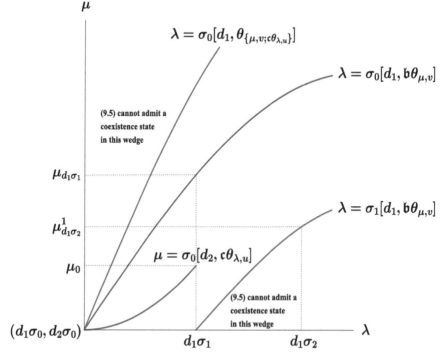

FIGURE 10.4: Some significant curves in the (λ, μ) plane.

Moreover, the component $\mathfrak{C}^+_{(\lambda,0,\theta_{\mu,v})}$ is bounded. Consequently, the associated global bifurcation diagram looks like those found by J. C. Eilbeck, J. E. Furter and J. López-Gómez [82] for the classical diffusive competing species model. As μ separates from $d_2\sigma_0$ and approximates μ_0, the previous behavior might change drastically, because the component of the set of coexistence states of (10.5) emanating from $(\lambda,\theta_{\lambda,u},0)$ at $\lambda := \lambda_\mu$, which will be denoted by $\mathfrak{C}^+_{(\lambda,\theta_{\lambda,u},0)}$, might be unbounded, as may the component $\mathfrak{C}^+_{(\lambda,0,\theta_{\mu,v})}$.

Actually, both components must be unbounded if one is unbounded. Indeed, if one of them is bounded, according to [149, Th. 7.2.2], it must link the two semi-trivial positive solutions. So,

$$\mathfrak{C}^+_{(\lambda,0,\theta_{\mu,v})} = \mathfrak{C}^+_{(\lambda,\theta_{\lambda,u},0)},$$

as in the classical situations described by J. C. Eilbeck, J. E. Furter and J. López-Gómez [82]. Figure 10.5 shows an admissible global bifurcation diagram of coexistence states when these components are unbounded.

Subsequently, we suppose that, in addition,

$$\mu_0 < \mu^1_{d_1\sigma_2} \tag{10.90}$$

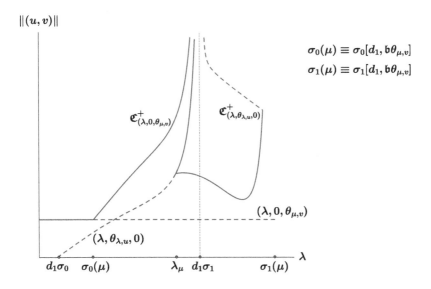

FIGURE 10.5: An admissible bifurcation diagram for $\mu \simeq \mu_0$, $\mu < \mu_0$.

as illustrated in Figure 10.4. It should be remembered that

$$d_1\sigma_2 = \sigma_1[d_1, \mathfrak{b}\theta_{\mu^1_{d_1\sigma_2},v}] = \lambda_1[-d_1\Delta + \mathfrak{b}\theta_{\mu^1_{d_1\sigma_2},v}, \Omega_{0,1}(d_1, \mathfrak{b}\theta_{\mu^1_{d_1\sigma_2},v})].$$

By Proposition 10.15, (10.5) cannot admit a coexistence state for

$$\lambda \geq \sigma_1[d_1, \mathfrak{b}\theta_{\mu,v}] = \lambda_1[-d_1\Delta + \mathfrak{b}\theta_{\mu,v}, \Omega_{0,1}(d_1, \mathfrak{b}\theta_{\mu,v})].$$

Pick $\mu_0 < \mu < \mu^1_{d_1\sigma_2}$ (see Figure 10.4). For such range of values of the parameter μ, (10.5) cannot admit a semi-trivial positive solution of the form $(\theta_{\lambda,u}, 0)$. Thus, by [149, Th. 7.2.2], the component $\mathfrak{C}^+_{(\lambda,0,\theta_{\mu,v})}$ is unbounded.

Moreover, by Proposition 10.15, whenever (10.5) has a coexistence state,

$$\lambda < \lambda_1[-d_1\Delta + \mathfrak{b}\theta_{\mu,v}, \Omega_{0,1}(d_1, \mathfrak{b}\theta_{\mu,v})]$$
$$< \lambda_1[-d_1\Delta + \mathfrak{b}\theta_{\mu^1_{d_1\sigma_2},v}, \Omega_{0,1}(d_1, \mathfrak{b}\theta_{\mu,v})]$$
$$= \lambda_1[-d_1\Delta + \mathfrak{b}\theta_{\mu^1_{d_1\sigma_2},v}, \Omega_{0,1}(d_1, \mathfrak{b}\theta_{\mu^1_{d_1\sigma_2},v})] = d_1\sigma_2$$

provided μ is sufficiently close to $\mu^1_{d_1\sigma_2}$ so that

$$\Omega_{0,1}(d_1, \mathfrak{b}\theta_{\mu,v}) = \Omega_{0,1}(d_1, \mathfrak{b}\theta_{\mu^1_{d_1\sigma_2},v}).$$

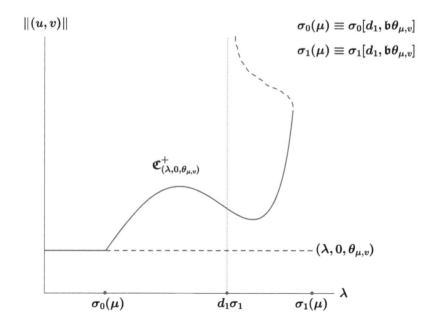

FIGURE 10.6: An admissible bifurcation diagram for $\mu_0 < \mu < \mu^1_{d_1\sigma_2}$.

Therefore, thanks to Theorem 10.12, the component $\mathfrak{C}^+_{(\lambda,0,\theta_{\mu,v})}$ blows up at $\lambda = d_1\sigma_1$. Consequently, it looks like Figure 10.6.

Thanks to Theorem 10.16, for sufficiently small $\mathfrak{c} \simeq 0$, the component $\mathfrak{C}^+_{(\lambda,0,\theta_{\mu,v})}$ is defined for values of λ as close as we wish to $\sigma_1[d_1, \mathfrak{b}\theta_{\mu,v}]$. Thinking of the special case $\mathfrak{c} = 0$ might help us understand what's going on. Therefore, the smaller $\mathfrak{c}_M$, the closer to $\sigma_1[d_1, \mathfrak{b}\theta_{\mu,v}]$ is the subcritical turning point of the component $\mathfrak{C}^+_{(\lambda,0,\theta_{\mu,v})}$ in Figure 10.6.

For larger values of μ, $\mu > \mu^1_{d_1\sigma_2}$, the component of coexistence states $\mathfrak{C}^+_{(\lambda,0,\theta_{\mu,v})}$ might blow up at some $d_1\sigma_j$ with $j \geq 2$. In such cases, the global bifurcation diagram of coexistence states represented in Figure 10.7 is admissible.

If $\mathfrak{C}^+_{(\lambda,0,\theta_{\mu,v})}$ blows up at $d_2\sigma_2$, then, by the multiplicity results of Section 10.6, the bifurcation diagram in Figure 10.7 occurs, because for each $\lambda \in (d_1\sigma_1, d_1\sigma_2)$, (10.5) possesses at least one coexistence state. Consequently, a further component of the set of coexistence states of (10.5), $\mathfrak{C}^+_{\text{isola}}$, should lie within that interval, as illustrated by Figure 10.7.

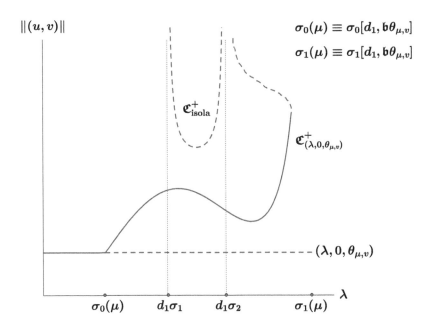

FIGURE 10.7: A possible bifurcation diagram for $\mu > \mu^1_{d_1\sigma_2}$.

Naturally, some numerical experiments are imperative for ascertaining the exact global bifurcation diagrams.

10.5   Strong maximum principle for quasi-cooperative systems

Throughout this section we will assume that

$$b(x) > 0 \quad \text{and} \quad c(x) > 0 \quad \text{for all } x \in \Omega. \tag{10.91}$$

If (u_0, v_0) is a coexistence state of (10.5), its linearized stability is given through the sign of the eigenvalues of the linearization of (10.5) at (u_0, v_0), i.e., by the signs of the τ's for which the linear boundary value problem

$$
\begin{cases}
\begin{pmatrix} -d_1\Delta & 0 \\ 0 & -d_2\Delta \end{pmatrix} \begin{pmatrix} u \\ v \end{pmatrix} = A \begin{pmatrix} u \\ v \end{pmatrix} + \tau \begin{pmatrix} u \\ v \end{pmatrix} & \text{in } \Omega, \\[2mm]
(u, v) = (0, 0) & \text{on } \partial\Omega.
\end{cases}
\tag{10.92}
$$

has a solution $(u, v) \neq (0, 0)$, where we have denoted

$$A = \begin{pmatrix} \alpha(x) & \beta(x) \\ \gamma(x) & \varrho(x) \end{pmatrix} \tag{10.93}$$

with

$$
\begin{aligned}
\alpha &:= \lambda - \mathfrak{a}f_1(\cdot, u_0) - \mathfrak{a}u_0\partial_u f_1(\cdot, u_0) - bv_0, \\
\varrho &:= \mu - \mathfrak{d}f_2(\cdot, v_0) - \mathfrak{d}v_0\partial_v f_2(\cdot, v_0) - cu_0, \\
\beta &:= -bu_0, \\
\gamma &:= -cv_0.
\end{aligned}
$$

Thanks to (10.91), the off-diagonal entries of the coupling matrix A are negative. This is why (10.92) is said to be of *quasi-cooperative* type. More generally, we will consider (10.92) with α, β, γ, $\varrho \in \mathcal{C}(\bar{\Omega})$ where

$$\beta(x) < 0 \quad \text{and} \quad \gamma(x) < 0 \quad \text{for almost all } x \in \Omega.$$

Subsequently, to simplify notations, we will set

$$\mathfrak{L} := \begin{pmatrix} -d_1\Delta & 0 \\ 0 & -d_2\Delta \end{pmatrix} - A = \begin{pmatrix} -d_1\Delta - \alpha(x) & -\beta(x) \\ -\gamma(x) & -d_2\Delta - \varrho(x) \end{pmatrix} \tag{10.94}$$

and suppose $p > N$. Given $u, v \in L^p(\Omega)$, it is said that $(u, v) \geq_q 0$ if $u \geq 0$ and $v \leq 0$; the q emphasizes the quasi-cooperative character of this ordering.

If, in addition, $u \neq 0$, or $v \neq 0$, it is said that $(u, v) >_q 0$. A couple $(u, v) \in W_0^{2,p}(\Omega) \times W_0^{2,p}(\Omega)$ is said to be *strongly positive* if $u(x) > 0$, $v(x) < 0$ for all $x \in \Omega$ and $\partial_n u(x) < 0$, $\partial_n v(x) > 0$ for all $x \in \partial\Omega$, where n stands for the outward unit normal to Ω at $x \in \partial\Omega$. In such case, we will simply write $(u, v) \gg_q 0$. Subsequently, we shall always refer to this order. Note that we are ordering $X_1 \times X_2$ with the cone $P := P_1 \times (-P_2)$ (see the beginning of Section 9.5). As P is a normal cone, according to [163, Th. 6.1] there exists a norm in $X_1 \times X_2$, $\|\cdot\|_{X_1 \times X_2}$, which is monotonic. This will be the norm used in Section 10.6 in the calculations of the topological indices.

Definition 10.17 *The operator $\mathfrak{L}$ defined by (10.94) is said to satisfy the strong maximum principle in Ω if $x \in W_0^{2,p}(\Omega) \times W_0^{2,p}(\Omega)$ and $\mathfrak{L}x >_q 0$ imply $x \gg_q 0$.*

Definition 10.18 *A function $\overline{x} \in W^{2,p}(\Omega) \times W^{2,p}(\Omega)$ is said to be a supersolution of $(\mathfrak{L}, \Omega)$ if $\mathfrak{L}\overline{x} \geq_q 0$ in Ω and $\overline{x} \geq_q 0$ on $\partial\Omega$. If, in addition, $\mathfrak{L}\overline{x} >_q 0$ in Ω, or $\overline{x} >_q 0$ on $\partial\Omega$, then it is said that $\overline{x}$ is a strict supersolution of $(\mathfrak{L}, \Omega)$.*

Adapting the proofs of Theorem 2.1 of J. López-Gómez and M. Molina-Meyer [130] and Theorem 6.3 of J. López-Gómez and J. C. Sabina de Lis [180, Th. 6.3], the next results follow readily. Note that we are adapting to a quasi-cooperative setting the main results of J. López-Gómez [163, Sect. 7] for the single equation.

Theorem 10.19 *The operator $\mathfrak{L}$, under homogeneous Dirichlet boundary conditions in Ω, possesses a unique eigenvalue to a positive eigenfunction, $\lambda_1[\mathfrak{L}, \Omega]$, called the principal eigenvalue of $\mathfrak{L}$ in Ω, or $(\mathfrak{L}, \Omega)$. The principal eigenvalue is algebraically simple. Consequently, the principal eigenfunction, $\Phi >_q 0$, is unique up to a positive multiplicative constant. Moreover, $\Phi \gg_q 0$ and any other eigenvalue τ of (10.92) satisfy $\operatorname{Re}\tau > \lambda_1[\mathfrak{L}, \Omega]$. Furthermore, for every $\kappa > -\lambda_1[\mathfrak{L}, \Omega]$, the operator*

$$(\mathfrak{L} + \kappa)^{-1} \in \mathcal{L}(L^p(\Omega) \times L^p(\Omega))$$

is compact and positive.

Actually, the last assertion is an immediate consequence of the next counterpart of Theorem 1.1.

Theorem 10.20 *The following conditions are equivalent:*

(a) $\lambda_1[\mathfrak{L}, \Omega] > 0$.

(b) $(\mathfrak{L}, \Omega)$ *admits a positive strict supersolution in $W^{2,p}(\Omega) \times W^{2,p}(\Omega)$.*

(c) $(\mathfrak{L}, \Omega)$ *satisfies the strong maximum principle.*

Actually, under any of these circumstances, the following result holds.

Theorem 10.21 *Suppose $\lambda_1[\mathfrak{L}, \Omega] > 0$. Then, any strict supersolution*

$$\overline{x} := (\overline{u}, \overline{v}) \in W^{2,p}(\Omega) \times W^{2,p}(\Omega)$$

of $(\mathfrak{L}, \Omega)$ must be strongly positive in Ω, in the sense that $\overline{u}(x) > 0$ and $\overline{v}(x) < 0$ for all $x \in \Omega$, $\partial_n \overline{u}(x) < 0$ for all $x \in (\overline{u})^{-1}(0) \cap \partial\Omega$ and $\partial_n \overline{v}(x) > 0$ for all $x \in (\overline{v})^{-1}(0) \cap \partial\Omega$.

From these results one can infer very easily the main monotonicity properties of $\lambda_1[\mathfrak{L}, \Omega]$, as in J. López-Gómez and M. Molina-Meyer [130] and H. Amann [12]; technical details are omitted here.

Throughout the rest of this chapter, we will also consider the inhomogeneous problem

$$\begin{cases} -d_1 \Delta u = \lambda u - \mathfrak{a}(x)\mathfrak{f}_1(x, u)u - \mathfrak{b}(x)uv & \text{in } \Omega, \\ -d_2 \Delta v = \mu v - \mathfrak{d}(x)\mathfrak{f}_2(x, v)v - \mathfrak{c}(x)uv & \\ (u, v) = (\varphi_u, \varphi_v) & \text{on } \partial\Omega, \end{cases} \quad (10.95)$$

with $\varphi_u, \varphi_v \in W^{1,p}(\partial\Omega)$.

Definition 10.22 *A function $\underline{x} = (\underline{u}, \underline{v}) \in W^{2,p}(\Omega) \times W^{2,p}(\Omega)$ is said to be a subsolution of (10.95) if*

$$-d_1 \Delta \underline{u} \leq \lambda \underline{u} - \mathfrak{a}(x)\mathfrak{f}_1(x, \underline{u})\underline{u} - \mathfrak{b}(x)\underline{u}\,\overline{v}$$
$$-d_2 \Delta \underline{v} \geq \mu \underline{v} - \mathfrak{d}(x)\mathfrak{f}_2(x, \underline{v})\underline{v} - \mathfrak{c}(x)\overline{u}\,\underline{v}$$

in Ω and $\underline{x} \leq_q (\varphi_u, \varphi_v)$ on $\partial\Omega$.

Similarly, a function $\overline{x} = (\overline{u}, \overline{v}) \in W^{2,p}(\Omega) \times W^{2,p}(\Omega)$ is said to be a supersolution of (10.95) if

$$-d_1 \Delta \overline{u} \geq \lambda \overline{u} - \mathfrak{a}(x)\mathfrak{f}_1(x, \overline{u})\overline{u} - \mathfrak{b}(x)\overline{u}\,\underline{v}$$
$$-d_2 \Delta \overline{v} \leq \mu \overline{v} - \mathfrak{d}(x)\mathfrak{f}_2(x, \overline{v})\overline{v} - \mathfrak{c}(x)\underline{u}\,\overline{v}$$

in Ω and $\overline{x} \geq_q (\varphi_u, \varphi_v)$ on $\partial\Omega$.

The following result can be easily obtained combining the previous results with H. Amann [11, Th. 9.4].

Theorem 10.23 *Suppose (10.95) possesses a subsolution $\underline{x} = (\underline{u}, \underline{v})$ and a supersolution $\overline{x} = (\overline{u}, \overline{v})$ such that $\underline{x} \leq_q \overline{x}$. Then (10.95) possesses a minimal solution $x_* = (u_*, v_*)$ and a maximal solution $x^* = (u^*, v^*)$ in the order interval $[\underline{x}, \overline{x}]$. If in addition $\varphi_u = \varphi_v = 0$, $\underline{u} > 0$ and $\overline{v} > 0$, then (10.5) has a coexistence state.*

Actually, most of the results of Chapter 1 remain valid for the system.

10.6 Multiplicity of coexistence states

Throughout this section, for every $\lambda > d_1\sigma_0$, we will denote by J_μ the set of $\mu \in \mathbb{R}$ for which (10.5) possesses a coexistence state.

Theorem 10.24 *Suppose* $\mathfrak{b} > 0$ *and* $\mathfrak{c} > 0$ *almost everywhere in* Ω *and* $\lambda \geq d_1\sigma_1$ *satisfies* $\lambda \neq d_1\sigma_j$, $1 \leq j \leq q_0$. *Then, some of the following alternatives occur:*

(a) $J_\mu = (\mu_\lambda, \infty)$, *where* $\lambda = \sigma_0[d_1, \mathfrak{b}\theta_{\mu_\lambda, v}]$, *or*

(b) $J_\mu = [\mu_*(\lambda), \infty)$ *for some* $\mu_*(\lambda) \in (\mu_\lambda^1, \mu_\lambda]$, *where* $\lambda = \sigma_1[d_1, \mathfrak{b}\theta_{\mu_\lambda^1, v}]$.

Proof: Thanks to Theorem 10.14,

$$(\mu_\lambda, \infty) \subset J_\mu \subset (\mu_\lambda^1, \infty).$$

Consequently, option (a) occurs when (10.5) does not admit a coexistence state for $\mu \leq \mu_\lambda$. It remains to show that if (10.5) has a coexistence state for some $\hat{\mu} < \mu_\lambda$, then it also admits a coexistence state for each $\mu \in [\hat{\mu}, \mu_\lambda]$, as this establishes option (b) easily.

Suppose (10.5) has a coexistence state for some $\hat{\mu} < \mu_\lambda$, $(\hat{u}, \hat{v})$. Then, $\hat{u} = \hat{v} = 0$ on $\partial\Omega$ and for every $\mu \in (\hat{\mu}, \mu_\lambda)$

$$-d_1\Delta\hat{u} = \lambda\hat{u} - \mathfrak{a}\mathfrak{f}_1(\cdot, \hat{u})\hat{u} - \mathfrak{b}\hat{u}\,\hat{v}$$
$$-d_2\Delta\hat{v} < \mu\hat{v} - \mathfrak{d}\mathfrak{f}_2(\cdot, \hat{v})\hat{v} - \mathfrak{c}\hat{u}\,\hat{v}$$

in Ω. Thus, $(\hat{u}, \hat{v})$ provides us with a strict supersolution of (10.5). Moreover,

$$-d_2\Delta\hat{v} < \mu\hat{v} - \mathfrak{d}\mathfrak{f}_2(\cdot, \hat{v})\hat{v}$$

and hence, by Lemma 1.8, $\hat{v} < \theta_{\mu, v}$. So, substituting this estimate in the $\hat{u}$-equation yields

$$-d_1\Delta\hat{u} = \lambda\hat{u} - \mathfrak{a}\mathfrak{f}_1(\cdot, \hat{u})\hat{u} - \mathfrak{b}\hat{u}\,\hat{v} > \lambda\hat{u} - \mathfrak{a}\mathfrak{f}_1(\cdot, \hat{u})\hat{u} - \mathfrak{b}\theta_{\mu, v}\hat{u}.$$

Thus, thanks again to Lemma 1.8, we find that

$$\hat{u} > \theta_{[\lambda, \mathfrak{b}\theta_{\mu, v}]}, \tag{10.96}$$

where $\theta_{[\lambda, \mathfrak{b}\theta_{\mu, v}]}$ stands for the unique positive solution of

$$(-d_1\Delta + \mathfrak{b}\theta_{\mu, v})u = \lambda u - \mathfrak{a}\mathfrak{f}_1(\cdot, u)u \quad \text{in} \quad \Omega, \qquad u|_{\partial\Omega} = 0.$$

Note that $\theta_{[\lambda, \mathfrak{b}\theta_{\mu, v}]} > 0$ is well defined, because $\mu_\lambda^1 < \mu < \mu_\lambda$ implies

$$\sigma_0[d_1, \mathfrak{b}\theta_{\mu, v}] < \lambda < \sigma_1[d_1, \mathfrak{b}\theta_{\mu, v}].$$

Set

$$\underline{u} := \theta_{[\lambda, \mathfrak{b}\theta_{\mu, v}]}, \qquad \underline{v} := \theta_{\mu, v}.$$

Then, since $\hat{v} < \theta_{\mu, v}$, it follows from (10.96) that

$$(\underline{u}, \underline{v}) <_q (\hat{u}, \hat{v}). \tag{10.97}$$

Moreover,

$$-d_1 \Delta \underline{u} = \lambda \underline{u} - \mathfrak{a} f_1(\cdot, \underline{u}) \underline{u} - \mathfrak{b} \, \underline{u} \, \underline{v},$$
$$-d_2 \Delta \underline{v} = \mu \underline{v} - \mathfrak{d} f_2(\cdot, \underline{v}) \underline{v} > \mu \underline{v} - \mathfrak{d} f_2(\cdot, \underline{v}) \underline{v} - \mathfrak{c} \, \underline{u} \, \underline{v}.$$

Thus, $(\underline{u}, \underline{v})$ is a subsolution of (10.5). Therefore, by (10.97), we find from Theorem 10.23 that (10.5) possesses a coexistence state for each $\mu \in [\hat{\mu}, \mu_\lambda)$.

To establish the existence of a coexistence state for $\mu = \mu_\lambda$, let (μ_n, u_n, v_n), $n \geq 1$, be a sequence of coexistence states such that $\mu_n < \mu_\lambda$, $n \geq 1$, and

$$\lim_{n \to \infty} \mu_n = \mu_\lambda.$$

As $\lambda \neq d_1 \sigma_j$ for all $1 \leq j \leq q_0$, according to Theorem 10.12, these coexistence states are bounded in $X_1 \times X_2$. Thus, by a rather standard compactness argument, there is a solution of (10.5), $(u_\lambda, v_\lambda) \in X_1 \times X_2$, such that along some subsequence, re-labeled by n,

$$\lim_{n \to \infty} \|u_n - u_\lambda\|_{X_1} = 0, \quad \lim_{n \to \infty} \|v_n - v_\lambda\|_{X_2} = 0.$$

Necessarily, $u_\lambda \geq 0$ and $v_\lambda \geq 0$. If $u_\lambda > 0$ and $v_\lambda > 0$, we are done.

Suppose $u_\lambda > 0$ and $v_\lambda = 0$. Then, $u_\lambda = \theta_{\lambda, u}$ and due to Theorem 4.1 we find that $d_1 \sigma_0 < \lambda < d_1 \sigma_1$, which contradicts $\lambda > d_1 \sigma_1$. So, this cannot occur.

Suppose $u_\lambda = 0$ and $v_\lambda = 0$. Then, by the uniqueness of the principal eigenvalue, we can infer from

$$-d_1 \Delta u_n = \lambda u_n - \mathfrak{a} f_1(\cdot, u_n) u_n - \mathfrak{b} u_n v_n, \qquad n \geq 1,$$

that

$$\lambda = \sigma_0[d_1, \mathfrak{a} f_1(\cdot, u_n) + \mathfrak{b} v_n], \qquad n \geq 1.$$

So, letting $n \to \infty$ yields $\lambda = d_1 \sigma_0$, which is impossible.

Suppose $u_\lambda = 0$ and $v_\lambda > 0$. Then, $v_\lambda = \theta_{\mu_\lambda, v}$ and, necessarily, $\mathfrak{C}^+_{(\mu, 0, \theta_{\mu, v})}$ bifurcates towards the left of μ_λ by the local uniqueness given by the main theorem of M. G. Crandall and P. H. Rabinowitz [59]. Consequently, if (10.5) does not admit a coexistence state for $\mu = \mu_\lambda$, then

$$\mathcal{P}_\mu \left(\mathfrak{C}^+_{(\mu, 0, \theta_{\mu, v})} \right) \subset (\mu_\lambda^1, \mu_\lambda),$$

because the μ projection of the component $\mathfrak{C}^+_{(\mu, 0, \theta_{\mu, v})}$ is an interval. But this contradicts Theorem 10.14. Therefore, (10.5) also possesses a coexistence state for $\mu = \mu_\lambda$. So, $[\hat{\mu}, \infty) \subset J_\mu$.

Finally, set $\mu_*(\lambda) := \inf J_\mu$. Thanks to the previous analysis, to complete the proof we must prove that $\mu_*(\lambda) \in J_\mu$. Let (μ_n, u_n, v_n), $n \geq 1$, be a sequence of coexistence states such that

$$\lim_{n \to \infty} \mu_n = \mu_*(\lambda), \qquad \mu_*(\lambda) < \mu_n < \mu_\lambda, \quad n \geq 1.$$

According to Theorem 10.12, these coexistence states are uniformly bounded in $X_1 \times X_2$. Thus, (10.5) admits a non-negative solution, $(\mu_*(\lambda), u_*, v_*) \in X_1 \times X_2$, such that along some subsequence, labeled again by n,

$$\lim_{n \to \infty} \|u_n - u_*\|_{X_1} = 0, \qquad \lim_{n \to \infty} \|v_n - v_*\|_{X_2} = 0.$$

Arguing as before, it becomes apparent that $u_* > 0$ and $v_* > 0$. $\qquad \square$

Adapting the arguments of the proof of Theorem 10.24, one can also show that J_μ is a bounded interval if $\lambda < d_1 \sigma_1$, as in the most classical competing species models dealt with in J. C. Eilbeck et al. [82] and J. E. Furter and J. López-Gómez [91]. Moreover, J_μ might shrink to a single point when both semi-trivial positive solutions are *neutrally stable* according to the shapes and the sizes of the function coefficients $\mathfrak{b}$ and $\mathfrak{c}$. Local bifurcation and singularity theory were the approaches adopted in these pioneering papers to establish some closely related features. Incidentally, re-elaborating on the results, techniques and ideas of J. C. Eilbeck, J. López-Gómez and J. E. Furter [82], E. N. Dancer [65] constructed some sharp examples establishing that some previous existence results by L. Li and R. Logan [134] and R. Logan and A. Ghoreishi [139] were incorrect as stated. Indeed, according to E. N. Dancer [65] (see pages 239 and 240):

"The main purpose of this paper is to show that for many smooth domains Ω (including some convex ones in all dimensions except two) there exists $a, d > \lambda_1$ such that (1) has no positive solution for $c = \bar{c}, b = \bar{b}$. This result is particularly interesting since there are two published proofs of the contrary result in the literature ([134], [139]). The errors in those proofs are incorrect degree calculations as in Lemma 4.5 in [139]...

It is necessary to have a copy of [82] available in reading this paper. "

Next, we will adapt the abstract theory of H. Amann in [11, Sect. 20], further developed by H. Amann and J. López-Gómez [13] and M. Delgado, J. López-Gómez and A. Suárez [72], to prove the following multiplicity which is the main result of this section; it deals with the situation described by Figure 10.3.

Theorem 10.25 *Under the assumptions of Theorem 10.24, suppose, in addition, that*

$$J_\mu = [\mu_*(\lambda), \infty) \quad \text{for some} \quad \mu_*(\lambda) \in (\mu_\lambda^1, \mu_\lambda). \tag{10.98}$$

Then, for every $\mu \in (\mu_*(\lambda), \mu_\lambda)$ the problem (10.5) has at least two coexistence states.

Condition (10.98) holds whenever the component $\mathfrak{C}^+_{(\mu,0,\theta_{\mu,v})}$ bifurcates towards the left from the solution branch $(\mu, 0, \theta_{\mu,v})$ at the critical value of the parameter $\mu = \mu_\lambda$. By Theorem 10.16, this occurs for sufficiently small $\mathfrak{c}_M$.

Proof of Theorem 10.25: Proof is based on the properties of the fixed point index in cones. First, we will establish that, for every $\mu \in (\mu_*(\lambda), \mu_\lambda)$, (10.5) has a minimal coexistence state. Indeed, according to the proof of Theorem 10.24, we already know that

$$(\underline{u}, \underline{v}) := (\theta_{[\lambda, b\theta_{\mu,v}]}, \theta_{\mu,v})$$

is a positive subsolution of (10.5) such that $(\underline{u}, \underline{v}) <_q (u, v)$ for all coexistence states (u, v). Suppose (10.5) has two coexistence states, (u_1, v_1) and (u_2, v_2). Then, the couple

$$(\overline{u}, \overline{v}) := (\max\{u_1, u_2\}, \min\{v_1, v_2\})$$

provides us with a supersolution satisfying

$$(\underline{u}, \underline{v}) <_q (\overline{u}, \overline{v}), \qquad (u_j, v_j) \in [(\underline{u}, \underline{v}), (\overline{u}, \overline{v})], \qquad j \in \{1, 2\}.$$

Thus, by Theorem 10.23, there is a coexistence state

$$(u_3, v_3) \in [(\underline{u}, \underline{v}), (\overline{u}, \overline{v})]$$

such that $(u_3, v_3) \leq_q (u_j, v_j)$, for each $j \in \{1, 2\}$. This gives the minimal coexistence state when (10.5) has finitely many coexistence states, and it shows that if (10.5) has infinitely many coexistence states but does not admit a minimal coexistence state, there exists a sequence of coexistence states, $(\tilde{u}_n, \tilde{v}_n)$, $n \geq 1$, such that

$$\lim_{n \to \infty} \tilde{v}_n = 0 \qquad \text{in } X_2.$$

As $\tilde{u}_n > \theta_{[\lambda, b\theta_{\mu,v}]}$ for all $n \geq 1$ and, due to Theorem 10.12, the coexistence states of (10.5) possess uniform a priori bounds, by a rather standard compactness argument it becomes apparent that (10.5) admits a semi-trivial positive solution of the form $(\tilde{u}, 0)$, which is impossible because we are imposing $\lambda > d_1\sigma_1$. This provides us with the minimal coexistence state, (u_μ, v_μ).

Next, we will make sure that (10.5) fits into the abstract setting of H. Amann [11]. Fix $\mu^1_\lambda < a < \mu_*(\lambda)$, $b > 0$, and consider the interval $I := [a, \mu_\lambda + b]$. By the existence of uniform a priori bounds in I, the constant K in (10.69) can be chosen sufficiently large so that, for every $\mu \in I$ and any non-negative solution (u, v) of (10.5),

$$\mathfrak{a}f_1(\cdot, u) + \mathfrak{b}v \leq \mathfrak{a}\frac{\partial f_1}{\partial u}(\cdot, u)u + \mathfrak{a}f_1(\cdot, u) + \mathfrak{b}v < \lambda + K,$$

$$\mathfrak{d}f_2(\cdot, v) + \mathfrak{c}u \leq \mathfrak{d}\frac{\partial f_2}{\partial v}(\cdot, v)v + \mathfrak{d}f_2(\cdot, v) + \mathfrak{c}u < \mu + K.$$

For this choice of K, the maps

$$u \mapsto [\lambda + K - \mathfrak{a}f_1(\cdot, u) - \mathfrak{b}v]u, \qquad v \mapsto [\mu + K - \mathfrak{d}f_2(\cdot, v) - \mathfrak{c}u]v,$$

are positive and increasing in u and v, respectively. Thus, for any pair of coexistence states, (u_1, v_1) and (u_2, v_2), with $(u_1, v_1) \leq_q (u_2, v_2)$, i.e., such that $u_1 \leq u_2$ and $v_1 \geq v_2$, one has that

$$[\lambda + K - \mathfrak{a}f_1(\cdot, u_1) - \mathfrak{b}v_1]u_1 \leq [\lambda + K - \mathfrak{a}f_1(\cdot, u_2) - \mathfrak{b}v_1]u_2$$
$$\leq [\lambda + K - \mathfrak{a}f_1(\cdot, u_2) - \mathfrak{b}v_2]u_2.$$

Similarly,

$$[\mu + K - \mathfrak{d}f_2(\cdot, v_1) - \mathfrak{c}u_1]v_1 \geq [\mu + K - \mathfrak{d}f_2(\cdot, v_2) - \mathfrak{c}u_1]v_2$$
$$\geq [\mu + K - \mathfrak{d}f_1(\cdot, v_2) - \mathfrak{c}u_2]v_2.$$

Consequently,

$$\mathcal{K}_{\lambda,\mu}(u_1, v_1) \leq_q \mathcal{K}_{\lambda,\mu}(u_2, v_2).$$

Moreover,

$$(u_1, v_1) <_q (u_2, v_2) \quad \Longrightarrow \quad \mathcal{K}_{\lambda,\mu}(u_1, v_1) \ll_q \mathcal{K}_{\lambda,\mu}(u_2, v_2),$$

because $\mathfrak{b}(x) > 0$ and $\mathfrak{c}(x) > 0$ for almost all $x \in \Omega$ and the operators $(-d_1\Delta + K)^{-1}$ and $(-d_2\Delta + K)^{-1}$ are strongly order preserving. Therefore, $\mathcal{K}_{\lambda,\mu}$ is compact and strongly order preserving for all $\mu \in I$.

Denote by B the unit ball of $X_1 \times X_2$ and, for every $\rho > 0$, let P_ρ be the positive part of ρB. Then, the fixed point index of $\mathcal{K}_{\lambda,\mu}$ in P_ρ is well defined for sufficiently large $\rho > 0$. Moreover, the next result holds.

Lemma 10.26 *Assume $\mu \in (\mu_\lambda^1, \mu_\lambda)$. Then, $(0,0)$ and $(0, \theta_{\mu,v})$ are isolated fixed points of $\mathcal{K}_{\lambda,\mu}$ in $P_1 \times P_2$ with*

$$i(\mathcal{K}_{\lambda,\mu}, (0,0)) = i(\mathcal{K}_{\lambda,\mu}, (0, \theta_{\mu,v})) = 0.$$

Moreover, $i(\mathcal{K}_{\lambda,\mu}, P_\rho) = 0$ for sufficiently large ρ.

Proof: Since $\mu < \mu_\lambda$, it is easily seen that $(0, \theta_{\mu,v})$ is linearly unstable. So, from J. López-Gómez [140, Le. 4.1], we have that $i(\mathcal{K}_{\lambda,\mu}, (0, \theta_{\mu,v})) = 0$. Moreover, by Lemma 13.1(ii) of H. Amann [11], $i(\mathcal{K}_{\lambda,\mu}, (0,0)) = 0$. The rest follows from the homotopy invariance of the index, since $(0,0)$ and $(0, \theta_{\mu,v})$ are the unique non-negative solutions of (10.5) if $\mu \in (a, \mu_*(\lambda))$. □

To ascertain the fixed point index of the minimal coexistence state, (u_μ, v_μ), which is the most delicate part of the proof, we need the following counterparts of Propositions 9.6, 9.7 and 9.14, whose proofs can be easily adapted from the proofs of these propositions

Lemma 10.27 *For every $\mu \in [\mu_*(\lambda), \mu_\lambda)$ the minimal coexistence state (u_μ, v_μ) of (10.5) is weakly stable, in the sense that $\lambda_1[\mathfrak{L}_\mu, \Omega] \geq 0$, where $\mathfrak{L}_\mu$ is the operator defined by (10.94) with $(u_0, v_0) = (u_\mu, v_\mu)$.*

Lemma 10.28 *The following assertions are true:*

(a) *Let $(\mu, u, v) = (\mu_0, u_0, v_0)$ be a coexistence state with $\lambda_1[\mathfrak{L}_{\mu_0}, \Omega] > 0$, where $\mathfrak{L}_{\mu_0}$ is the operator defined by (10.94) with $A(x)$ given by (10.93). Then, there exist $\varepsilon > 0$ and a differentiable map*

$$(u, v) : (\mu_0 - \varepsilon, \mu_0 + \varepsilon) \to P_1 \times P_2$$

such that $(u(\mu_0), v(\mu_0)) = (u_0, v_0)$ and $(\mu, u(\mu), v(\mu))$ is a coexistence state for each $\mu \in (\mu_0 - \varepsilon, \mu_0 + \varepsilon)$. Moreover, the mapping $\mu \mapsto (u(\mu), v(\mu))$ is decreasing and there exists a neighborhood $\mathcal{Q}_0$ of (μ_0, u_0, v_0) in $\mathbb{R} \times X_1 \times X_2$ such that the unique solutions of (10.5) in $\mathcal{Q}_0$ are those of the form $(u(\mu), v(\mu))$ with $\mu \in (\mu_0 - \varepsilon, \mu_0 + \varepsilon)$.

(b) *Suppose $\lambda_1[\mathfrak{L}_{\mu_0}, \Omega] = 0$ and let $\Phi > 0$ denote a principal eigenfunction of $\lambda_1[\mathfrak{L}_{\mu_0}, \Omega] = 0$. Then, there exist $\varepsilon > 0$ and a differentiable map*

$$(\mu, u, v) : (-\varepsilon, \varepsilon) \to \mathbb{R} \times P_1 \times P_2$$

such that $(\mu(0), u(0), v(0)) = (\mu_0, u_0, v_0)$ and $(\mu(s), u(s), v(s))$ is a coexistence state for each $s \in (-\varepsilon, \varepsilon)$. Moreover,

$$\mu(s) = \mu_0 + \hat{\mu}(s), \quad (u(s), v(s)) = (u_0, v_0) + s\Phi + (\hat{u}(s), \hat{v}(s)),$$

where $\hat{\mu}(s) = 0(s)$, $\hat{u}(s) = o(s)$ and $\hat{v}(s) = o(s)$ as $s \to 0$. In addition, there exists a neighborhood, $\mathcal{Q}_0$, of (μ_0, u_0, v_0) in $\mathbb{R} \times X_1 \times X_2$ such that the unique solutions of (10.5) in $\mathcal{Q}_0$ are of the form $(\mu(s), u(s), v(s))$ with $s \in (-\varepsilon, \varepsilon)$. Furthermore,

$$\operatorname{sgn} \mu'(s) = -\operatorname{sgn} \lambda_1[\mathfrak{L}_{\mu(s)}], \tag{10.99}$$

where $\mathfrak{L}_{\mu(s)}$ is the operator defined in (10.94) with $A(x)$ given by (10.93) and $(u_0, v_0) = (u(s), v(s))$.

In the context of Lemma 10.28(a), differentiating with respect to μ yields

$$\mathfrak{L}_\mu \begin{pmatrix} u'(\mu) \\ v'(\mu) \end{pmatrix} = \begin{pmatrix} 0 \\ v(\mu) \end{pmatrix} <_q 0.$$

Thus, since $\lambda_1[\mathfrak{L}_\mu, \Omega] > 0$, it follows from Theorem 10.20 that

$$(u'(\mu), v'(\mu)) \ll_q 0.$$

Therefore, $\mu \mapsto u(\mu)$ is decreasing, whereas $\mu \mapsto v(\mu)$ is increasing.

Similarly, in the context of Lemma 10.28(b), differentiating with respect to the parameter s shows that

$$\mathfrak{L}_{\mu(s)} \begin{pmatrix} u'(s) \\ v'(s) \end{pmatrix} = \mu'(s) \begin{pmatrix} 0 \\ v(s) \end{pmatrix}.$$

Thus, since

$$(u'(s), v'(s)) \sim \Phi \gg_q 0,$$

(10.99) holds from Theorem 10.20 taking into account that $(0, v(s)) <_q 0$.

We are ready to ascertain the fixed point index of the minimal positive solution, (μ, u_μ, v_μ). According to Lemma 10.27,

$$\lambda_1[\mathfrak{L}_\mu, \Omega] \geq 0.$$

Suppose $\lambda_1[\mathfrak{L}_\mu, \Omega] > 0$. The Schauder formula implies that

$$i(\mathcal{K}_{\lambda,\mu}, (u_\mu, v_\mu)) = 1.$$

Thus, according to Lemma 10.26, (10.5) admits a further coexistence state, as stated in Theorem 10.25.

Suppose $\lambda_1[\mathfrak{L}_\mu, \Omega] = 0$ and let $(\mu(s), u(s), v(s))$ be the curve of coexistence states whose existences are guaranteed by Lemma 10.28(b), with

$$(\mu(0), u(0), v(0)) = (\mu, u_\mu, v_\mu).$$

Since $\Phi >_q 0$, $s \mapsto (u(s), v(s))$ is strictly increasing. Hence, if $\mu(s) = \mu$ for some $s \neq 0$, then (10.5) possesses two coexistence states: $(u(0), v(0)) = (u_\mu, v_\mu)$ and $(u(s), v(s))$. Thus, without loss of generality, to complete the proof of Theorem 10.25 we can assume that

$$\mu(s) \neq \mu \qquad \text{for all } 0 < |s| < \varepsilon. \tag{10.100}$$

In such case, we claim that

$$\mu(s) > \mu \qquad \text{for all } s \in (-\varepsilon, 0). \tag{10.101}$$

Indeed, if $\mu_1 := \mu(s_1) < \mu$ for some $s_1 < 0$, since $s \to (u(s), v(s))$ is increasing and $\mu \mapsto (u_\mu, v_\mu)$ is non-increasing, we find that

$$(u(s_1), v(s_1)) <_q (u(0), v(0)) = (u_\mu, v_\mu) \leq_q (u_{\mu_1}, v_{\mu_1}), \tag{10.102}$$

where (u_{μ_1}, v_{μ_1}) stands for the minimal coexistence state at $\mu = \mu_1$. One should take into account that (u_{μ_1}, v_{μ_1}) is a supersolution of (10.5) if $\mu_1 < \mu$. As (10.102) contradicts the minimality of (u_{μ_1}, v_{μ_1}), (10.101) holds. Moreover, by (10.100), either $\mu(s) < \mu$ for all $s \in (0, \varepsilon)$ or $\mu(s) > \mu$ for all $s \in (0, \varepsilon)$. Subsequently, we will deal with each of these situations separately.

Case (i): Suppose $\mu(s) > \mu$ for all $s \in (0, \varepsilon)$. Then, since $[\mu_*(\lambda), \mu] \subset J_\mu$, there exists a sequence of coexistence states, (μ_n, u_n, v_n), $n \geq 1$, such that

$$\lim_{n \to \infty} \mu_n = \mu, \qquad \mu_n < \mu, \quad n \geq 1.$$

By the existence of a priori bounds, we can extract a subsequence, re-labeled by n, such that

$$\lim_{n \to \infty} (u_n, v_n) = (u_0, v_0)$$

for some non-negative solution (u_0, v_0). Adapting the arguments of the proof of Theorem 10.24, it is easy to see that (μ, u_0, v_0) is a coexistence state. Moreover, by the uniqueness obtained as an application of Lemma 10.28(b), we find that $(\mu_n, u_n, v_n) \notin \mathcal{Q}_0$ for all $n \geq 1$, because $\mu_n < \mu < \mu(s)$. Thus, $(\mu, u_0, v_0) \notin \mathcal{Q}_0$ and hence $(\mu, u_0, v_0) \neq (\mu, u_\mu, v_\mu)$. Therefore, (10.5) has two coexistence states, as required.

Case (ii): Suppose

$$\mu(s) < \mu \qquad \forall\, s \in (0, \varepsilon). \tag{10.103}$$

Then, owing to Lemma 10.28(b), (μ, u_μ, v_μ) is an isolated solution of (10.5) and the index $i(\mathcal{K}_{\lambda,\mu}, (u_\mu, v_\mu))$ is well defined. According to Lemma 10.26, to complete the proof of Theorem 10.25 it suffices to prove that

$$i(\mathcal{K}_{\lambda,\mu}, (u_\mu, v_\mu)) = 1. \tag{10.104}$$

By (10.103), $\mu'(s_1) < 0$ for some $s_1 \in (0, \varepsilon)$. Thus, by (10.99),

$$\lambda_1[\mathcal{L}_{\mu(s_1)}, \Omega] > 0.$$

Consequently, owing to Theorem 10.19, $(\mu(s_1), u(s_1), v(s_1))$ is exponentially asymptotically stable and hence by the Schauder formula,

$$i(\mathcal{K}_{\lambda,\mu(s_1)}, (u(s_1), v(s_1))) = 1. \tag{10.105}$$

On the other hand, since $(\mu(s_1), u(s_1), v(s_1))$ is non-degenerate and the mapping $s \to (u(s), v(s))$ is increasing, there exists $\delta > 0$ with the property that (10.5) cannot admit a coexistence state in

$$[\mu(s_1) - \delta, \mu(s_1)] \times \partial(P_{\rho_1} \setminus \overline{P}_{\rho_2})$$

where

$$\rho_1 := \|(u(s_1), v(s_1))\|_{X_1 \times X_2} - \delta, \qquad \rho_2 := \|(u_\mu, v_\mu)\|_{X_1 \times X_2} - \delta.$$

Indeed, the fact that (10.5) cannot admit a coexistence state (μ, u, v) with

$$\mu \in [\mu(s_1) - \delta, \mu(s_1)] \quad \text{and} \quad \|(u, v)\|_{X_1 \times X_2} = \rho_1$$

follows easily from the uniqueness obtained as an application of Lemma 10.28(b), by the monotonicity of the norm $\| \cdot \|_{X_1 \times X_2}$, as in the proof of [13, Th. 7.4]. That (10.5) cannot admit a coexistence state (μ, u, v) with

$$\mu \in [\mu(s_1) - \delta, \mu(s_1)] \quad \text{and} \quad \|(u,v)\|_{X_1 \times X_2} = \rho_2$$

follows easily from the fact that the minimal coexistence state, (ξ, u_ξ, v_ξ), satisfies

$$\|(u_\xi, v_\xi)\|_{X_1 \times X_2} \geq \|(u_\mu, v_\mu)\| > \rho_2$$

for all $\xi \in [\mu(s_1) - \delta, \mu(s_1)]$ because $\mu(s_1) < \mu$. Moreover, thanks again to the uniqueness obtained as an application of Lemma 10.28(a), $\delta > 0$ can be shortened, if necessary, so that (10.5) cannot admit a coexistence state in $P_{\rho_1} \setminus \overline{P}_{\rho_2}$ at $\mu = \mu(s_1) - \delta$. For this choice, by the homotopy invariance of the index, we find that

$$i(\mathcal{K}_{(\lambda,\mu(s_1))}, P_{\rho_1} \setminus \overline{P}_{\rho_2}) = 0. \tag{10.106}$$

Finally, for sufficiently small $\delta > 0$, we set

$$\rho := \|(u(s_1), v(s_1))\|_{X_1 \times X_2} + \delta.$$

According to (10.105) and (10.106),

$$i(\mathcal{K}_{(\lambda,\mu(s_1))}, P_\rho \setminus \overline{P}_{\rho_2}) = 1. \tag{10.107}$$

Moreover, by the monotonicity of $(u(s), v(s))$ and the uniqueness given by Lemma 10.28(b), (10.5) cannot admit a coexistence state in

$$[\mu(s_1), \mu] \times \partial(P_\rho \setminus P_{\rho_2}).$$

Combining this property with (10.107) yields (10.104) and ends the proof. $\quad\square$

10.7 Dynamics of (1.1) when $\mathfrak{b} = 0$

We begin by characterizing the existence of coexistence and metacoexistence states in the special case $\mathfrak{b} = 0$. Then, we shall analyze their attractive character with respect to the classical positive solutions of (10.1). To state the main result we need to introduce some notations. Throughout this section, for every $1 \leq j \leq q_0$, we denote by X_j^∞ the set of functions $\Phi \in C^\nu(\Omega_j)$, $\Phi \geq 0$, such that

$$\Phi = 0 \quad \text{on} \quad \partial\Omega_j \cap \partial\Omega \quad \text{and} \quad \lim_{\text{dist}(x, \partial\Omega_j \cap \partial\Omega) \downarrow 0} \Phi(x) = \infty$$

where Ω_j is defined in (4.3), and for every $1 \leq j \leq q_0$ and $\Phi \in X_j^\infty$ we denote

$$\Phi_n := \min\{\Phi, n\} \quad \text{for all integer } n \geq 1.$$

Obviously,

$$\Phi_n \in C^\nu(\bar{\Omega}_j; \mathbb{R}_+) \quad \text{and} \quad \Phi_n \leq \Phi_{n+1} \quad \text{for all } n \geq 1.$$

Thus, for every $V \in C^\nu(\bar{\Omega}_j)$, the sequence of principal eigenvalues

$$\lambda_1[-d_2\Delta + V + \mathfrak{c}\Phi_n, \Omega_j], \qquad n \geq 1,$$

is non-decreasing. Moreover, it is bounded above. Indeed, pick $x \in \Omega_-$ and $R > 0$ with $\bar{B}_R(x) \subset \Omega_-$. Then, since $\Phi \in C^\nu(\bar{B}_R(x))$, there exists $n_0 \in \mathbb{N}$ such that $\Phi_n = \Phi$ in $\bar{B}_R(x)$ for all $n \geq n_0$. Therefore, by monotonicity with respect to the domain,

$$\lambda_1[-d_2\Delta + V + \mathfrak{c}\Phi_n, \Omega_j] \leq \lambda_1[-d_2\Delta + V + \mathfrak{c}\Phi, B_R(x)], \qquad n \geq n_0.$$

Consequently, the limit

$$\lambda_1[-d_2\Delta + V + \mathfrak{c}\Phi, \Omega_j] := \lim_{n \to \infty} \lambda_1[-d_2\Delta + V + \mathfrak{c}\Phi_n, \Omega_j]$$

is well defined in $\mathbb{R}$. Using these notations the next result holds.

Theorem 10.29 *Suppose $\mathfrak{b} = 0$ in Ω. Then, the following properties hold:*

(a) *The problem (10.5) possesses a coexistence state if, and only if,*

$$d_1\sigma_0 < \lambda < d_1\sigma_1 \quad \text{and} \quad \mu > \sigma_0[d_2, \mathfrak{c}\theta_{\lambda,u}],$$

where $\theta_{\lambda,u}$ stands for the unique positive solution of (10.7).

(b) *Suppose $1 \leq j \leq q_0$. Then, (10.5) has a metacoexistence state, $(\mathfrak{M}_u, \mathfrak{M}_v)$, supported in Ω_j, as discussed by Definition 10.1, if, and only if,*

$$\lambda < d_1\sigma_{j+1} \quad \text{and} \quad \mu > \lambda_1[-d_2\Delta + \mathfrak{c}\mathfrak{M}^{\min}_{[\lambda,\Omega_j]}, \Omega_j], \tag{10.108}$$

where $\mathfrak{M}^{\min}_{[\lambda,\Omega_j]}$ is the minimal metasolution of (10.7) supported in Ω_j. Moreover, $\mathfrak{M}_v$ can be chosen so that $\mathfrak{M}_v = 0$ on $\partial\Omega_j$.

As usual, when $j = q_0$ we are taking $\sigma_{q_0+1} \equiv \infty$ and $\Omega_{q_0} = \Omega_-$. In such case, it should be noted that (10.108) is not imposing any restriction on the size of λ, but only on μ. Our proof of Theorem 10.29 relies on the following result.

Proposition 10.30 *Suppose $1 \leq j \leq q_0$ and $\Phi \in X_j^\infty$. Then, the problem*

$$\begin{cases} (-d_2\Delta + \mathfrak{c}\Phi)v = \mu v - \partial f_2(\cdot, v)v & \text{in } \Omega_j, \\ v = 0 & \text{on } \partial\Omega_j, \end{cases} \tag{10.109}$$

admits a weak bounded positive solution if, and only if,

$$\mu > \lambda_1[-d_2\Delta + \mathfrak{c}\Phi, \Omega_j].$$

Moreover, any weak bounded positive solution, v, of (10.109) satisfies $v \in C^{2+\nu}(\Omega_j) \cap C(\bar{\Omega}_j)$, and is unique if it exists.

Note that $\lambda_1[-d_2\Delta + \mathfrak{c}\Phi, \Omega_j]$ is the value defined just before the statement of Theorem 10.29 with $V = 0$. To prove Proposition 10.30, we need the next lemma.

Lemma 10.31 *Consider the linear eigenvalue problem*

$$\begin{cases} (-d_2\Delta + V + \mathfrak{c}\Phi)v = \mu v & in \ \ \Omega_j, \\ v = 0 & on \ \ \partial\Omega_j, \end{cases} \tag{10.110}$$

where $V \in \mathcal{C}^\nu(\bar{\Omega}_j)$. Then,

(a) *$\mu(V) := \lambda_1[-d_2\Delta + V + \mathfrak{c}\Phi, \Omega_j]$ is the unique value of μ for which (10.110) admits a weak positive solution in $H_0^1(\Omega_j)$. Moreover, for $\mu = \mu(V)$ there exists a weak solution $\varphi_j \in H_0^1(\Omega_j) \cap L^\infty(\Omega_j)$ such that $\varphi_j(x) > 0$ almost everywhere in Ω_j.*

(b) *If $V < W$, then $\mu(V) < \mu(W)$.*

Proof of Lemma 10.31: Consider the truncations

$$\Phi_n := \min\{\Phi, n\}, \qquad n \geq 1,$$

and let $\varphi_{j,n} > 0$, $n \geq 1$, denote the (unique) principal eigenfunction of

$$\lambda_{1,n,j} := \lambda_1[-d_2\Delta + V + \mathfrak{c}\Phi_n, \Omega_j]$$

such that

$$\|\varphi_{j,n}\|_{L^\infty(\Omega_j)} = 1, \qquad n \geq 1.$$

Then,

$$(-d_2\Delta + V)\varphi_{j,n} \leq (-d_2\Delta + V + \mathfrak{c}\Phi_n)\varphi_{j,n} = \lambda_{1,n,j}\varphi_{j,n}$$

for every $n \geq 1$. Thus, there exists a constant $C > 0$ such that

$$(-d_2\Delta + V)\varphi_{j,n} \leq C\varphi_{j,n} \qquad \text{for all } n \geq 1.$$

Consequently, by Lemma 10.13, there exist a subsequence of $\varphi_{j,n}$, $n \geq 1$, re-labeled by n, and a function $\varphi_j \in H_0^1(\Omega_j) \cap L^\infty(\Omega_j)$, $\varphi_j \neq 0$, such that

$$\lim_{n\to\infty} \varphi_{j,n} = \varphi_j \quad \text{weakly in } H_0^1(\Omega_j) \text{ and in } L^p(\Omega_j) \text{ for all } p > 1. \tag{10.111}$$

Necessarily, $\varphi_j > 0$, because $\varphi_{j,n} > 0$ for all $n \geq 1$, and, owing to (10.111), it is easily seen that φ_j is a weak solution of (10.110) for $\mu = \mu(V)$. Moreover, thanks to the weak Harnack inequality, $\varphi_j(x) > 0$ almost everywhere (a.e.) in Ω_j. To complete the proof of Part (a) we will argue by contradiction. Suppose $\mu_1 \neq \mu_2$ are two values of μ for which (10.110) admits a weak positive solution, ψ_1 and ψ_2, respectively. Then, thanks to the weak Harnack inequality, $\psi_k(x) >$

0 a.e. in Ω_j for $k \in \{1, 2\}$. Pick an arbitrary test function $\eta \in C_0^\infty(\Omega_j)$. Then, $\eta \psi_k \in H_0^1(\Omega_j)$ for $k \in \{1, 2\}$ and hence,

$$d_2 \int_{\Omega_j} \langle \nabla(\eta \psi_2), \nabla \psi_1 \rangle + \int_{\Omega_j} \eta(V + \mathfrak{c}\Phi)\psi_1 \psi_2 = \mu_1 \int_{\Omega_j} \eta \psi_1 \psi_2,$$

$$d_2 \int_{\Omega_j} \langle \nabla(\eta \psi_1), \nabla \psi_2 \rangle + \int_{\Omega_j} \eta(V + \mathfrak{c}\Phi)\psi_1 \psi_2 = \mu_2 \int_{\Omega_j} \eta \psi_1 \psi_2.$$

Thus, subtracting these two identities yields

$$d_2 \int_{\Omega_j} \langle \nabla \eta, \psi_2 \nabla \psi_1 - \psi_1 \nabla \psi_2 \rangle = (\mu_1 - \mu_2) \int_{\Omega_j} \eta \psi_1 \psi_2. \qquad (10.112)$$

As (10.112) holds true for all $\eta \in C_0^\infty(\Omega_j)$, it becomes apparent that

$$(\mu_1 - \mu_2) \int_{\Omega_j} \psi_1 \psi_2 = 0$$

and hence $\mu_1 = \mu_2$, which is a contradiction. This ends the proof of Part (a).

To prove Part (b), let φ_j and ψ_j denote two weak positive solutions of (10.110) associated to $\mu = \mu(V)$ and $\mu = \mu(W)$, respectively. Then, for every test function $\eta \in C_0^\infty(\Omega_j)$,

$$d_2 \int_{\Omega_j} \langle \nabla \eta, \psi_j \nabla \varphi_j - \varphi_j \nabla \psi_j \rangle + \int_{\Omega_j} \eta(V - W)\varphi_j \psi_j = (\mu(V) - \mu(W)) \int_{\Omega_j} \eta \varphi_j \psi_j$$

and hence

$$\int_{\Omega_j} (V - W)\varphi_j \psi_j = (\mu(V) - \mu(W)) \int_{\Omega_j} \varphi_j \psi_j.$$

As $V < W$, the left hand side of this identity is negative. Therefore, $\mu(V) < \mu(W)$, as claimed.   □

Proof of Proposition 10.30: Let $1 \leq j \leq q_0$ and $\Phi \in X_j^\infty$ be, and suppose (10.109) has a weak bounded positive solution, v. Then,

$$(-d_2 \Delta + \mathfrak{d}\mathfrak{f}_2(\cdot, v) + \mathfrak{c}\Phi)v = \mu v$$

and hence by Lemma 10.31,

$$\mu = \lambda_1[-d_2 \Delta + \mathfrak{d}\mathfrak{f}_2(\cdot, v) + \mathfrak{c}\Phi, \Omega_j] > \lambda_1[-d_2 \Delta + \mathfrak{c}\Phi, \Omega_j].$$

To show the converse, suppose

$$\mu > \lambda_1[-d_2 \Delta + \mathfrak{c}\Phi, \Omega_j].$$

Then, by definition, there exists $n_0 \in \mathbb{N}$ such that

$$\mu > \lambda_1[-d_2 \Delta + \mathfrak{c}\Phi_n, \Omega_j] \qquad \text{for all } n \geq n_0.$$

Consequently, by Remark 5.3 and Theorem 2.2, for every $n \geq n_0$, the problem

$$\begin{cases} (-d_2\Delta + \mathfrak{c}\Phi_n)v = \mu v - \mathfrak{d}\mathfrak{f}_2(\cdot, v)v & \text{in} \quad \Omega_j, \\ v = 0 & \text{on} \quad \partial\Omega_j, \end{cases} \tag{10.113}$$

has a unique positive (strong) solution, v_n. As $\Phi_{n+1} \geq \Phi_n$,

$$(-d_2\Delta + \mathfrak{c}\Phi_{n+1})v_n \geq (-d_2\Delta + \mathfrak{c}\Phi_n)v_n = \mu v_n - \mathfrak{d}\mathfrak{f}_2(\cdot, v_n)v_n$$

and hence by Lemma 1.8,

$$v_n \geq v_{n+1} \qquad \text{for all} \ \ n \geq n_0. \tag{10.114}$$

On the other hand, for each $n \geq n_0$,

$$-d_2\Delta v_n \leq (-d_2\Delta + \mathfrak{c}\Phi_n)v_n = \mu v_n - \mathfrak{d}\mathfrak{f}_2(\cdot, v_n)v_n \leq \mu v_n.$$

By (10.114), it follows from Lemma 10.13 that there exists $v \in H_0^1(\Omega_j) \cap L^\infty(\Omega_j)$ such that

$$\lim_{n\to\infty} v_n = v \quad \text{weakly in} \ \ H_0^1(\Omega_j) \ \ \text{and strongly in} \ \ L^p(\Omega_j) \ \ \forall\, p > 1.$$

Also thanks to (10.114), it becomes apparent that $v \in \mathcal{C}(\bar{\Omega}_j)$ and that $v = 0$ on $\partial\Omega_j$. Naturally, it is easy to see that $v \geq 0$ is a weak solution of (10.109). By interior elliptic regularity, $v \in C^{2+\nu}(\Omega_j)$. To complete the proof of the existence, it suffices to show that $v > 0$. On the contrary, suppose $v = 0$. Then, setting

$$\hat{v}_n := \frac{v_n}{\|v_n\|_{L^\infty(\Omega_j)}}, \qquad n \geq n_0,$$

we again have that

$$-d_2\Delta\hat{v}_n \leq \mu\hat{v}_n \quad \text{in} \ \ \Omega_j$$

for all $n \geq n_0$ and hence, once again by Lemma 10.13, there exists $\varphi \in H_0^1(\Omega_j) \cap L^\infty(\Omega_j)$ such that along some subsequence, re-labeled by n,

$$\lim_{n\to\infty} \hat{v}_n = \varphi \quad \text{weakly in} \ \ H_0^1(\Omega_j) \ \ \text{and strongly in} \ \ L^p(\Omega_j) \ \ \forall\, p > 1.$$

Necessarily $\varphi > 0$, because $\|\hat{v}_n\|_{L^\infty(\Omega_j)} = 1$ for all $n \geq n_0$. Moreover, dividing the v_n equation by $\|v_n\|_{L^\infty(\Omega_j)}$ and letting $n \to \infty$, it becomes apparent that φ provides us with a weak bounded positive solution of

$$(-d_2\Delta + \mathfrak{c}\Phi)\varphi = \mu\varphi,$$

which contradicts Lemma 10.31. Therefore, $v > 0$. The uniqueness of v can be demonstrated from Lemma 10.31 by adapting the proof of the uniqueness in Theorem 1.7. $\quad\square$

Proof of Theorem 10.29: By Remark 5.3, Part (a) is an easy consequence from Theorem 4.1, because $\mathfrak{b} = 0$. To prove Part (b) pick $1 \leq j \leq q_0$ and suppose (10.5) possesses a meta-coexistence state, $(\mathfrak{M}_u, \mathfrak{M}_v)$, supported in Ω_j such that $\mathfrak{M}_v = 0$ on $\partial\Omega_j$. Then, $\mathfrak{M}_u$ is a metasolution of (10.7) supported in Ω_j. Consequently, thanks to Theorem 4.5, $\lambda < d_1\sigma_{j+1}$. Moreover, $\mathfrak{M}_u \in X_j^\infty$ and $\mathfrak{M}_v$ solves the problem

$$\begin{cases} (-d_2\Delta + c\mathfrak{M}_u)v = \mu v - \mathfrak{d}f_2(\cdot, v)v & \text{in} \quad \Omega_j, \\ v = 0 & \text{on} \quad \partial\Omega_j. \end{cases}$$

Thus, we find from Proposition 10.30 that

$$\mu > \lambda_1[-d_2\Delta + c\mathfrak{M}_u, \Omega_j] \geq \lambda_1[-d_2\Delta + c\mathfrak{M}_{[\lambda,\Omega_j]}^{\min}, \Omega_j],$$

because $\mathfrak{M}_u \geq \mathfrak{M}_{[\lambda,\Omega_j]}^{\min}$. Therefore, the second estimate of (10.108) holds.

Conversely, suppose (10.108). Then, by Theorem 4.5, (10.7) possesses a minimal metasolution $\mathfrak{M}_{[\lambda,\Omega_j]}^{\min} \in X_j^\infty$ supported in Ω_j. Moreover, owing to Proposition 10.30, the singular problem

$$\begin{cases} (-d_2\Delta + c\mathfrak{M}_{[\lambda,\Omega_j]}^{\min})v = \mu v - \mathfrak{d}f_2(\cdot, v)v & \text{in} \quad \Omega_j, \\ v = 0 & \text{on} \quad \partial\Omega_j, \end{cases}$$

has a positive solution. The proof is complete. $\square$

The relevance of the coexistence and meta-coexistence states constructed in Theorem 10.29 relies on the fact that they provide us with the asymptotic behavior of the positive solutions of (10.1) in the special case when $\mathfrak{b} = 0$. Indeed, the next result is satisfied. The notations introduced in the previous sections will be kept.

Theorem 10.32 *Suppose $\mathfrak{b} = 0$ and $u_0, v_0 \in C_0(\bar\Omega)$ satisfy $u_0 > 0$ and $v_0 > 0$. Let $(u(x,t), v(x,t))$ denote the unique classical solution of (10.1). Then, the following properties are satisfied:*

(a) *If $d_1\sigma_0 < \lambda < d_1\sigma_1$ and $\mu \leq \sigma_0[d_2, c\theta_{\lambda,u}]$, then*

$$\lim_{t\uparrow\infty} \|u(\cdot, t) - \theta_{\lambda,u}\|_{L^\infty(\Omega)} = 0, \quad \lim_{t\uparrow\infty} \|v(\cdot, t)\|_{L^\infty(\Omega)} = 0.$$

(b) *If $d_1\sigma_0 < \lambda < d_1\sigma_1$ and $\mu > \sigma_0[d_2, c\theta_{\lambda,u}]$, then*

$$\lim_{t\uparrow\infty} \|u(\cdot, t) - \theta_{\lambda,u}\|_{L^\infty(\Omega)} = 0, \quad \lim_{t\uparrow\infty} \|v(\cdot, t) - \theta_{\{\mu,v;c\theta_{\lambda,u}\}}\|_{L^\infty(\Omega)} = 0.$$

(c) *If there exists $j \in \{1, ..., q_0\}$ such that*

$$d_1\sigma_j \leq \lambda < d_1\sigma_{j+1} \quad \text{and} \quad \mu \leq \lambda_1[-d_2\Delta + c\mathfrak{M}_{[\lambda,\Omega_j]}^{\min}],$$

then

$$\mathfrak{M}_{[\lambda,\Omega_j]}^{\min} \leq \liminf_{t\uparrow\infty} u(\cdot, t) \leq \limsup_{t\uparrow\infty} u(\cdot, t) \leq \mathfrak{M}_{[\lambda,\Omega_j]}^{\max} \qquad (10.115)$$

and

$$\lim_{t\uparrow\infty} \|v(\cdot, t)\|_{L^\infty(\Omega)} = 0.$$

(d) *If there exists $j \in \{1, ..., q_0\}$ such that*

$$d_1 \sigma_j \leq \lambda < d_1 \sigma_{j+1} \quad and \quad \mu > \lambda_1[-d_2\Delta + \mathfrak{c}\mathfrak{M}^{\min}_{[\lambda,\Omega_j]}],$$

then (10.115) *holds, and*

$$\Theta_{\{\mu,v;\mathfrak{c}\mathfrak{M}^{\max}_{[\lambda,\Omega_j]}\}} \leq \liminf_{t\uparrow\infty} v(\cdot,t) \leq \limsup_{t\uparrow\infty} v(\cdot,t) \leq \Theta_{\{\mu,v;\mathfrak{c}\mathfrak{M}^{\min}_{[\lambda,\Omega_j]}\}}$$

where, given any $\Phi \in X_j^\infty$, $\Theta_{\{\mu,v;\mathfrak{c}\Phi\}}$ stands for the extension by zero to Ω of the unique positive solution of

$$\begin{cases} (-d_2\Delta + \mathfrak{c}\Phi)v = \mu v - \mathfrak{d}\mathfrak{f}_2(\cdot, v)v & in \ \Omega_j, \\ v = 0 & on \ \partial\Omega_j, \end{cases}$$

whose existence is guaranteed by Proposition 10.30.

The proof of Theorem 10.32 follows readily by combining Theorem 5.2 and Remark 5.3 with the next result whose proof is omitted. The technical details can be reconstructed easily from the parabolic maximum principle. At this stage of the book, the proofs should be routine.

Theorem 10.33 *Under the same assumptions of Theorem 10.32, assume that $d_1\sigma_j \leq \lambda < d_1\sigma_{j+1}$ for some $j \in \{1, ..., q_0\}$, and let $u(x,t)$ denote the positive solution of the parabolic counterpart of* (10.7). *Lastly, let $v := v(x,t)$ be the (unique) positive solution of*

$$\begin{cases} \frac{\partial v}{\partial t} - d_2\Delta v + \mathfrak{c}u(x,t)v = \mu v - \mathfrak{d}(x)\mathfrak{f}_2(x,v)v & in \ \Omega \times (0,\infty), \\ v = 0 & on \ \partial\Omega \times (0,\infty), \\ v(\cdot,0) = v_0 & in \ \Omega. \end{cases}$$

Then

$$\Theta_{\{\mu,v;\mathfrak{c}\mathfrak{M}^{\max}_{[\lambda,\Omega_j]}\}} \leq \liminf_{t\uparrow\infty} v(\cdot,t) \leq \limsup_{t\uparrow\infty} v(\cdot,t) \leq \Theta_{\{\mu,v;\mathfrak{c}\mathfrak{M}^{\min}_{[\lambda,\Omega_j]}\}}.$$

10.8 Existence of meta-coexistence states

This section reveals that the necessary and sufficient conditions given in Section 10.7 for a meta-coexistence state in the special case $\mathfrak{b} = 0$ are actually sufficient when $\mathfrak{b}(x) > 0$ for all $x \in \Omega$.

Theorem 10.34 *Suppose there exists $j \in \{1, ..., q_0\}$ such that*

$$\lambda < d_1\sigma_{j+1} \quad and \quad \mu > \lambda_1[-d_2\Delta + \mathfrak{c}\mathfrak{M}^{\min}_{[\lambda,\Omega_j]}, \Omega_j]. \tag{10.116}$$

Then, (10.5) *possesses a meta-coexistence state $(\mathfrak{M}_u, \mathfrak{M}_v)$ supported in Ω_j with $\mathfrak{M}_v = 0$ on $\partial\Omega_j$.*

Condition (10.116) is optimal because, due to Theorem 10.29, it is necessary and sufficient if $\mathfrak{b} = 0$. When

$$\mathfrak{M}^{\max}_{[\lambda,\Omega_j]} = \mathfrak{M}^{\min}_{[\lambda,\Omega_j]} \equiv \mathfrak{M}_{[\lambda,\Omega_j]}$$

is non-degenerate, the curve

$$\mu = \lambda_1[-d_2\Delta + \mathfrak{c}\mathfrak{M}_{[\lambda,\Omega_j]}, \Omega_j], \qquad \lambda < d_1\sigma_{j+1},$$

provides us with the set of values of (λ, μ) for which bifurcation to meta-coexistence states, $(\mathfrak{M}_u, \mathfrak{M}_v)$, supported in Ω_j with $\mathfrak{M}_v = 0$ on $\partial\Omega_j$ from the semi-trivial metasolution $(\mathfrak{M}_{[\lambda,\Omega_j]}, 0)$ occurs. As this bifurcation can be subcritical, (10.5) can indeed have a meta-coexistence state for

$$\mu < \lambda_1[-d_2\Delta + \mathfrak{b}\mathfrak{M}_{[\lambda,\Omega_j]}, \Omega_j].$$

Thus, in general (10.116) is not necessary for the existence of meta-coexistence states supported in Ω_j with $\mathfrak{M}_v = 0$ on $\partial\Omega_j$.

Proof of Theorem 10.34: Suppose (10.116) holds for some $j \in \{1, ..., q_0\}$ and consider the nonlinear boundary value problems

$$\begin{cases} -d_1\Delta u = \lambda u - \mathfrak{a}\mathfrak{f}_1(\cdot, u)u - \mathfrak{b}uv \\ -d_2\Delta v = \mu v - \mathfrak{d}\mathfrak{f}_2(\cdot, v)v - \mathfrak{c}uv \end{cases} \quad \text{in } \Omega_j, \\ u = n \quad \text{on } \partial\Omega_j \cap \Omega, \qquad (10.117) \\ u = 0 \quad \text{on } \partial\Omega_j \cap \partial\Omega, \\ v = 0 \quad \text{on } \partial\Omega_j, $$

where $n \in \mathbb{N}$, $n \geq 1$. For each integer $n \geq 1$, let $\Theta_{[\lambda,\Omega_j,n]}$ denote the unique positive solution of

$$\begin{cases} -d_1\Delta u = \lambda u - \mathfrak{a}\mathfrak{f}_1(\cdot, u)u & \text{in } \Omega_j, \\ u = n & \text{on } \partial\Omega_j \cap \Omega, \\ u = 0 & \text{on } \partial\Omega_j \cap \partial\Omega, \end{cases}$$

whose existence and uniqueness are guaranteed by Theorem 4.5. The solutions satisfy

$$\Theta_{[\lambda,\Omega_j,n]} \leq \Theta_{[\lambda,\Omega_j,n+1]} \leq \mathfrak{M}^{\min}_{[\lambda,\Omega_j]}$$

for all $n \geq 1$ and, actually, thanks to Theorem 4.7, by elliptic regularity,

$$\lim_{n\to\infty} \|\Theta_{[\lambda,\Omega_j,n]} - \mathfrak{M}^{\min}_{[\lambda,\Omega_j]}\|_{C^{2+\nu}(\Omega_j)} = 0.$$

Let $\varphi > 0$ be a principal eigenfunction of

$$\mu_j := \lambda_1[-d_2\Delta + \mathfrak{c}\mathfrak{M}^{\min}_{[\lambda,\Omega_j]}, \Omega_j] < \mu,$$

whose existence is guaranteed by Lemma 10.31. We claim that there exists $\varepsilon_0 > 0$ such that for each $0 < \varepsilon < \varepsilon_0$ and $n \in \mathbb{N}$, $n \geq 1$, the couple

$$(\overline{u}_n, \overline{v}) := (\Theta_{[\lambda,\Omega_j,n]}, \varepsilon\varphi)$$

provides us with a supersolution of (10.117). Indeed, by definition,

$$-d_1\Delta\bar{u}_n = \lambda\bar{u}_n - \mathfrak{a}\mathfrak{f}_1(\cdot,\bar{u}_n)\bar{u}_n \geq \lambda\bar{u}_n - \mathfrak{a}\mathfrak{f}_1(\cdot,\bar{u}_n)\bar{u}_n - \mathfrak{b}\bar{u}_n\bar{v},$$

because $\mathfrak{b}\bar{u}_n\bar{v} \geq 0$. Moreover, $\bar{u}_n = n$ on $\partial\Omega_j \cap \Omega$ and $\bar{u}_n = 0$ on $\partial\Omega_j \cap \partial\Omega$. Thus, since $\varepsilon\varphi = 0$ on $\partial\Omega_j$, to prove the previous claim it suffices to show that there exists $\varepsilon_0 > 0$ such that for each $\varepsilon \in (0,\varepsilon_0)$ and $n \in \mathbb{N}$, $n \geq 1$, the following estimate holds:

$$-d_2\Delta(\varepsilon\varphi) \leq \mu\varepsilon\varphi - \mathfrak{d}\mathfrak{f}_2(\cdot,\varepsilon\varphi)\varepsilon\varphi - \mathfrak{c}\Theta_{[\lambda,\Omega_j,n]}\varepsilon\varphi. \tag{10.118}$$

Since (10.118) can be written in the form

$$(-d_2\Delta + \mathfrak{c}\mathfrak{M}_{[\lambda,\Omega_j]}^{\min})(\varepsilon\varphi) + \mathfrak{c}(\Theta_{[\lambda,\Omega_j,n]} - \mathfrak{M}_{[\lambda,\Omega_j]}^{\min})\varepsilon\varphi \leq \mu\varepsilon\varphi - \mathfrak{d}\mathfrak{f}_2(\cdot,\varepsilon\varphi)\varepsilon\varphi,$$

and $\Theta_{[\lambda,\Omega_j,n]} \leq \mathfrak{M}_{[\lambda,\Omega_j]}^{\min}$ for all $n \geq 1$, it becomes apparent that (10.118) holds provided

$$\mu_j\varepsilon\varphi \leq \mu\varepsilon\varphi - \mathfrak{d}\mathfrak{f}_2(\cdot,\varepsilon\varphi)\varepsilon\varphi. \tag{10.119}$$

As (10.119) is equivalent to

$$\mu_j \leq \mu - \mathfrak{d}\mathfrak{f}_2 g(\cdot,\varepsilon\varphi),$$

$\mu_j < \mu$ and $\mathfrak{f}_2(\cdot,0) = 0$, there exists $\varepsilon_0 > 0$ such that (10.119) holds for all $\varepsilon \in (0,\varepsilon_0)$, which ends the proof of the claim above. Note that ε_0 can be chosen sufficiently small so that

$$\varepsilon\varphi < \theta_{\mu,v}, \qquad 0 < \varepsilon < \varepsilon_0, \tag{10.120}$$

where $\theta_{\mu,v}$ is the unique positive solution of (10.8).

On the other hand, it is easily seen that

$$(\underline{u}_0,\underline{v}_0) := (0,\theta_{\mu,v})$$

provides us with a subsolution of (10.117). Fix $\varepsilon \in (0,\varepsilon_0)$. Thanks to (10.120),

$$(\underline{u}_0,\underline{v}_0) = (0,\theta_{\mu,v}) <_q (\bar{u}_n,\bar{v}) := (\Theta_{[\lambda,\Omega_j,n]},\varepsilon\varphi)$$

for all $n \geq 1$. Thus, thanks to Theorem 10.23, the problem (10.117) with $n = 1$ has a minimal solution, (u_1,v_1), in the interval $[(\underline{u}_0,\underline{v}_0),(\bar{u}_1,\bar{v})]$. As $0 \leq u_1$ and $u_1 = 1$ on $\partial\Omega_j \cap \Omega$, $u_1 > 0$. Also $v_1 > 0$, because $v_1 \geq \bar{v} = \varepsilon\varphi$. Similarly, $(\underline{u}_1,\underline{v}_1) := (u_1,v_1)$ provides us with a subsolution of (10.117) for $n = 2$ such that

$$(\underline{u}_0,\underline{v}_0) \leq_q (\underline{u}_1,\underline{v}_1) \leq_q (\bar{u}_1,\bar{v}) \leq_q (\bar{u}_2,\bar{v}).$$

Consequently, by Theorem 10.23, we find that (10.117) with $n = 2$ has a minimal solution, (u_2,v_2), in the interval $[(u_1,v_1),(\bar{u}_2,\bar{v})]$. In particular, $(u_1,v_1) \leq (u_2,v_2)$. By a simple induction argument, it is apparent that, for every $n \geq 1$, the problem (10.117) has a component-wise positive solution (u_n, v_n) such that

$$(u_n, v_n) \leq_q (u_{n+1}, v_{n+1}) \leq_q (\mathfrak{M}^{\min}_{[\lambda, \Omega_j]}, \varepsilon\varphi) \qquad \text{for all } n \geq 1.$$

By monotonicity, the point-wise limits

$$u_\infty := \lim_{n\to\infty} u_n \quad \text{and} \quad v_\infty := \lim_{n\to\infty} v_n$$

are well defined. Moreover, $u_\infty > 0$ and $v_\infty > 0$. Finally, combining the Lebesgue theorem on dominated convergence with some classical elliptic estimates, it is seen easily that indeed (u_∞, v_∞) is a meta-coexistence state of (10.5) supported in Ω_j with $v_\infty = 0$ on $\partial\Omega_j$. This ends the proof.   □

10.9   Comments on Chapter 10

The non-spatial model

$$\begin{cases} u' = \lambda u - Au^2 - Buv \\ v' = \mu v - Dv^2 - Buv \end{cases} \tag{10.121}$$

with $A > 0$, $B > 0$, $C > 0$ and $D > 0$, was introduced by V. Volterra [229] and independently by A. J. Lotka [185]. The first general results for its diffusive counterpart,

$$\begin{cases} \partial_t u - d_1 \Delta u = \lambda u - Au^2 - Buv & \text{in } \Omega \times (0, \infty), \\ \partial_t v - d_2 \Delta v = \mu v - Dv^2 - Buv & \\ u = v = 0 & \text{on } \partial\Omega \times (0, \infty), \\ u(\cdot, 0) = u_0 \geq 0 \text{ and } v(\cdot, 0) = v_0 \geq 0 & \text{in } \Omega, \end{cases} \tag{10.122}$$

were given by J. Blat and K. J. Brown [29], where the global bifurcation theorem of P. H. Rabinowitz [214] was invoked to construct the first global continua of coexistence states in the context of reaction-diffusion systems. Indeed, J. Blat and K. J. Brown [29] established the existence of a component of coexistence states linking the two semi-trivial positive solutions. This methodology had been previously used by J. Cushing [63] in the periodic Lotka–Volterra non-spatial model.

Some time later, these pioneering findings were substantially sharpened and generalized to cover wide classes of spatially heterogeneous competing species models by the author in [142], [149], where the global unilateral theory of P. H. Rabinowitz [214] was revised and updated to get the most general existence result for two species systems, Theorem 7.2.2 of [149], which is valid for a very general class of nonlinear elliptic spatially heterogeneous systems:

$$\begin{cases} \mathfrak{L}_1 u = \lambda u + f(x, u)u + F(x, u, v)uv & \text{in } \Omega, \\ \mathfrak{L}_2 v = \mu v + g(x, v)v + G(x, u, v)uv & \\ \mathfrak{B}_1 u = \mathfrak{B}_2 v = 0 & \text{on } \partial\Omega, \end{cases} \tag{10.123}$$

where $\mathfrak{L}_k$, $k = 1, 2$, are linear second order elliptic operators like (1.48) and $\mathfrak{B}_1$ and $\mathfrak{B}_2$ are two general boundary operators of the mixed general type (1.49). Theorem 7.2.2 of [149] has been the main bifurcation theorem used to show the existence of coexistence states in this chapter.

Other closely related papers analyzing the effects of spatial heterogeneities on the dynamics of competing species models are by J. López-Gómez [143], V. Hutson at al. [120] and J. López-Gómez and J. C. Sabina [180]. The authors analyzed the effects of vanishing the coefficient functions $\mathfrak{b}(x)$ and $\mathfrak{c}(x)$ on the dynamics of (10.1). The patches where $\mathfrak{b}(x)$ vanishes are protection zones for the individuals of the species u, while the components of $\mathfrak{c}^{-1}(0)$ are protection zones for those of the species v. When the protection zones of each of the competitors can maintain the species in isolation, there is permanence independent of the intensity of the competition. Consequently, as observed often in empirical studies, the paradigm inherent to the principle of competitive exclusion is false in heterogeneous habitats. The reader is sent to J. López-Gómez [145], S. Cano-Casanova and J. López-Gómez [36] and J. López-Gómez and M. Molina-Meyer [166], [168]–[172] for refinements and a detailed discussion on the implications of these features in population dynamics. Undoubtedly, [143], [120] and [180] constitute a paradigm of the discussion by R. S. Cantrell and C. Cosner in Section 6.2.1 of [43], *"How Spatial Segregation May Facilitate Coexistence"*.

The model (10.1) was introduced in J. López-Gómez [153], whose materials have been considerably polished in this chapter. It was the first time that the carrying capacity of one of the competitors was allowed to be infinity on some patches of the inhabiting area. Almost simultaneously, Y. Du submitted [77], where the same problem in the very special case when Ω_0 is connected was dealt with.

From reading Chapter 10, it becomes apparent that the simplest way to overcome most of the technical difficulties when Ω_0 has an arbitrary number of components, as has been always the case throughout this book, is assuming that Ω_0 is connected. Naturally, if Ω_0 consists of a single component, then the hierarchies of the protection zones, according to their relative sizes measured by the underlying principal eigenvalues, cannot change as the parameters λ and μ increase, as occurs when Ω_0 has an arbitrary number of components. Most precisely, when Ω_0 is connected, the analysis carried out in Section 10.2 does not make any sense, while in the case of two or more components it is a categorical imperative. It actually comes as a "bolt out of the blue".

Bibliography

[1] S. Alama and G. Tarantello, On semilinear elliptic equations with indefinite nonlinearities, *Calc. Var. Part. Diff. Eqns.* **1** (1993), 439–475.

[2] S. Alama and G. Tarantello, Elliptic problems with nonlinearities indefinite in sign, *J. Funct. Anal.* **141** (1996), 159–215.

[3] S. Alarcón, G. Díaz, R. Letelier and J. M. Rey, Expanding the asymptotic explosive boundary behaviour of large solutions to a semilinear elliptic equation, *Nonl. Anal.* **72** (2010), 2426–2443.

[4] S. Alarcón, G. Díaz and J. M. Rey, The influence of sources terms on the boundary behaviour of the large solutions of quasilinear elliptic equations; the power-like case, *Z. Angew. Math. Phys.* DOI 10.1007/s00033-012-0253-5, 2012.

[5] S. Alarcón, J. García-Melián and A. Quaas, Keller-Osserman type conditions for some elliptic problems with gradient terms, *J. Diff. Eqns.* **252** (2012), 886–914.

[6] D. Aleja and J. López-Gómez, Some paradoxical effects of advection on a class of diffusive equations in Ecology, *Disc. Cont. Dyn. Systems B* **19** (2014) 10, 3031–3056.

[7] D. Aleja and J. López-Gómez, Concentration through large advection, *J. Diff. Eqns.* **257** (2014), 3135–3164.

[8] D. Aleja and J. López-Gómez, Dynamics of a class of advective-diffusive equations in ecology, *Adv. Nonl. Studies* **15** (2015), 557–585.

[9] P. Álvarez-Caudevilla and J. López-Gómez, Metasolutions in cooperative systems, *Nonl. Anal. R.W.A.* **9** (2008), 1119–1157.

[10] H. Amann, On the existence of positive solutions of nonlinear elliptic boundary value problems, *Ind. Univ. Math. J.*, **21** (1971), 125–146.

[11] H. Amann, Fixed point equations and nonlinear eigenvalue problems in ordered Banach spaces, *SIAM Review* **18** (1976), 620–709.

[12] H. Amann, Maximum principles and principal eigenvalues, in *Ten Mathematical Essays on Approximation in Analysis and Topology* (J. Ferrera,

J. López-Gómez and F. R. Ruiz del Portal Eds.), pp. 1–60, Elsevier B. V., Amsterdam, 2005.

[13] H. Amann and J. López-Gómez, A priori bounds and multiple solutions for superlinear indefinite elliptic problems, *J. Diff. Eqns.* **146** (1998), 336–374.

[14] A. Ambrosetti and J. L. Gámez, Branches of positive solutions for some semilinear Schrödinger equations, *Math. Z.* **224** (1997), 347–362.

[15] C. Aneda and G. Porru, Boundary estimates for solutions of weighted semilinear elliptic equations, *Discrete Contin. Dyn. Syst.* **32** (2012), 3801-3817.

[16] C. Bandle and M. Marcus, Sur les solutions maximales de problémes elliptiques nonlinéaires: bornes isopérimetriques et comportement asymptotique, *C. R. Acad. Sci. Paris 1* **311** (1990), 91–93.

[17] C. Bandle and M. Marcus, Large solutions of semilinear elliptic equations: Existence, uniqueness and asymptotic behavior, *J. D'Analysis Math.* **58** (1992), 9–24.

[18] C. Bandle and M. Marcus, On second order effects in the boundary behavior of large solutions of semilinear elliptic problems, *Diff. Int. Eqns.* **11** (1998), 23–34.

[19] J. Bao, X. Ji and H. Li, Existence and nonexistence theorem for entire subsolutions of k-Yamabe type equations, *J. Diff. Eqns.* **253** (2012), 2140–2160.

[20] P. Baras and L. Cohen, Complete blow-up after $T_{\max}$ for the solution of a semilinear heat equation, *J. Funct. Anal.* **71** (1987), 142–174.

[21] F. Belgacem and C. Cosner, The effects of dispersal along environmental gradients on the dynamics of populations in heterogeneous environments, *Can. Appl. Math. Quart.* **3** (1995), 379–397.

[22] N. Belhaj Rhouma, A. Drissi and W. Sayeb, Existence and asymptotic behavior of boundary blow-up weak solutions for problems involving the p-Laplacian, *J. Partial Differ. Equ.* **26** (2013), 172–192.

[23] H. Berestycki, I. Capuzzo-Dolcetta and L. Nirenberg, Superlinear indefinite elliptic problems and nonlinear Liouville theorems, *Top. Meth. Nonl. Anal.* **4** (1994), 59–78.

[24] H. Berestycki, I. Capuzzo-Dolcetta and L. Nirenberg, Variational methods for indefinite superlinear homogeneous elliptic problems, *Nonl. Diff. Eqns. Appl.* **2** (1995), 553–572.

[25] H. Berestycki and P. L. Lions, Some applications of the method of super and subsolutions. Bifurcation and nonlinear eigenvalue problems (Proc., Session, Univ. Paris XIII, Villetaneuse, 1978), pp. 16-41, Lecture Notes in Mathematics 782, Springer, Berlin, 1980.

[26] M. D. Bertness and R. M. Callaway, Positive interactions in communities, *Trends in Ecology and Evolution* **9** (1994), 191–193.

[27] M. Bertsch and R. Rostamian, The principle of linearized stability for a class of degenerate diffusion equations, *J. Diff. Eqns.* **57** (1985), 373–405.

[28] L. Bieberbach, $\Delta u = e^u$ und die automorphen Funktionen, *Math. Annalen* **77** (1916), 173–212.

[29] J. Blat and K. J. Brown, Bifurcation of steady state solutions in predator-prey and competition systems, *Proc. Royal Soc. Edinburgh* **97A** (1984), 21–34.

[30] M. M. Boureanu, Uniqueness of singular radial solutions for a class of quasilinear problems, *Bull. Belg. Math. Soc. Simon Stevin* **16** (2009), 665-685.

[31] H. Brézis and L. Oswald, Remarks on sublinear elliptic equations, *Nonl. Anal.* **10** (1986), 55–64.

[32] H. Browder, Reflection on the future of mathematics, *Notices of the AMS* **49** (2002), 658–662.

[33] R. M. Callaway and L. R. Walker, Competition and facilitation: A synthetic approach to interactions in plant communities, *Ecology* **78** (1997), 1958–1965.

[34] S. Cano-Casanova, Decay rate at infinity of the positive solutions of a generalized class of Thomas-Fermi equations, 8th AIMS Conference, *Discrete Contin. Dyn. Syst.* Suppl. Vol. I (2011), 240–249.

[35] S. Cano-Casanova and J. López-Gómez, Properties of the principal eigenvalues of a general class of non-classical mixed boundary value problems, *J. Diff. Eqns.* **178** (2002), 123–211.

[36] S. Cano-Casanova and J. López-Gómez, Permanence under strong aggressions is possible, *Ann. I. H. Poincaré Anal. Nonlin.* **20** (2003), 999–1041.

[37] S. Cano-Casanova and J. López-Gómez, Varying boundary conditions in a general class of elliptic problems of mixed type, *Nonl. Anal.* **55** (2003), 47–72.

[38] S. Cano-Casanova and J. López-Gómez, Varying domains in a general class of sublinear elliptic problems, *El. J. Diff. Eqns.* **74** (2004), 1–41.

[39] S. Cano-Casanova and J. López-Gómez, Existence, uniqueness and blow-up rate of large solutions for a canonical class of one-dimensional problems on the half line, *J. Diff. Eqns.* **244** (2008), 3180–3203.

[40] S. Cano-Casanova and J. López-Gómez, Blow-up rates of radially symmetric large solutions, *J. Math. Anal. Appl.* **352** (2009), 166–174.

[41] S. Cano-Casanova, J. López-Gómez and M. Molina-Meyer, Permanence through spatial segregation in heterogeneous competition, in *Proceedings of the 9th IEEE International Conference on Methods and Models in Automation and Robotics*, Miedzyzdroje, pages 123–130. University of Szczecin, Poland, 2003.

[42] S. Cano-Casanova, J. López-Gómez and M. Molina-Meyer, Isolas: Compact solution components separated away from a given equilibrium curve, *Hiroshima Math. J.* **34** (2004), 177-199.

[43] R. S. Cantrell and C. Cosner, *Spatial Ecology via Reaction-Diffusion Equations*, Wiley, New York, 2003.

[44] L. Chen, Y. Chen and D. Luo, Boundary blow-up solutions for a cooperative system involving the *p*-Laplacian, *Ann. Polon. Math.* **109** (2013), 297–310.

[45] Y. Chen and M. Wang, Large solutions for quasilinear elliptic equation with nonlinear gradient term, *Nonl. Anal. Real World Appl.* **12** (2011), 455–463.

[46] Y. Chen and M. Wang, Boundary blow-up solutions for elliptic equations with gradient terms and singular weights: Existence, asymptotic behaviour and uniqueness, *Proc. Roy. Soc. Edinburgh Sect. A* **141** (2011), 717–737.

[47] Y. Chen and M. Wang, Boundary blow-up solutions for *p*-Laplacian elliptic equations of logistic type, *Proc. Roy. Soc. Edinburgh Sect. A* **142** (2012), 691–714.

[48] Y. Chen and M. Wang, Boundary blow-up solutions of *p*-Laplacian elliptic equations with a weakly superlinear nonlinearity, *Nonl. Anal. Real World Appl.* **14** (2013), 1527–1535.

[49] Y. Chen, P. Y. H. Pang and M. Wang, Blow-up rates and uniqueness of large solutions for elliptic equations with nonlinear gradient term and singular or degenerate weights, *Manuscripta Math.* **141** (2013), 171–193.

[50] Y. Chen and Y. Zhu, Large solutions for a cooperative elliptic system of *p*-Laplacian equations, *Nonl. Anal.* **73** (2010), 450–457.

[51] Y. Chen, Y. Zhu and R. Hao, Large solutions with a power nonlinearity given by a variable exponent for p-Laplacian equations, *Nonl. Anal.* **110** (2014), 130–140.

[52] M. Chuaqui, C. Cortázar, M. Elgueta, C. Flores, J. García-Melián, On an elliptic problem with boundary blow-up and a singular weight: The radial case, *Proc. Roy. Soc. Edinburgh* **133A** (2003), 1283–1297.

[53] M. Chuaqui, C. Cortázar, M. Elgueta, J. García-Melián, Uniqueness and boundary behaviour of large solutions to elliptic problems with singular weights, *Comm. Pure Appl. Anal.* **3** (2004), 653–662.

[54] F. C. Cirstea and V. Radulescu, Existence and uniqueness of blow-up solutions for a class of logistic equations, *Comm. Contemp. Math.* **4** (2002), 559–586.

[55] F. C. Cirstea and V. Radulescu, Uniqueness of the blow-up boundary solution of logistic equation with absorbtion, *C. R. Acad. Sci. Paris 1* **335** (2002), 447–452.

[56] F. C. Cirstea and V. Radulescu, Asymptotics for the blow-up boundary solution of the logistic equation with absorption, *C. R. Acad. Sci. Paris 1* **336** (2003), 231–236.

[57] F. C. Cirstea and V. Radulescu, Solutions with boundary blow-up for a class of nonlinear elliptic problems, *Houston J. Math.* **29** (2003), 821–829.

[58] J. H. Connell, On the prevalence and relative importance of interspecific competition: evidence from field experiments, *Amer. Natur.* **122** (1983), 661–696.

[59] M. G. Crandall and P. H. Rabinowitz, Bifurcation from simple eigenvalues, *J. Funct. Anal.* **8** (1971), 321–340.

[60] M. G. Crandall and P. H. Rabinowitz, Bifurcation, pertubation from simple eigenvalues and linearized stability, *Arch. Rat. Mech. Anal.* **52** (1973), 161–180.

[61] R. Courant and D. Hilbert, *Methods of Mathematical Physics Vol. I*, Wiley, New York, 1962.

[62] R. Cui, J. Shi and B. Wu, Strong Allee effect in a diffusive predator-prey system with a protection zone, *J. Diff. Eqns.* **256** (2014), 108–129.

[63] J. Cushing, Two species competition in a periodic environment, *J. Math. Biol.* **10** (1980), 385–390.

[64] E. N. Dancer, The effects of domain shape on the number of positive solutions of certain nonlinear equations II, *J. Diff. Eqns.* **87** (1990), 316–339.

[65] E. N. Dancer, A counterexample on competing species equations, *Diff. Int. Eqns.* **9** (1996), 239–246.

[66] E. N. Dancer and Y. Du, Effect of certain degeneracies in the predator-prey model, *SIAM J. Math. Anal.* **34** (2002), 292–314.

[67] E. N. Dancer and J. López-Gómez, Semiclassical analysis of general second order elliptic operators on bounded domains, *Trans. Amer. Math. Soc.* **352** (2000), 3723–3742.

[68] D. Daners and P. Koch-Medina, *Abstract Evolution Equations, Periodic Problems and Applications*, Longman Scientific, Harlow, Essex, 1992.

[69] Ch. Darwin, *On the Origin of Species by Means of Natural Selection*, John Murray, London, 1859.

[70] M. A. del Pino, Positive solutions of a semilinear elliptic equation on a compact manifold, *Nonlinear Anal.* **22** (1994), 1423–1430.

[71] M. A. del Pino and R. Letelier, The influence of domain geometry in boundary blow-up elliptic problems, *Nonl. Anal.* **48** (2002), 897–904.

[72] M. Delgado, J. López-Gómez and A. Suárez, On the symbiotic Lotka–Volterra model with diffusion and transport effects, *J. Diff. Eqns.* **160** (2000), 175–262.

[73] M. Delgado, J. López-Gómez and A. Suárez, Nonlinear versus linear diffusion: From classical solutions to metasolutions, *Adv. Diff. Eqns.* **7** (2002), 1101–1124.

[74] M. Delgado, J. López-Gómez and A. Suárez, Characterizing the existence of large solutions for a class of sublinear problems with nonlinear diffusion, *Adv. Diff. Eqns.* **7** (2002), 1235–1256.

[75] M. Delgado, J. López-Gómez and A. Suárez, Singular boundary value problems of a porous media logistic equation, *Hiroshima J. Maths.* **34** (2004), 57–80.

[76] G. Díaz and R. Letelier, Explosive solutions of quasilinear elliptic equations: Existence and uniqueness, *Nonl. Anal.* **20** (1993), 97–125.

[77] Y. Du, Effects of a degeneracy in the competition model: Classical and generalized steady-state solutions, *J. Diff. Eqns.* **181** (2002), 92–132.

[78] Y. Du, Spatial patterns for population models in a heterogeneous environment, *Taiwanese J. Math.* **8** (2004), 155–182.

[79] Y. Du, *Order Structure and Topological Methods in Nonlinear Partial Differential Equations*, World Scientific Publishing, Singapore, 2006.

[80] Y. Du and Q. Huang, Blow-up solutions for a class of semilinear elliptic and parabolic equations, *SIAM J. Math. Anal.* **31** (1999), 1–18.

[81] S. Dumont, L. Dupaigne, O. Goubet and V. Radulescu, Back to the Keller-Osserman condition for boundary blow-up solutions, *Adv. Nonl. Studies* **7** (2007), 271–298.

[82] J. C. Eilbeck, J. E. Furter and J. López-Gómez, Coexistence in the competition model with diffusion, *J. Diff. Eqns.* **107** (1994), 96–139.

[83] C. Faber, Beweis das unter allen homogenen Membranen von gleicher Fläche und gleicher Spannung die kreisdörmige den tiefsten Grundton gibt, *Sitzungsber. Bayer. Akad. der Wiss. Math. Phys.* (1923), 169–171.

[84] G. Feltrin and F. Zanolin, Multiple positive solutions for a superlinear problem: A topological approach, *J. Diff. Eqns.*, in press.

[85] H. Feng and C. Zhong, Boundary behavior of solutions for the degenerate logistic type elliptic problem with boundary blow-up, *Nonl. Anal.* **73** (2010), 3472–3478.

[86] P. Feng, Remarks on large solutions of a class of semilinear elliptic equations, *J. Math. Anal. Appl.* **356** (2009), 393–404.

[87] A. Fick, Ueber Diffusion, *Annalen der Physik* **XCII** (1855), 59–86.

[88] R. A. Fisher, The wave of advances of advantageous genes, *Ann. Eugen.* **7** (1937), 355–369.

[89] J. B. J. Fourier, *Théorie Analytique de la Chaleur*, Firmin Didot, Paris, 1822.

[90] J. M. Fraile, P. Koch-Medina, J. López-Gómez and S. Merino, Elliptic eigenvalue problems and unbounded continua of positive solutions of a semilinear equation, *J. Diff. Eqns.* **127** (1996), 295–319.

[91] J. E. Furter and J. López-Gómez, On the existence and uniqueness of co-existence states for the Lotka-Volterra competition model with diffusion and spatially dependent coefficients, *Nonl. Anal.* **25** (1995), 363–398.

[92] J. E. Furter and J. López-Gómez, Diffusion mediated permanence problem for an heterogeneous Lotka–Volterra competition model, *Proc. Roy. Soc. Edinburgh* **127A** (1997), 281–336.

[93] M. Gaudenzi, P. Habets and F. Zanolin, Positive solutions of superlinear boundary value problems with singular indefinite weight, *Comm. Pure Appl. Anal.* **2** (2003), 403–414.

[94] M. Gaudenzi, P. Habets and F. Zanolin, An example of superlinear problem with multiple positive solutions, *Atti Sem. Fis. Univ. Modena* **LI** (2003), 259–272.

[95] J. García-Melián, Nondegeneracy and uniqueness for boundary blow-up elliptic problems, *J. Diff. Eqns.* **223** (2006), 208–227.

[96] J. García-Melián, Uniqueness for boundary blow-up problems with continuous weights, *Proc. Amer. Math. Soc.* **135** (2007) 2785–2793.

[97] J. García-Melián, Boundary behavior for large solutions of elliptic equations with singular weights, *Nonl. Anal.* **67** (2007) 818–826.

[98] J. J. García-Melián, Uniqueness of positive solutions for a boundary value problem, *J. Math. Anal. Appns.* **360** (2009), 530–536.

[99] J. J. García-Melián, Multiplicity of positive solutions to boundary blow-up elliptic problems with sign-changing weights, *J. Funct. Anal.* **261** (2011), 1775–1798.

[100] J. García-Melián, R. Gómez-Reñasco, J. López-Gómez and J. C. Sabina de Lis, Point-wise growth and uniqueness of positive solutions for a class of sublinear elliptic problems where bifurcation from infinity occurs, *Arch. Rat. Mech. Anal.*, **145** (1998), 261–289.

[101] J. García-Melián, R. Letelier and J. C. Sabina de Lis, Uniqueness and asymptotic behaviour for solutions of semilinear problems with boundary blow-up, *Proc. Amer. Math. Soc.* **129** (2001), 3593–3602.

[102] J. García-Melián, C. Morales-Rodrigo, J. D. Rossi and A. Suárez, Nonnegative solutions to an elliptic problem with nonlinear absorption and a nonlinear incoming flux on the boundary, *Ann. Mat. Pura Appl.* **187** (2008), 459–486.

[103] D. Gilbarg and N. S. Trudinger, *Elliptic Partial Differential Equations of Second Order*, Springer, Berlin, 2001.

[104] M. Golubitsky and D. G. Shaeffer, *Singularity and Groups in Bifurcation Theory*, Springer, Berlin, 1985.

[105] R. Gómez-Reñasco, *The Effect of Varying Coefficients in Semilinear Elliptic Boundary Value Problems: From Classical Solutions to Metasolutions*, Ph.D. dissertation, University of La Laguna, Tenerife, Spain, March 1999.

[106] R. Gómez-Reñasco and J. López-Gómez, The effect of varying coefficients on the dynamics of a class of superlinear indefinite reaction diffusion arising in population dynamics, (unpublished) preprint 1999.

[107] R. Gómez-Reñasco and J. López-Gómez, The effect of varying coefficients on the dynamics of a class of superlinear indefinite reaction diffusion equations, *J. Diff. Eqns.* **167** (2000), 36–72.

[108] R. Gómez-Reñasco and J. López-Gómez, The uniqueness of the stable positive solution for a class of superlinear indefinite reaction diffusion equations, *Diff. Int. Eqns.*, **14** (2001), 751–768.

[109] R. Gómez-Reñasco and J. López-Gómez, On the existence and numerical computation of classical and non-classical solutions for a family of elliptic boundary value problems, *Nonlinear Anal.* **48** (2002), 567–605.

[110] G. Green, *An Essay of the Applications of Mathematical Analysis to the Theory of Electricity and Magnetism*, Bromley House, Nottingham, 1828.

[111] J. Hadamard, Mémoires sur le problème d'analyse relatif à l'equilibre des plaques élastiques encastrées, *Mém. Acad. Sci. Paris* **39** (1908), 128–259.

[112] D. Henry, *Geometric Theory of Semilinear Parabolic Equations*, Lectures Notes in Mathematics 840, Springer, Berlin, 1981.

[113] E. Hopf, A remark on linear elliptic differential equations of the second order, *Proc. Amer. Math. Soc.* **3** (1952), 791–793.

[114] S. Huang and Q. Tian, Boundary blow-up rates of large solutions for elliptic equations with convection terms, *J. Math. Anal. Appl.* **373** (2011), 30–43.

[115] S. Huang, Q. Tian and Y. Mi, General uniqueness results and blow-up rates for large solutions of elliptic equations, *Proc. Roy. Soc. Edinburgh Sect. A* **142** (2012), 825–837.

[116] S. Huang, Q. Tian, S. Zhang and J. Xi, A second-order estimate for blow-up solutions of elliptic equations, *Nonl. Anal.* **74** (2011), 2342–2350.

[117] S. Huang, Q. Tian, S. Zhang, J. Xi and Z. Fan, The exact blow-up rates of large solutions for semilinear elliptic equations, *Nonl. Anal.* **73** (2010), 3489–3501.

[118] S. Huang, W. T. Li, Q. Tian and C. Mu, Large solution to nonlinear elliptic equation with nonlinear gradient terms, *J. Diff. Eqns.* **251** (2011), 3297–3328.

[119] G. E. Hutchinson, *The Ecological Theater and the Evolutionary Play*, Yale University Press, New Haven, 1965.

[120] V. Hutson, J. López-Gómez, K. Mischaikow and G. Vickers, Limit behaviour for a competing species problem with diffusion, *World Scientific Series Anal. Appl.* **4** (1995), 343–358.

[121] X. Ji and J. Bao, Necessary and sufficient conditions on solvability for Hessian inequalities, *Proc. Amer. Math. Soc.* **138** (2010), 175–188.

[122] T. Kato, *Perturbation Theory for Linear Operators*, Springer, Berlin, 1995.

[123] J. B. Keller, On solutions of $\Delta u = f(u)$, *Comm. Pure Appl. Math.* **X** (1957), 503–510.

[124] A. N. Kolmogorov, I. G. Petrovsky and N. S. Piskunov, Etude de l'équation de la diffusion avec croissance de la quantité de matière et son application à un problème biologique, *Bull. Math. Univ. d'Etat à Moscow (Sèr. Int.)* **A 1** (1937), 1–129.

[125] V. A. Kondratiev, and V. A. Nikishin, Asymptotics, near the boundary, of a solution of a singular boundary value problem for a semilinear elliptic equation, *Diff. Eqns.* **26** (1990), 345–348.

[126] E. Krahn, Über eine von Rayleigh formulierte Minimaleigenschaft des Kreises, *Math. Ann.* **91** (1925), 97–100.

[127] O. A. Ladyzenskaja, V. A. Solonnikov and N. N. Uraltzeva, *Linear and Quasilinear Equations of Parabolic Type*, American Mathematical Society, Providence, 1968.

[128] M. Langlais, and D. Phillips, Stabilization of solutions of nonlinear and degenerate evolution equations, *Nonl. Anal.* **9** (1985), 321–333.

[129] A. C. Lazer and P. J. McKenna, On a problem of L. Bieberbach and H. Rademacher, *Nonl. Anal.* **21** (1993), 327–335.

[130] A. C. Lazer and P. J. McKenna, A singular elliptic boundary value problem, *Appl. Math. Comp.* **65** (1994), 183–194.

[131] A. C. Lazer and P. J. McKenna, Asymptotic behaviour of solutions of boundary blow-up problems, *Diff. Int. Eqns.*, **7** (1994), 1001–1019.

[132] W. Lei and M. Wang, Existence of large solutions of a class of quasilinear elliptic equations with singular boundary, *Acta Math. Hungar.* **129** (2010), 81–95.

[133] Y. Liang, Q. Zhang and C. Zhao, On the boundary blow-up solutions of $p(x)$-Laplacian equations with gradient terms, *Taiwanese J. Math.* **18** (2014), 599–632.

[134] L. Li and R. Logan, Positive solutions to general elliptic Lotka–Volterra models, *Diff. Int. Eqns.* **4** (1991), 817–834.

[135] H. Li, P. Y. H. Pang and M. Wang, Boundary blow-up solutions for logistic-type porous media equations with nonregular source, *J. Lond. Math. Soc.* **80** (2009), 273–294.

[136] H. Li, P. Y. H. Pang and M. Wang, Boundary blow-up of a logistic-type porous media equation in a multiply connected domain, *Proc. Roy. Soc. Edinburgh Sect. A* **140** (2010), 101–117.

[137] C. Liu and Z. Yang, Boundary blow-up quasilinear elliptic problems with nonlinear gradient terms, *Complex Var. Elliptic Eqns.* **57** (2012), 687–704.

[138] C. Loewner and L. Nirenberg, Partial differential equations invariant under conformal or projective transformations, in: L. V. Ahlfors, I. Kra, B. Maskit, L. Nirenberg (Eds.), *Contributions to Analysis*, pp. 245–272, Academic Press, New York, 1974.

[139] R. Logan and A. Ghoreishi, Positive solutions of a class of biological models in a heterogeneous environment, *Bull. Aust. Math. Soc.* **44** (1991), 79–94.

[140] J. López-Gómez, Positive periodic solutions of Lotka–Volterra reaction-diffusion systems, *Diff. Int. Eqns.* **5** (1992), 55–72.

[141] J. López-Gómez, On linear weighted boundary value problems, in *Partial Differential Equations; Models in Physics and Biology*, pages 188–203, Akademie Verlag, Berlin, 1994.

[142] J. López-Gómez, Nonlinear eigenvalues and global bifurcation: Application to the search of positive solutions for general Lotka–Volterra reaction diffusion systems with two species, *Diff. Int. Eqns.* **7** (1994), 1427–1452.

[143] J. López-Gómez, Permanence under strong competition, *World Scientific Series Anal. Appl.* **4** (1995), 473–488.

[144] J. López-Gómez, The maximum principle and the existence of principal eigenvalues for some linear weighted boundary value problems, *J. Diff. Eqns.* **127** (1996), 263–294.

[145] J. López-Gómez, Strong competition with refuges, *Nonl. Anal.* **30** (1997), 5167–5178.

[146] J. López-Gómez, On the existence of positive solutions for some indefinite superlinear elliptic problems, *Comm. Part. Diff. Eqns.* **22** (1997), 1787–1804.

[147] J. López-Gómez, Large solutions, metasolutions, and asymptotic behaviour of the regular positive solutions of sublinear parabolic problems, *El. J. Diff. Eqns.* **Conf. 05** (2000), 135–171.

[148] J. López-Gómez, Varying bifurcation diagrams of positive solutions for a class of indefinite superlinear boundary value problems, *Trans. Amer. Math. Soc.* **352** (2000), 1825–1858.

[149] J. López-Gómez, *Spectral Theory and Nonlinear Functional Analysis*, Research Notes in Mathematics 426, Chapman & Hall/CRC, Boca Raton, 2001.

[150] J. López-Gómez, Approaching metasolutions by classical solutions, *Diff. Int. Eqns.* **14** (2001), 739–750.

[151] J. López-Gómez, Metasolutions, *Bol. Soc. Esp. Math. Appl.* **24** (2003), 59–90.

[152] J. López-Gómez, Classifying smooth supersolutions in a general class of elliptic boundary value problems, *Adv. Diff. Eqns.* **8** (2003), 1025–1042.

[153] J. López-Gómez, Coexistence and metacoexistence states in competing species models, *Houston J. Math.* **29** (2003), 485–538.

[154] J. López-Gómez, The boundary blow-up rate of large solutions, *J. Diff. Eqns.* **195** (2003), 25–45.

[155] J. López-Gómez, Dynamics of parabolic equations: From classical solutions to metasolutions, *Diff. Int. Eqns.* **16** (2003), 813–828.

[156] J. López-Gómez, Varying stoichiometric exponents I: Classical steady-states and metasolutions, *Adv. Nonl. Studies* **3** (2003), 327–354.

[157] J. López-Gómez, Global existence versus blow-up in superlinear indefinite parabolic problems, *Sci. Math. Jap.* **e-2004** (2004), 449–472.

[158] J. López-Gómez, Global existence versus blow-up in superlinear indefinite parabolic problems, *Sci. Math. Jap.* **61** (2005), 493–517.

[159] J. López-Gómez, Uniqueness of large solutions for a class of radially symmetric elliptic equations, *Spectral Theory and Nonlinear Analysis with Applications to Spatial Ecology* (S. Cano-Casanova, J. López-Gómez and C. Mora-Corral, Eds.), 75–110, World Scientific, Singapore, 2005.

[160] J. López-Gómez, Metasolutions: Malthus versus Verhulst in population dynamics: A dream of Volterra, in *Handbook of Differential Equations, Stationary Partial Differential Equations* (M. Chipot and P. Quittner eds.), pages 211–309, Elsevier, Amsterdam, 2005.

[161] J. López-Gómez, Optimal uniqueness theorems and exact blow-up rates of large solutions, *J. Diff. Eqns.* **224** (2006), 385–439.

[162] J. López-Gómez, Uniqueness of radially symmetric large solutions, *Disc. Cont. Dynam. Systems*, Proceedings of Sixth AIMS Conference, Supplement 2007, 677–686.

[163] J. López-Gómez, *Linear Second Order Elliptic Operators*, World Scientific, Singapore, 2013.

[164] J. López-Gómez and L. Maire, Uniqueness of large positive solutions for a class of radially symmetric cooperative systems, submitted.

[165] J. López-Gómez and M. Molina-Meyer, The maximum principle for cooperative weakly coupled elliptic systems and some applications, *Diff. Int. Eqns.* **7**, (1994), 383–398.

[166] J. López-Gómez and M. Molina-Meyer, Singular perturbations in economy and ecology: The effect of strategic symbiosis in random competitive environments, *Adv. Math. Sci. Appl.* **14** (2004), 87–107.

[167] J. López-Gómez and M. Molina-Meyer, Bounded components of positive solutions of abstract fixed point equations: mushrooms, loops and isolas, *J. Diff. Eqns.* **209** (2005), 416–441.

[168] J. López-Gómez and M. Molina-Meyer, The competitive exclusion principle versus biodiversity through segregation and further adaptation to spatial heterogeneities, *Th. Pop. Biol.* **69** (2006), 94–109.

[169] J. López-Gómez and M. Molina-Meyer, In the blink of an eye, in *Progress in Nonlinear Partial Differential Equations and Their Applications 64*, pages 291–327, Birkhäuser, Basel, 2005.

[170] J. López-Gómez and M. Molina-Meyer, Cooperation and competition, strategic alliances, and the Cambrian explosion, in *Spectral Theory and Nonlinear Analysis with Applications to Spatial Ecology*, pages 111–126, World Scientific, Singapore, 2005.

[171] J. López-Gómez and M. Molina-Meyer, Superlinear indefinite systems: beyond Lotka–Volterra models, *J. Diff. Eqns.* **221** (2006), 343–411.

[172] J. López-Gómez and M. Molina-Meyer, Biodiversity through coopetition, *Disc. Cont. Dyn. Syst. B* **8(1)** (2007), 187–205.

[173] J. López-Gómez, M. Molina-Meyer and A. Tellini, The uniqueness of the linearly stable positive solution for a class of super-linear indefinite problems with non-homogeneous boundary conditions, *J. Diff. Eqns.*, **255** (2013), 503–523.

[174] J. López-Gómez, M. Molina-Meyer and A. Tellini, Intricate bifurcation diagrams for a class of one-dimensional superlinear indefinite problems of interest in population dynamics, *Disc. Cont. Dyn. Syst.* Supplement 2013, 515–524.

[175] J. López-Gómez, M. Molina-Meyer and A. Tellini, Complex dynamics caused by facilitation in competitive environments within polluted habitat patches, *Eur. J. Appl. Math.* **25** (2014), 213–229.

[176] J. López-Gómez, M. Molina-Meyer and A. Tellini, Spiraling bifurcation diagrams in superlinear indefinite problems, *Disc. Cont. Dyn. Syst.* **35** (2015), 1561–1588.

[177] J. López-Gómez and P. Quittner, Complete energy blow-up in indefinite superlinear parabolic problems, *Disc. Cont. Dyn. Syst.* **14** (2006), 169–186.

[178] J. López-Gómez and P. H. Rabinowitz, Nodal solutions for a class of degenerate boundary value problems, *Adv. Nonl. Studies*, **15** (2015), 253-288.

[179] J. López-Gómez and P. H. Rabinowitz, The effects of spatial heterogeneities on some multiplicity results, *Disc. Cont. Dyn. Syst.*, **36** (2016), in press.

[180] J. López-Gómez and J. C. Sabina de Lis, Coexistence states and global attractivity for some convective diffusive competing species models, *Trans. Amer. Math. Soc.* **347** (1995), 3797–3833.

[181] J. López-Gómez and J. C. Sabina de Lis, First variations of principal eigenvalues with respect to the domain and point-wise growth of positive solutions for problems where bifurcation from infinity occurs, *J. Diff. Eqns.* **148** (1998), 47–64.

[182] J. López-Gómez and A. Suárez, Combining fast, linear and slow diffusion, *Topol. Meth. Nonl. Anal.* **23** (2004), 275–300.

[183] J. López-Gómez and A. Tellini, Generating an arbitrarily large number of isolas in a superlinear indefinite problem, *Nonl. Anal.* **108** (2014) 223–248.

[184] J. López-Gómez, A. Tellini and F. Zanolin, High multiplicity and complexity of the bifurcation diagrams of large solutions for a class of superlinear indefinite problems, *Comm. Pure Appl. Anal.* **13** (2014), 1–73.

[185] A. J. Lotka, The growth of mixed populations: Two species competing for a common food supply, *J. Washington Acad. Sci.* **22** (1932), 461–469.

[186] A. Lunardi, Analytic semigroups and optimal regularity in parabolic problems, in *Progress in Nonlinear Partial Differential Equations and Their Applications*, Birkhäuser, Berlin, 1995.

[187] Th. R. Malthus, *An Essay on the Principle of Population*, St. Paul's Church Yard, London, 1798.

[188] M. Marcus and L. Véron, Uniqueness and asymptotic behavior of solutions with boundary blow-up for a class of nonlinear elliptic equations, *Ann. Inst. Henri Poincaré* **14** (1997), 237–274.

[189] M. Marcus and L. Véron, The boundary trace and generalized boundary value problem for semilinear elliptic equations with coercive adsorption, *Comm. Pure Appl. Math.* **56** (2003), 689–731.

[190] V. Maric and M. Tomic, Asymptotics of solutions of a generalized Thomas-Fermi equation, *J. Diff. Eqns.* **35** (1980), 36–44.

[191] M. Marras and G. Porru, Estimates and uniqueness for boundary blow-up solutions of p-Laplace equations, *Electron. J. Diff. Eqns.* **119** (2011), 10.

[192] J. L. Mawhin, The legacy of P. F. Verhulst and V. Volterra in population dynamics, in *The First 60 Years of Nonlinear Analysis of Jean Mawhin*, pages 147–160, World Scientific, Singapore, 2004.

[193] J. L. Mawhin, D. Papini and F. Zanolin, Boundary blow-up for differential equations with indefinite weight, *J. Diff. Eqns.* **188** (2003), 33–51.

[194] L. Mi and B. Liu, Second order expansion for blow-up solutions of semilinear elliptic problems, *Nonl. Anal.* **75** (2012), 2591–2613.

[195] M. Molina-Meyer and F. R. Prieto-Medina, A collocation-spectral method to solve the bi-dimensional degenerate diffusive logistic equation with spatial heterogeneities in circular domains, preprint, May 2014.

[196] C. Mu, S. Huang, Q. Tian and L. Liu, Large solutions for an elliptic system of competitive type: existence, uniqueness and asymptotic behavior, *Nonl. Anal.* **71** (2009), 4544–4552.

[197] J. D. Murray, *Mathematical Biology*, Springer, Berlin, 1989.

[198] S. Nakamori and K. Takimoto, Uniqueness of boundary blowup solutions to k-curvature equation, *J. Math. Anal. Appl.* **399** (2013), 496–504.

[199] L. Nirenberg, A strong maximum principle for parabolic equations, *Comm. Pure Appl. Math.* **6** (1953), 167–177.

[200] A. Okubo and S. A. Levin, *Diffusion and Ecological Problems: Modern Perspectives*, Interdisciplinary Applied Mathematics 14, Springer, New York, 2001.

[201] O. A. Oleinik, On properties of some boundary problem for equations of elliptic type, *Math. Sbornik, N.S.* **30** (1952), 695–702.

[202] R. Osserman, On the inequality $\Delta u \geq f(u)$, *Pacific J. Math.* **7** (1957), 1641–1647.

[203] T. Ouyang, On the positive solutions of semilinear equations $\Delta u + \lambda u - hu^p = 0$ on the compact manifolds, *Trans. Amer. Math. Soc.*, **331** (1992), 503–527.

[204] T. Ouyang, On the positive solutions of semilinear equations $\Delta u + \lambda u - hu^p = 0$ on the compact manifolds, Part II, *Indiana Univ. Math. J.* **40** (1991), 1083–1141.

[205] T. Ouyang and Z. Xie, The uniqueness of blow-up for radially symmetric semilinear elliptic equations, *Nonl. Anal.* **64** (2006), 2129–2142.

[206] T. Ouyang and Z. Xie, The exact boundary blow-up rate of large solutions for semilinear elliptic problems, *Nonl. Anal.* **68** (2008), 2791–2800.

[207] R. Pearl and L. L. Reed, On the rate of growth of the population of the United States since 1790 and its mathematical representation, *Proc. Nat. Acad. Sci.* **6** (1920), 275–288.

[208] O. Perron, Eine neue Behandlung der Randwertaufgabe für $\Delta u = 0$, *Mat. Z.* **18** (1923), 42–54.

[209] M. Picone, Sui valori eccenzionali di un pammetro da cui dipende un'equazione differenziale ordinaria del secondo ordine, *Ann. Sc. Nor. Sup. Pisa* **11** (1910), 1–141.

[210] M. W. Protter and H. F. Weinberger, *Maximum Principles in Differential Equations*, Prentice-Hall, Englewood Cliffs, 1967.

[211] F. I. Pugnaire (Editor), *Positive Plant Interactions and Community Dynamics*, CRC Press, Boca Raton, 2010.

[212] A. Quatelet, *Sur l'homme et le développement de ses facultés. Essai de physique sociale*, Bacheller, Paris, 1835.

[213] P. Quittner and P. Souplet, *Superlinear Parabolic Problems. Blow-up, Global Existence and Steady States*, Birkhäuser, Basel, 2007.

[214] P. H. Rabinowitz, Some global results for nonlinear eigenvaue problems, *J. Funct. Anal.* **7** (1971), 487–513.

[215] P. H. Rabinowitz, *Minimax Methods in Critical Point Theory with Applications to Differential Equations*, Regional Conference Series in Mathematics 65, American Mathematical Society, Providence, 1988.

[216] H. Rademacher, Einige besondere Probleme partieller Differentialgleichungen, in *Die Differential und Integralgleichungen der Mechanik und Physik I*, (P. Frank and R. von Mises, eds.) pages 838–845, Rosenberg, New York, 1943.

[217] V. Radulescu, Singular phenomena in nonlinear elliptic problems: From boundary blow-up solutions to equations with singular nonlinearities, in *Handbook of Differential Equations: Stationary Partial Differential Equations*, Vol. 4, pages 483–591, Elsevier, Amsterdam, 2007.

[218] D. Repovs, Asymptotics for singular solutions of quasilinear elliptic equations with an absorption term, *J. Math. Anal. Appl.* **395** (2012), 78–85.

[219] M. B. Saffo, Invertebrates in endosymbiotic associations, *Amer. Zool.* **32** (1992), 557–565.

[220] D. Sattinger, *Topics in Stability and Bifurcation Theory*, Lecture Notes in Mathematics 309, Springer, Heildelberg, 1973.

[221] T. Shibata, Spectral asymptotics for inverse nonlinear Sturm-Liouville problems, *Electron. J. Qual. Theory Diff. Eqns.* **58** (2009), 1–18.

[222] T. W. Shoener, Field experiments on interspecific competition, *Amer. Natur.* **122** (1983), 240–285.

[223] E. Shrödinger, *The Physical Aspect of the Living Cell*, Cambridge University Press, Cambridge, 1944.

[224] B. Simon, Semiclassical analysis of low lying eigenvalues, I. Non-degenerate minima: Asymptotic expansions, *Ann. Inst. Henri Poincaré A*, **XXXVIII** (1983), 12–37.

[225] S. D. Taliaferro, Asymptotic behavior of solutions of $y'' = \varphi(t)y^{\lambda}$. *J. Math. Anal. Appl.* **86** (1978), 95–34.

[226] S. D. Taliaferro, Asymptotic behavior of solutions of $y'' = \varphi(t)f(y)$, *SIAM J. Math. Anal.* **12** (1981), 853-865.

[227] A. Tellini, *Positive Blow-up Solutions for a Class of Nonlinear Differential Equations*, Tesi di Laura, Udine, July 2010.

[228] P. F. Verhulst, Notice sur la loi que la population suit dans son accroissement, *Corr. Math. Phys.* **10** (1838), 113–125.

[229] V. Volterra, *Lecons sur la théorie mathématique de la lutte pour la vie*, Gauthier-Vilars, Paris, 1931.

[230] L. Véron, Semilinear elliptic equations with uniform blow up on the boundary, *J. D'Analyse Math.* **59** (1992), 231–250.

[231] M. Wang and L. Wei, Existence and boundary blow-up rates of solutions for boundary blow-up elliptic systems, *Nonl. Anal.* **71** (2009), 2022–2032.

[232] W. Wang, H. Gong and S. Zheng, Asymptotic estimates of boundary blow-up solutions to the infinity Laplace equations, *J. Diff. Eqns.* **256** (2014), 3721–3742.

[233] Y. Wang and M. Wang, The blow-up rate and uniqueness of large solution for a porous media logistic equation, *Nonl. Anal. Real World Appl.* **11** (2010), 1572–1580.

[234] L. Wei, The existence of large solutions of semilinear elliptic equations with negative exponents, *Nonl. Anal.* **73** (2010), 1739–1746.

[235] L. Wei and M. Wang, Existence and estimate of large solutions for an elliptic system, *Nonl. Anal.* **70** (2009), 1096–1104.

[236] L. Wei and J. Zhu, The existence and blow-up rate of large solutions of one-dimensional p–Laplacian equations, *Nonl. Anal. Real World Appl.* **13** (2012), 665–676.

[237] J. L. Wulff, Clonal organisms and the evolution of mutualism. In Jackson, J.B.C., Buss, L.W., Cook, R.E. (Eds.), *Population Biology and Evolution of Clonal Organisms*, pages 437–466, Yale University Press, New Haven, 1985.

[238] Z. Xie, Uniqueness and blow-up rate of large solutions for elliptic equation $-\Delta u = \lambda u - b(x)h(u)$, *J. Diff. Eqns.* **247** (2009), 344–363.

[239] Z. Xie and C. Zhao, Blow-up rate and uniqueness of singular radial solutions for a class of quasi-linear elliptic equations, *J. Diff. Eqns.* **252** (2012), 1776–1788.

[240] H. Yamabe, On a deformation of Riemaniann structures on compact manifolds, *Osaka Math. J.* **12** (1960), 21–37.

[241] H. Yang and Y. Chang, On the blow-up boundary solutions of the Monge-Ampère equation with singular weights, *Comm. Pure Appl. Anal.* **11** (2012), 697–708.

[242] Q. Zhang, Existence, nonexistence and asymptotic behavior of boundary blow-up solutions to $p(x)$-Laplacian problems with singular coefficient, *Nonl. Anal.* **74** (2011), 2045-2061.

[243] Q. Zhang, X. Liu and Z. Qiu [243], On the boundary blow-up solutions of $p(x)$-Laplacian equations with singular coefficient, *Nonl. Anal.* **70** (2009), 4053–4070.

[244] Q. Zhang, Y. Wang and Z. Qiu, Existence of solutions and boundary asymptotic behavior of $p(r)$-Laplacian equation multi-point boundary value problems, *Nonl. Anal.* **72** (2010), 2950-2973.

[245] Q. Zhang and C. Zhao, Existence, uniqueness and blow-up rate of large solutions of quasi-linear elliptic equations with higher order and large perturbation, *J. Partial Differ. Eqns.* **26** (2013), 226-250.

[246] Z. Zhang, The asymptotic behaviour of solutions with boundary blow-up for semilinear elliptic equations with nonlinear gradient terms, *Nonl. Anal.* **62** (2005), 1137–1148.

[247] Z. Zhang, The asymptotic behaviour of solutions with blow-up at the boundary for semilinear elliptic problems, *J. Math. Anal. Appl.* **308** (2005), 532–540.

[248] Z. Zhang, Boundary blow-up elliptic problems with nonlinear gradient terms, *J. Diff. Eqns.* **228** (2006), 661–684.

[249] Z. Zhang, Boundary blow-up elliptic problems with nonlinear gradient terms and singular weights, *Proc. Roy. Soc. Edinburgh Sect. A* **138** (2008), 1403–1424.

[250] Z. Zhang, A boundary blow-up elliptic problem with an inhomogeneous term, *Nonl. Anal.* **68** (2008), 3428–3438.

[251] Z. Zhang, Boundary behavior of large solutions to semilinear elliptic equations with nonlinear gradient terms, *Nonl. Anal.* **73** (2010), 3348–3363.

[252] Z. Zhang, The second expansion of large solutions for semilinear elliptic equations, *Nonl. Anal.* **74** (2011), 3445–3457.

[253] Z. Zhang, Y. Ma, L. Mi and X. Li, Blow-up rates of large solutions for elliptic equations, *J. Diff. Eqns.* **249** (2010), 180–199.

[254] Z. Zhang and L. Mi, Blow-up rates of large solutions for semilinear elliptic equations, *Comm. Pure Appl. Anal.* **10** (2011), 1733–1745.

Index